A VENOMOUS LIFE

Professor Struan Sutherland is well known for his work on animal toxins. He is Director of the Australian Venom Research Unit, Department of Pharmacology, University of Melbourne.

In 1967 he founded the Immunology Research Department at the Commonwealth Serum Laboratories. In 1980 his team developed the first successful antivenom for bites by the dreaded Sydney Funnel-web spider. Other achievements were the discovery of the new pressure-immobilisation type of first aid and the development of venom detection kits.

Struan Keith Sutherland was born in Sydney and brought up in Bendigo, Victoria. He studied medicine at the University of Melbourne and received his MBBS in 1960. From 1962–65 he served as a surgeon lieutenant in the RAN, sailing on HMAS *Voyager* and HMAS *Melbourne* with shore stints at Flinders Naval Depot and Darwin.

In the course of a distinguished and tumultuous career he became FRCPA and FRACP, received his MD and DSc from the University of Melbourne, was awarded the AMA Prize for Medical Research and the James Cook Medal from the Royal Society of NSW.

He has also been suspended from duties twice at CSL and on numerous occasions jeopardised his career for the sake of his patients and his resesearch. Always for him the patient came first.

By the same author:

Family Guide to Dangerous Animals and Plants of Australia
Venomous Creatures of Australia
*Australian Animal Toxins: The creatures, their toxins and
   care of the poisoned patient*
Take Care! Poisonous Australian Animals
Hydroponics for Everyone

A Venomous Life

S. K. SUTHERLAND
MD, DSc, FRCPA, FRACP

First Published in Australia in 1998 by
Hyland House Publishing Pty Ltd
Hyland House
387-389 Clarendon Street
South Melbourne
Victoria 3205

© Struan Sutherland 1998

This book is copyright. Apart from any fair dealing for the purposes of private study, research, criticism or review, as permitted under the Copyright Act, no part may be reproduced by any process without written permission. Enquiries should be addressed to the publisher.

National Library of Australia
Cataloguing-in-publication data:

Sutherland, Struan K. (Struan Keith), 1936-.
   A venomous life.

   Bibliography.
   Includes index.
   ISBN 1 86447 026 7.

   1. Sutherland, Struan K. (Struan Keith), 1936-. 2. Toxicologists - Australia - Biography. I. Title

615.90092

Typeset in Bembo 11/12pt by Hyland House
Printed by Brown Prior Anderson, Melbourne, Australia

Contents

List of Illustrations vii
Acknowledgements ix
Introduction xi

PART ONE—BACKGROUND AND EARLY LIFE,
1936–1965   1
A Severely Pruned Family Tree   2

1   Soldiers, Sailors, Balloonists, Doctors, Bankers and
Other Assorted Ancestors   3
2   Growing Up   21
3   Uncles   51
4   A Student and Hospital Intern in Melbourne,
1954–1961   69
5   The Navy Lark—RAN, 1962–1965   107

PART TWO—A RESEARCH CAREER, 1966–1998   159

Chronology of Main Events 1966–1998   160

6   Snakes, Spiders and Test Tubes   163
7   The Sydney Funnel-web Spider   179

8	Improving the Lot of Victims of Snake Bite and Other Bites and Stings	219
9	Getting into Trouble at CSL	249
10	Into the Wilderness then Fighting the Odds	301
11	Australian Venom Research Unit, July 1994 to December 1998	349

Appendix: First Aid for Venomous Bites and Stings   369
Index   379

List of Illustrations

Facing Page

My paternal grandfather, Hector McKenzie Sutherland	18
Minnie, Hector's wife	18
My dad as Australia's youngest bank manager in Darwin, 1910	18
My mother, Dorothy Hunt	19
My dad after becoming a father in 1930	19
My dad as a troop, 1917	19
My maternal grandfather, Frederick Knight Hunt, with Grandma Hunt	50
Dad's 1918 diary showing damage done by the Kaiser	50
My great-great grandfather, James Glaisher, ballooning and in big trouble, in 1862	51
The author as an infant	82
First day at school	82
Camp Hill State School in 1987	82
The author as a young man	83
Ready for a family picnic, with sisters Diana and Rosemary	83
A good sized one-teacher class of 1946, showing the author and other pupils	83
Uncle Angus, my dad's favourite brother	114
Uncle Ian looking cheerful	114
Dr Jim Sharland, a fine Bendigo doctor and close friend of the author's father	114
Dr Keith Kerr resting in the bush near Bendigo	114
Uncle Charles and Aunt Marjory, in London, c. 1953	114
The author as a horseman	115
Three pals—the author, Dick Harvey and Alan Kerr	115
Outside 'St Richards', Parkville, 1956. The author, Bob Dalgleish and Barry Thompson	115
The author feeling the tropical heat	146
Steaming along behind HMAS *Melbourne*	146

Facing Page

HMAS *Voyager* doing 'wheelies'	146
Three pretty boys in the Philippines. the author, Mike Hudson and Richard Carpendale	146
The Red-back spider	147
The Blue-ringed octopus, the most dangerous octopus in the world	147
The Black House or Window spider	147
Male Sydney Funnel-web spider	147
The White-tailed spider	147
Wedding Day with my first wife, Wendy	147
The author's second wife, Megan	178
John and Susie, 1989. John has just been admitted to the Bar.	178
Mrs Marjory Davey at her farewell, June 1994	179
Drs Gabrielle Hawdon, Steven Pincus, James Tibballs, Ken Winkel, Anna Young after AVRU meeting, 15 June 1998	179
The author with Alan Kerr, then Administrator of Norfolk Island, 1996	179
Some of the ship's company in Hong Kong, with Captain Wells and the author	210
HMAS *Voyager*	210
Analysis of human serum by immunoelectrophoresis	211
Technician Moira Whigley dissecting a Blue-ringed octopus, c. 1968	211
Dissection of Blue-ringed octopus	211
Anne Miller cautiously 'milking' a female Sydney Funnel-web spider	242
John going to work with his dad, c. 1970	242
Immunology Research, May 1979	242
The author with the founder of the Australian Reptile Park, the late Eric Worrell, 1981	243
Successful collaborators: the author with Dr Alan Duncan and Dr James Tibballs	243
Two dedicated right-hand men, Alan Coulter and Rodney Harris, 1978	274
The author demonstrating the pressure-immobilisation first aid for snake bite, with Erin Lovering as 'patient', c. 1979	274
Dr Bill Lane, Director of CSL 1966-74	275
Publisher's Note: The author would have liked to have photographs of more people from his CSL days but the management of CSL declined permission.	306
Dr John Trinca from the author's collection	306
Dr Norman Coles from the author's collection	306
Mr Viv Davey from the author's collection	306
Mr Merv Hinton from the author's collection	306
Professor James Angus, Department Head and chief mentor of AVRU	307
Ayse Berke	307

Acknowledgements

My children, John and Susie, for their consistent tolerance, support and especially affection, the last of which is returned in full.

Dr Gabrielle Hawdon and Susie Kennewell whose beady eyes scrutinised all chapters. Their collective expertise reduced inconsistencies, medical gaffs, sexism and open invitations for litigation. They were excellent sounding boards. John Sutherland also contributed to this process.

Lee White and Anne Godden who independently assessed the first draft in 1997. Their reports were highly critical but filled with positive suggestions—including a massive re-write. Crestfallen, I followed their guidelines but have not regained my former high opinion of my writing skills.

My sister Rosemary for her enthusiastic help, especially with the severely pruned family tree.

My old ship-mate Commander Chris Bolton RAN (Rtd) for checking Chapter 5 and making encouraging noises.

The following dear friends read various chapters and made useful suggestions: Verna Cook, Alan Coulter, Frank Crowley, Alan Kerr, Lesley Lane, the Reverend John Macmillan, Dr John Meyer, Joyce Newman and Pamela Sharland-Normark.

The support of my colleagues is greatly appreciated. Professors John Pearn and John Williamson, both fine teachers and true believers, have been outstanding.

The typing skills of Marjory Davey, Pauline Embury, Joanne Cook and especially Vanessa Tresidder.

My distinguished editor Anne Godden, Al Knight, Susie Godden, Nerissa Greenfield, Najiye Nihat and others of Hyland House Publishing.

Dr Stephen Smith, my caring physician, who is so keen to see this and other works finished that he once apologised for the interruption of his house call!

Special thanks to my little earthbound firecracker, to Joyce, and for God who created Brighton Beach.

Introduction

For some time I have been wondering how to start this tale. Yesterday I received some information which seems to fit the bill, since it may affect both the quality of the work and its chances of completion.

I'd been having trouble with writing and other fine movements like doing up buttons. Whereas before my pen flew across the page, now it barely crawled. My writing, always small at the best of times, now resembled the meanderings of a drunken ant dipped in ink.

A few months before this Bob the Neurologist had decided that I had Parkinson's Disease and given me a large bottle of pills which he said would have me scribbling away in no time. They didn't, so he ordered an MRI (Magnetic Resonance Imaging). I dutifully went along and spent a lonely half hour tucked away in a great big machine having my brain scanned. I felt a bit like hand luggage being checked at an airport, only the process was much slower.

Yesterday Bob looked as though he needed cheering up as he ushered me into his office. He inquired if I'd seen my MRI. 'Yes,' I replied. 'The technician showed me some of his work. The shots were fantastic but seeing my eyeballs in cross-section was quite revolting.' I had been pleased to see that there seemed to be actually plenty of grey matter and my brain had not shrunk to the size of a walnut. But Bob smiled weakly and said, 'Stru, it's not Parkinson's Disease. I'm afraid it's a bit worse. It's Striato Nigral Degeneration or SND for short.' He explained that whereas the malfunctioning area in the brain in Parkinson's Disease is localised, with SND it is widespread. This diagnosis tied everything up neatly; the failure to respond to therapy and the rapid progression of symptoms, particularly deterioration of writing and speech.

Although Parkinson's Disease had been 'invented' by James Parkinson way back in 1817, SND had only mooched into the medical literature sometime after I had graduated in 1960. Clear-cut cases like mine were rare in Bob's experience; he only knew of one other in Melbourne. Looking at the widespread patches of iron pigmentation on my brain scan, which indicated non-functioning areas, I was amazed that overall I was still ticking over reasonably well. However, a few specific questions disclosed that in eighteen months or so I would be somewhat 'uninteresting company'. Bob's advice was to slow down which was a bit of a joke because that was my basic problem.

As doctors do, the conversation then continued as though I 'the patient' had left the room. We agreed that he should start backing out from University life and the 24-hour Advisory Service for doctors, before speech and movement, etc., became an embarrassment. The patient's brain would also be of interest to a distinguished Melbourne Neuropathologist in due course. Armed with a copy of Bob's newest textbook and a print-out of relevant abstracts, I went off trying to memorise the name of my disease. I left him looking a bit more cheerful: 'Just think Bob, at least my brain scan would make a great slide to show at an International Meeting.'

Helped by a tape-recorder two equally important but separate writing challenges face me at the moment. The first is a revision of a 1983 textbook *Australian Animal Toxins*. Long overdue, there is much material in notes and in my head which is potentially useful to doctors and should not be lost. As a positive step, next week I will ask my good friend Dr James Tibballs if he will be my co-author.

The second is this autobiography which presents a long awaited opportunity to present my side of various skirmishes with 'seniors and betters'. Hopefully it will lead to a better understanding of why I took certain actions. To restore some balance, I am aware that bias will no doubt creep into the narrative. After all, an autobiography is an ideal vehicle for self-promotion and subtle shafting of one's enemies!

But best of all, my autobiography provides the opportunity to record glimpses of many fine people who made my life all the richer.

Struan Sutherland
May 1996

The author will not mind if a reader occasionally skips through a paragraph to a subject of greater interest. Hopefully, few will fling the book away in despair. Keep the First Aid section or give it to a friend.

In memoriam:

Ayse Berke (1974–1998)

Part 1

Background and Early Life, 1936–1965

Severely Pruned Family Tree Showing Relations Mentioned in Text

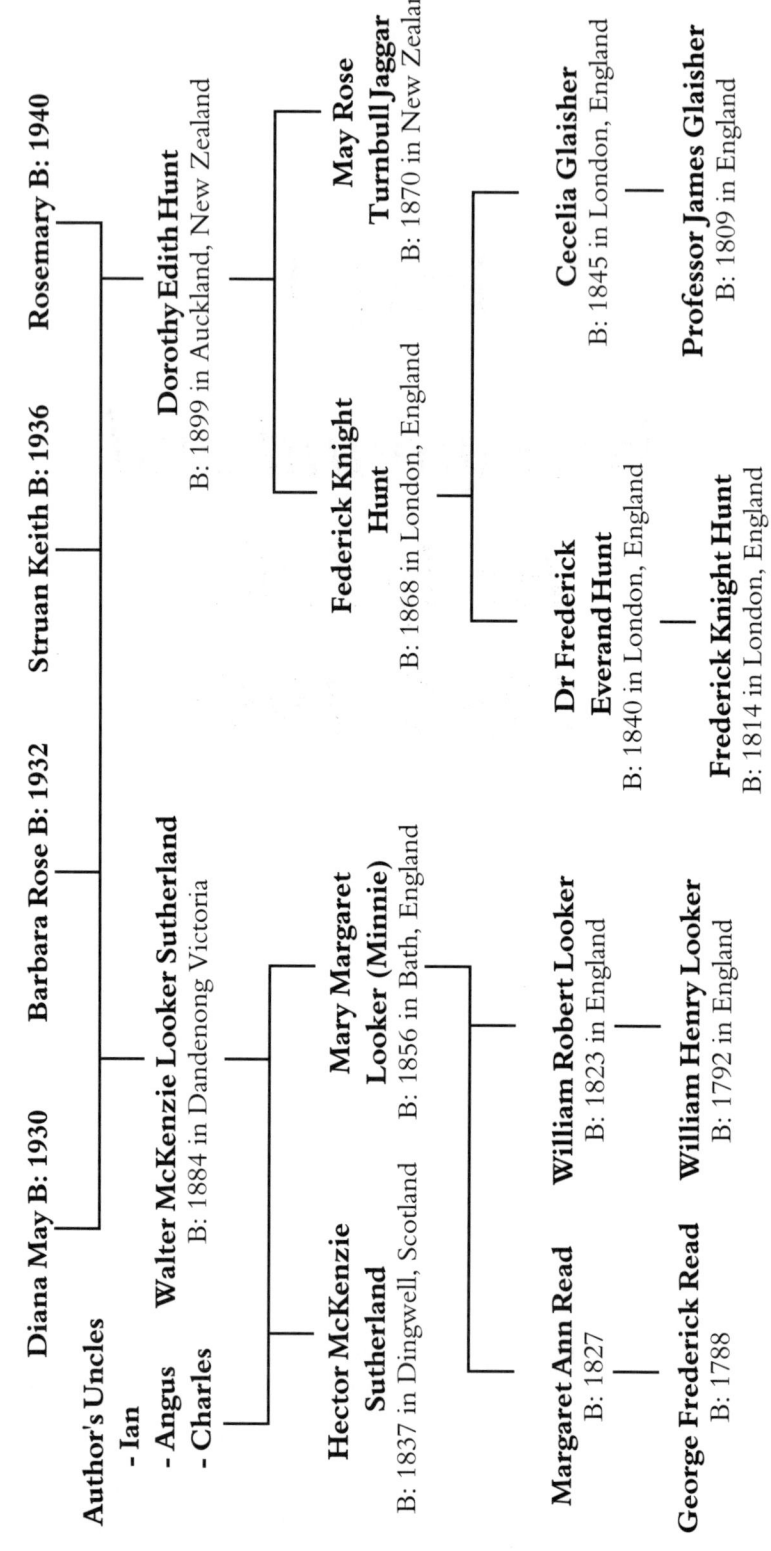

Soldiers, Sailors, Balloonists, Doctors, Bankers and Other Assorted Ancestors 1

'People will not look forward to posterity, who never looked backward to their ancestors.'

Edmund Burke (1729–1797)

Prologue	3
The Old Tin Trunk	4
Grandfather Sutherland	5
Minnie, my Paternal Grandmother and her Fantastically Fertile Family	7
My Father, including a bit about World War I	9
My Mother's Family	12
Tales my mother told me and the hazards of hot-air ballooning	14
Emigrating to New Zealand: an arduous experience	14
Mother's memoirs	16

Prologue

'Is he paid for?' was my three-year-old sister Barbara's rather ominous reaction to my birth. In retrospect, my appearance on 17 June 1936 at 1 Shellcove Road, Neutral Bay, Sydney, was greeted rather coolly by my female relatives. Even the timing counted against me. My emergence at precisely 8.20 p.m. interrupted mother's favourite radio serial. When saying her prayers the following evening, my eldest sister, Diana, aged six, got quite personal. After thanking God for a baby brother, presumably on her parents' instructions, she was heard to observe, *'But God he is ugly, oh so ugly!'*

My father, as usual, appears to have kept his thoughts to himself. He was chief accountant at the Sydney office of the Commercial Bank of Australia and was, at fifty-two, unknowingly at the peak of his career. My mother, who was thirty-six, was not to know she would never see her beloved New Zealand again.

They would have been flabbergasted if told they would shortly commence a collective stint of eighty-five years in the Victorian provincial city of Bendigo. As will be seen, Bendigo offered their son opportunities for activities neither available nor permissible in Neutral Bay. Furthermore, the tranquility of country life afforded him favourable circumstances to reflect and puzzle over all manner of things.

My first name was chosen by my father because he liked it. It was left to me to track down its origins which I did rather publicly in 1970. I was zooming through the tiny village of Struan on the Isle of Skye in a packed tourist bus. Our guide broadcast details of the special significance of the next few seconds to me then stuffed up my moment of glory by observing that Struan meant 'little trickle'! Later he sucked up to me saying 'bubbling brook' was an alternative, but I remained miffed.

My second name commemorates (if that is the correct word) Dad's cousin, Keith Looker, who was killed at Gallipoli. Keith's mother had his medals beaten into a table napkin ring for me which appeared on special occasions. To this day I wish she had just passed the medals over but I guess she thought she did the right thing.

The next bit of personal information I would like to impart to the reader is that I have been married three times and at least two of these arrangements brought me great happiness and inspiration for a significant period. The first to Wendy was blessed with two children, John and Susie, of whom we are very proud. The four grandchildren also seem to be coming on nicely. Hereafter, details of my personal life will only be introduced when they are highly relevant to events being described. Thus, little will be said about my second wife, Megan, and even less about my last wife.

The Old Tin Trunk

I'd like now to delve into some family background. Family histories are generally dull and of little interest to the living unless money is involved—that's a fact. Accepting this, I use the excuse that my parents' forebears exerted a strong influence which affected my upbringing and therefore they should be considered part of my tale.

Once when mother was skiting about a famous relative of hers, Dad took his pipe out of his mouth and in a quiet aside to me said 'She doesn't mention the ones that were hanged'. If any were it's been well covered up, because sadly I can find no convict forebears. Although there is no Saucy Sally Sutherland giving the officers of the First Fleet the hots, I have found a few ancestors who did most interesting things. Some of them I'd love to meet, or at least observe from a distance. A heavily pruned family tree on page 2 may help to identify them.

I'll start with my grandparents whom, although I unfortunately never met, I got to know well via an old tin trunk which sits in the back shed. This trunk is nothing much to look at, but its dark and rusted exterior has staunchly protected its special contents for many years. The inside of the trunk was painted a brilliant blue by hands long dead and is in perfect condition. Each time I open the trunk it reminds me of Tutankhamen's tomb. The contents spook me a bit as every item has been precious and sometimes very private to someone.

For forty years I have added to this collection. Sometimes when rummaging through the effects of deceased relatives, old family documents and bundles of letters, etc., have caught my eye. If no-one else wanted them they ended up in the old tin trunk. Other relatives might have been engaged in more lucrative pursuits like fossicking for a bundle of stirling silver Aunty had tucked away somewhere safe from burglars. However, the bits and pieces I collected were nearly always more fascinating than silver spoons.

'Don't read other people's letters!' so I was taught. But is it okay when they have been dead for fifty or a hundred years? If you have letters you want no-one else to read destroy them now, otherwise some inquisitive prat of a relative may delve into them a hundred years hence. Toss out those uncomplimentary school reports even though they were penetratingly correct. If a love letter shows both you and the writer in the most impeccable light, expressing sentiments which would bring tears to Jane Austen's eyes, then keep it!

Photographs left for posterity should be labelled. There are so many nameless faces, often velvet-framed daguerreotypes, in the trunk. These often sad unidentifiable faces staring out of the past can produce a feeling of melancholy. In fact, to be frank, looking at pictures and reading letters by the long-dead can be a little depressing. It can also be very time consuming and turn the researcher into a bore. Best to be avoided.

Many of the letters, particularly those prior to 1850, I have left unread. They take hours to decipher as often, to make maximum use of the paper, the writer swings it through 90° and continues over the earlier writing. However, some things cannot be ignored.

For example, when taking out some specific Sutherland documents a little envelope fell onto the table. It was labelled 'Dear Little Willy's hair cut 5th October, 1894'. Inside was a little sheath of ash-blonde hair tied with pink ribbon. I found out that Willy had died the previous day, aged 6 years, during an epidemic of meningitis in Dandenong. Meditating over this little package halted writing for a while. We might start with Willy's Dad, who was my paternal grandfather.

Grandfather Sutherland

My grandfather Hector McKenzie Sutherland was born in Dingwall, Scotland, in May 1837. He was very Scottish.

In 1866 at the age of twenty-seven Hector decided to migrate to the colonies to join his brother and a number of other relatives in Melbourne.

He sailed on the famous steamship *Great Britain* which was the world's first ocean going propeller-driven ship.

For the technical minded she had an overall length of 322 feet, a beam of 51 feet and was originally rigged as a 6 masted schooner using sail where possible to save coal. The *Great Britain* joined the England to Australia run in 1852 and her maiden voyage was a record 83 days. (By the time Hector came aboard the average was 60 days.) She was the largest and most luxurious ship afloat and some 630 passengers were carried in four classes with an enormous crew of 137. Later she spent 50 years anchored in Stanley Harbour in the Falkland Islands as a store ship. In 1970 she was towed back to Bristol from where she had been launched 127 years before. The *Great Britain* has now been fully restored thanks to British ship-lovers and is a tourist attraction in Bristol, England.

Hector's treasured documents range from pre-1850 letters from his brother Charles in Australia to an impassioned note from his son Angus pleading for permission to join up during World War I. In the midst of Hector's bits and pieces I found a tough little envelope labelled 'Great Britain Papers'. I'm pleased to report he came out in style: cabin 10, lower saloon, and paid £63 for the privilege—which is big bickies in today's terms.

If the bill of fare for Day 59 was anything to go by Hector would have arrived a plump little Scotsman. The choice included saddles of mutton and jelly, boiled turkey with oyster sauce, tripe cutlets and onions, rissoles of fowls or corned pork and pease pudding.

The unfortunate steerage passengers, who paid £15, had quite a different menu and had to provide their own beds, bedding and utensils, including a keg to hold 3 gallons of water. They were issued a weekly scale of provisions which included 1 lb of preserved meats and soups, 3 1/2 lbs of biscuits, 1/2 lb of fresh or preserved potatoes and, fortunately, 6 ozs of lime juice. Small stores such as ham, bacon and jams could be bought from the purser 'at moderate prices'. Perhaps in case any steerage passengers got any subversive ideas the ship's regulations also stated that 'Gunpowder, Lucifer matches and Combustibles of every description, strictly prohibited from being taken on board by any Passenger'.

From all reports Hector was pretty dour and kept out of trouble. He hated the sound of bagpipes and when he swore no-one could understand what he said. When Dad told me about this I wondered how they knew he was swearing.

Hector joined the Commercial Bank and in 1868 he was appointed Bank Manager of the new Dandenong branch and remained in that position for the next thirty-seven years.

Fortunately, his standing must have been high at the time of the land boom crash when most banks failed and unemployment rocketed. A public meeting was held by the inhabitants of Dandenong and surrounding districts on Friday, 14 April 1893 with the aim 'To express sympathy with H. M. Sutherland, JP, Manager of Commercial Bank, Dandenong, in his present unmerited troubles'.

When my grandfather Hector retired in 1902 an impressive collection was taken up by the local citizens and he was presented with an illustrated address, a fine Cutler desk and other items. In the desk, the canny Scot stored away a detailed list of the donors and their donations.

For forty years he was an Honorary Justice of the Peace and sat on the bench regularly until a week prior to his death in 1918. Hector's obituaries describe him as a Scot/pioneer banker and a good citizen.

Minnie, my Paternal Grandmother, and her Fantastically Fertile Family

Hector's somewhat austere lifestyle and singleminded devotion to the bank changed quite suddenly one evening in 1880. The forty-three-year-old Hector was asked to keep an eye on Miss Minnie Looker, aged twenty-four, during her brief stay in Dandenong. The story is he liked her shapely ankle as she alighted from the carriage and even he must have managed to smile at least once, because they were duly married on 17 August 1881.

Whereas available records of Sutherland family history are sparse, Minnie's make up for this deficiency. There were some spectacular breeding pairs. Minnie was the fifth of twelve children born to Margaret and William Looker and Minnie's mother, Margaret Ann Read, born in 1827, was one of fifteen children born to the wives of Captain George Frederick Read (1788–1860).

George was the sort of ancestor one would really liked to meet. He went to sea at the age of eleven, most probably with the East India Company. After serving as an officer in that company's war time service, he became master and part owner of the Brig *Lynx*.

George visited Hobart in 1808 and again in 1812 when, because of a food shortage, his cargo was comandeered. This action, as well as having his ship placed on rations, was said to have greatly irritated him. The *Lynx* is thought to have been the first merchant vessel brought through the Torres Straits. The ship's log in Sydney's Mitchell Library records voyages all over the place including Calcutta and Batavia (Jakarta). Also, he didn't neglect the west, calling at the Swan River in 1816.

In 1817 George decided to live ashore in Sydney. He became a merchant and obtained a grant of land. However, he was affected by asthma in Sydney so he went to Hobart with a recommendation from Governor Macquarie and shortly afterwards obtained a grant of 800 acres of land with government servants. He must have been a live-wire because he was highly successful as a merchant, pastoralist and banker and was described as 'one of the most important men in the colonies' formative years'. Owner of several ships, he traded for years between other parts of Australia, India, China and the East Indies. He had four large properties, including Redland's where the family established the famous salmon ponds. George was the founding proprietor of the Van Dieman's Land Bank and its first Managing Director (1827–49). At Melbourne's first land sale in 1837 he bought some city sections in Queen Street. Very impressive but not a sniff of his fortune trickled down to my generation because the Reads all bred at a furious rate and my grandmother was only one of twenty-five or so grandchildren. He died in 1860.

My grandmother Minnie's father, William Robert Looker (1823–1900) came from pretty fertile stock, too, being one of nine children. His father, William Henry Looker (1792–1872) served at the age of twenty as an officer at the battle of Waterloo and at his death was one of the last surviving officers of that historical engagement. In 1835 he was appointed Deputy Commissary-General of Van Diemen's Land, a post he held until 1850.

William Robert Looker was born in London and was working with his father in the Commissariat when he married Margaret Ann Read in 1848 at St John's, Hobart town. After the elder Looker had returned to England in 1850 the young couple moved to Victoria and by 1854 they had purchased six Victorian properties totalling 64 000 acres. In 1855 they sold up and went to England with their four children. William Senior now owned a business in Pall Mall supplying 'the needs of the empire's officers' (the mind boggles). He was elected a Fellow of the Royal Geographic Society in 1861. He also gained four more grandchildren over the next few years. The first one was our Minnie (baptised Mary Margaret) who was born in Bath in 1856. In 1866 the whole tribe came back to live at Edgewood, a property in Oakleigh, Melbourne. By 1871 Minnie had eleven brothers and sisters. W. R. Looker also took up properties in New South Wales and Queensland and in 1877 established the firm of W. R. Looker and Sons, Auctioneers and Stock Station Agents.

Thus, by the time Hector helped Minnie out of her carriage his future father-in-law was doing pretty well and Minnie had hundreds of Read and Looker relatives.

My paternal grandparents, Minnie and Hector, soon settled down and produced Sutherlands. First was Ian McKenzie in 1882, followed by my Dad Walter, then Barbara, Angus (actually Hector Aeneas), Margaret Flora and finally Charles. Little Willy came along in the midst of these but was lost during an epidemic.

In 1884 Hector purchased a property which had a frontage of over a quarter of a mile on to the main Melbourne–Dandenong Road. He built Novar which was later described as a 'modern villa' of ten rooms with the large verandah and outbuildings usual for that period. Dad showed me the house in 1960 when it had been divided into three flats. I was quite impressed when we approached it by Sutherland Road and then drove past Novar Post Office.

Minnie died in 1936 shortly after I was born. I remember my mother one day waving a little singlet at me from the rag cupboard and saying that Granny Sutherland had died when she was knitting it for me. This gave me an uneasy feeling that I was implicated in her death.

My Father, including a bit about World War I

My father, Walter McKenzie Looker Sutherland, was born in Dandenong on 2 June 1884 and died in Bendigo on 12 October 1973. I was very fond of him but until his funeral I was unaware of how many others felt the same.

18th June 1901

Dear Sir,

I beg to apply for an appointment in your bank as a probationer. I am 17 years of age and I am about to leave school. My teacher and my father can testify as to my fitness and good conduct.

Yours respectfully,
Walter Sutherland.

So wrote my father to the General Manager of the Commercial Bank in Melbourne. No doubt it was written a number of times before Hector approved the final letter. Walter got the job and served the bank for forty-seven years during which time he had less than fourteen days sick leave.

By 1912 he was the youngest Bank Manager in Australia. This might sound impressive but the bank was in Darwin. The experience seemed to have been an enjoyable one despite the isolation, lack of refrigeration, etc. The photos of the period are real 'outpost of the empire' stuff: pith-helmet, white tropical rig and worried looking aborigines.

Walter enlisted in 1917 despite the experiences of his recently returned younger brother Angus (page 53). He saw action in France from July 1918 serving in the 46th Battalion AIF. His father Hector had died early in April but judging by Dad's diary it was two months before he heard the news. Reading the correspondence I am moved by the affectionate closeness of the family and the clarity with which they expressed their feelings. As happened years later when my sister Barbara was killed in England (page 90), letters to and from the deceased person continued to arrive long after the death.

The Germans made serious attempts to kill Dad but his luck held out. Although, like me and my son John, he could at times be stunningly accurate with a rifle, it's hard to imagine him shooting to kill. He was such a polite and basically gentle person. Rarely were his war experiences ever mentioned but there is no doubt he had been deeply affected. Like many veterans he had occasional nightmares all his life. He was a tireless and generous worker for Legacy. I remember being miffed as a twelve-year-old when I discovered that he was providing pocket money to a War Widow's son at double my meagre rate! Dad seemed somewhat bemused by my 'charity begins at home' approach and promptly found me a Saturday morning job as a chemist's delivery boy as a solution to the problem.

One of the old tin trunk's most treasured items is Dad's war diary. It was wounded quite severely by shrapnel and the damaged leather cover gives off a pleasant fragrance. Now my most precious possession, I was unaware of its existence until a few years ago.

In his neat copperplate writing he pencilled a day by day account of what a Private went through. The keeping of such a diary was illegal for security reasons but many of the entries merely record how far he and his mates marched and how hungry they were. Let's look at some examples:

Saturday August 24, 1918:
On fatigue this morning. In the afternoon went to Corbie with Bourke who called in on a friend of his (Lieut Freer) who treated us to stout. At Corbie saw great traffic transport and several batches of German prisoners. Captured Hun gun too big for bridge and almost fell into canal. Greatly enjoyed watching two caterpillar tractors pull it out of danger, in doing so they stood on its muzzle.

Monday September 16, 1918:
Had rough bath—change of u/clothing. All ready to go forward at 6.P.M. Reached trench at 8.P.M. and was shelled all night in poorly sheltered trench—4 killed. Tunic blown up and hence damage to this diary.

Tuesday September 17, 1918:
Moved to some better dugouts at rear of trench. Blankets are being called in. Looks like a hop over soon. Everything ready for action. Breakfast to be at 2.A.M.

Wednesday September 18, 1918:
Marched out through terrific barrage at 5.10.A.M. and advanced 6,000 yds through shell fire. Held by M.G. fire until 10.30 when fresh barrage silenced enemy and we gained our objective.

Thursday September 19, 1918:
Holding outpost trench ... Manned fritz M.G.

Friday September 20, 1918:
Ditto.

Saturday September 21, 1918:
Relieved at 9.P.M. Got out of trench but when 2 miles out aeroplane dropped 7 bombs near us. Marched about 14 kilo to Tincourt reaching there at 4.A.M. Sunday morning.

Monday September 23, 1918:
Awakened at 5.30 by shrapnel close by. Two horses killed but no other harm done.

Wednesday September 25, 1918:
Spent in getting mud off clothes and gear. Handed in mutilated pay books. (*A laconic note.*)

Saturday September 28, 1918:
Marched to Pissy a village 3 kilo, and had first bath since coming out of trenches. Got wet through coming back.

Wednesday November 6, 1918:
Cabled mother 'happy returns twenty-fifth well'. Cost 17.30 francs. (*Explanation—birthday greetings*)

Monday November 11, 1918:
Amnesty signed at 11 o'clock today. Great rejoicing in Fluey. Aussies rang the Church bells from 3 to 6.

Sunday November 17, 1918:
Big Church parade. Very cold, ground beginning to freeze.

Saturday November 23, 1918:
Marched on for about 27 kilo no food from 6.30 until 6.P.M. Many fell out.

The diary recalls deaths of friends as late as 27 September 1918.

In 1919 Alf L'Hotellier published a book of poetry which included the poem 'My Comrades'. This poem was dedicated to Dad and his friend 'Con'. It is a fair poem and includes the lines

> *They were soldiers of Australia,*
> *Loyal in the strictest sense,*
> *For they'd thrown up good positions,*
> *For a paltry sixty pence ...*
> *But I never heard them grumble once,*

> *My comrades 'Wal' and 'Con'*
> *And when the final battle's*
> *being fought and truly won,*
> *They'll be foremost in the fighting,*
> *Will my comrades 'Wal' and 'Con'.*

Dad got back safe and sound to Melbourne in late 1919 and was given a number of relieving managers jobs by the bank. During this time he helped to fund his younger brother Charles' medical course at the University of Melbourne.

He had a lucky escape when he was relieving manager of the Camberwell branch. The day after he left, the resident manager returned and was shot dead by a couple of thieves who were later executed.

In March 1920 there was a flurry of correspondence between Dad and the Chief Inspector of the bank. The inspector pointed out that Dad had been passed over for promotion a number of times recently because of his desire to remain in Victoria. Presumably this was to be near his family.

After being pointedly asked whether he would accept an appointment 'in any of the other states and in the Dominion of New Zealand', he replied he would be prepared to proceed to those places with the provision that 'I would prefer to, at present, avoid residence in the tropics'. Two weeks later he was appointed accountant at the Auckland branch with a £50 pay rise, bringing his annual salary to £330.

He set off with a light heart, confident that it would be more pleasurable than the trenches of France. His return to Australia some years later was far more complicated—he had a wife and two kids in tow.

My Mother's Family

At the beginning of the twentieth century my mother, Dorothy Edith Hunt, was two-months-old. Born on 26 October 1899 she had a good run, lasting until 22 May 1992. She was fair complexioned, red haired, a little above normal height and a very alert person. Like my late sister Diana she had a remarkable memory. Mother had an opinion on everything and a modicum of good ol' fashion Victorian bigotry which at times could be startling. She admired Bob Menzies and Gough Whitlam. Ten years before he became Prime Minister she had a budgerigar named Johnny Howard because, she explained, it never stops talking.

Mother was very proud of her family background. Her father, Frederick Knight Hunt (1868–1945) was for twenty-five years the stipendiary magistrate and city coroner for Auckland and the surrounding area. Obituaries describe him as 'one of the most colourful legal personalities in the dominion and informal, shrewd and noted for explosive wit on the bench'. Known as 'Freddie' Hunt he had wider powers than an English county court judge. Things moved very quickly when he was presiding and sometimes information was extracted from a witness when they were only halfway to the witness box.

Freddie must have known his law because the supreme court never had an occasion to change one of his sentences. His verdicts were often accompanied by comments that probed social sores and called attention to facts that might otherwise have gone unnoticed. He was involved in many famous cases such as the Cooper 'Baby farming' Case which he allowed my mother to attend. Mrs Cooper had been paid to raise a number of orphaned or abandoned infants. The profitability of this venture was increased by the murdering and burying of many of her charges. Mother never forgot the sight of rows of shoeboxes filled with the bones of tiny babies.

An extract from Freddie's unpublished autobiography recalls a ghastly murder.

> One of the most horrible crimes in Mr Hunt's experience occurred in Christchurch. Two nice little Scotch girls had just arrived in New Zealand which they had heard about favourably. They were school mistresses but being unable to find work in their calling, they got jobs as waitresses at a Boarding House. One of them received overtures from a 'sheep farmer', who was really a crook just out of jail. He was always after her so that she finally married him, writing to her people that she had married a sheep farmer.
>
> The poor girl had little to be glad of, for the man proved himself to be a terrible creature and she was compelled to leave him. This was not to be the end of his attentions, for he still persecuted her, writing letters asking her to meet him. She decided to meet him in a park, and there he certainly fulfilled his appointment, bringing an axe with him with which he killed the Scotch lassie. He then poisoned himself.
>
> The most revolting part of this tragedy is, perhaps, the strange way the murderer left a complete record of his intentions regarding the girl. He wrote that he was looking forward to seeing the fear in her eyes as he carried out his designs, which were fully set forth in papers found on him afterwards.

Freddie Hunt was a splendid source of copy for court reporters. As one of her earliest motorists and a coroner he was especially interested in road safety and framed New Zealand's right-hand rule. On the other hand, Dad, who got on famously with him, recalled being startled by the way he would occasionally flaunt the law. Once the pair arrived at a railway station in a hurry and Freddie parked illegally, barking 'Police!, Police!' at the startled railway attendant. I can just see my little Dad looking over his shoulder as he followed in the wake of the imperious Freddie.

Tales my mother told me and the hazards of hot-air ballooning

Freddie had been born in London. He was the first son of Dr Frederick Everard Hunt (1840–1900) who married Cecilia Glaisher (1845–1932). Cecilia had had chilblains and consulted her local doctor; obviously they got on like a house on fire. There was opposition from Cecilia's family because, although Frederick was a member of the Royal College of Surgeons, in those days doctors generally were not of high social standing because they worked for a living. Dr Hunt's father (Frederick Knight Hunt, 1814–54) was also a doctor but became a journalist and as editor of the Daily News gave Charles Dickens a job.

Freddie's father had been educated at University College School and Guys Hospital but Cecilia's dad outranked him academically. Professor James Glaisher (1809–1903) was the mainstay of the Royal Meteorological Society and a famous balloonist, having in 1862 reached the greatest height yet recorded. He and Henry Coxwell passed the 26 000 feet mark and both lost consciousness due to lack of oxygen. All good books on the history of ballooning have an etching showing my great-great-grandfather unconscious in the basket while his plucky assistant manages to open a valve with his teeth, his frost-bitten blackened hands being useless, before he too falls unconscious. Some of the technicians I have worked with over the years would think nothing had changed. They do all the work and the boss gets the credit!

Apart from aeronautics, Professor Glaisher was active in astronomy and the mathematical sciences, and he must have dabbled in photography because in 1887 the Fellows of the Royal Photographic Society commissioned a bust of him. In fact two were made. One is now held by the Royal Meteorological Society and the other is currently looked after by my younger sister Rosemary. At home we always kept a party hat on Jimmy (as my Dad called him) as this made him look a little more cheerful.

Emigrating to New Zealand: an arduous experience

After Cecilia and Dr Hunt, my mother's paternal grandparents, were married he practised at St Ives in Cornwall, which was about as far as the couple could get from relatives and yet still be in England. A glance at the family tree on page 2 may be helpful. In 1880 they set off with their seven children to New Zealand in the *Edwin Fox*. The family bible records that Mary Cecilia, aged six years, died at sea on 24 January. Mary died of fever, as did two other children during this voyage which was described as one of hardship. It is a doctor's nightmare to lose a patient who is also a close family member. The *Edwin Fox* was built of teak in 1853 and this was her second-last passenger carrying voyage. Like other teak vessels conditions were somewhat primitive compared with the newer ships. (Grandfather Sutherland had a far more comfortable and faster voyage to Melbourne.)

The Hunts were amongst the twenty salon passengers, there being twelve second class and seventy-seven steerage passengers, who arrived in New Zealand on 3 May 1880.

Although this 1880 voyage had been uncomfortable, it was nothing to earlier trips to New Zealand made by the *Edwin Fox*. On the first voyage, the crew got fighting drunk and was unfit for duty when a strong gale hit the ship. To make matters worse, the vessel sprang a leak but was kept afloat by relays of passengers manning the pumps. Whilst helping, the ship's doctor was impaled on a metal rod and killed.

A miraculous event occurred during this crisis when a young girl was washed overboard but luckily swept back by the next wave. She was snatched to safety by her father and seemed none the worse for her unintentional swim. When the ship eventually reached dry land all the crew were arrested and returned to England, where they received six months hard labour. The passengers then had to wait for another six weeks before the ship was seaworthy and there was a serious outbreak of scarlet fever which resulted in a number of deaths and a period of quarantine before disembarking in New Zealand. Her third voyage to New Zealand was longer than expected because of a collision and a drowning. Five other deaths and six births also occurred on the voyage so the books were balanced.

Everything considered, Freddie Hunt had one of the better trips to New Zealand on the *Edwin Fox*.

Dr Hunt promptly set up a practice in Christchurch. Faded photographs of his surgery and dispensary show it well stocked both with thick books and an interesting collection of jars and bottles of medicine. He was a good doctor and founded the Christchurch Medical Society. There were some touching tributes by patients following his sudden death aged fifty-nine in 1900.

For example a memorial card bore a touching poem and the following:

Contributed by a few of his lady patients of the Woolston and Lynwood Districts as a mark of their appreciation of his skill and kindness during the many years he practised amongst them.

The press concluded his obituary by saying 'his kindness of heart made him very popular'. Cecilia lived on for many years and my mother remembered her with great affection.

On 7 January 1889, my maternal grandfather Doctor Hunt's son, also called Freddie, then a young barrister, married May Rose Turnbull Jaggar and my mother appeared in October. May Rose was the youngest of eleven children and her parents were the first Headmaster and Headmistress of the Timaru High School. Henceforth May Rose will be known as Grandma Hunt. She died in 1944.

Of the relatives I have been telling you about Grandma Hunt is low on my meeting list for the next life. She comes across as a hard selfish person who became formidable as time passed.

Mother's memoirs

About 1960 my mother settled down at her old typewriter and belted out a draft autobiography. Some excerpts are informative.

Her parents first met at a Ball and mother reported as follows: 'Father had the supper dance with mother and was so nervous in her presence that he spilt the sweet he was handing her, to wit, some form of jelly, on the floor.'

Dorothy wrote extensively about her school days, which she enjoyed, taking the discipline in her stride. Of St Margaret's School, Christchurch, in 1909 she writes: 'St Margaret brought us up in a very strict manner, no more than 3 girls to walk together at a time and only two talk. No-one must be seen outside the gate without gloves.'

And on etiquette: 'We were taught how to sit down gracefully, arise from a chair silently and how to nod to people in the street etc.'

On a lighter note mother writes: 'I ran into bad trouble for listening to adult conversations. When my parents were talking at the end of the day I often heard my father saying "Oh, I saw Tom W. today. His wife is not well, she is in the family way." I had not a clue what it meant but Aunt Hepsie came to stay and was in bed one day not feeling very well when I brightly said, "Perhaps you are in the family way." My Aunt being a confirmed spinster was furious and mother after smoothing things over, soundly boxed my ears. I was sent to my bedroom in tears—it was so unfair I was only trying to be helpful.'

'One Christmas our chinaman who supplied our vegetables gave us a huge packet of crackers. Mother said: "We must do something with these." At that time we had two lavatories, the outside one being known as Father's Dub. While one morning we waited till father was comfortably settled on his "Dub" when mother who had a spade with a long handle thereupon placed the crackers (truth to tell we were all frightened of them), lit them and pushed the spade under the door—there was a great explosion and smoke poured out everywhere—we were so afraid we had killed Father. After quite a considerable time he emerged out of the smoke and walked quickly away but we saw the smile on his face. We were crippled up with laughter and for many years when we talked about it we would start laughing. Poor man he must have got a fright.'

'Father was very proud of his moustache which was a ginger colour with long handle-bars greased at the end. Mother longed to have him with just a small moustache but he would not hear of it—so one afternoon when he was asleep mother crept up and snipped off one of his handle-bars. There was a slight storm but in the end father laughed. Father was very good tempered and thereafter the moustache was a small one.' *(The above two incidents might be grounds for justifiable homicide.)*

Mother left school to work in her father's office which was desperately short staffed as World War I dragged on. She enjoyed working there and her responsibilities were increasing until her mother saw an advertisement for a typist at the Auckland branch of the Commercial Bank. Grandma Hunt successfully applied for this better paid job on behalf of her daughter and then increased her board! Mother had been working there for several years when the accountant whom she disliked intensely was replaced by a Mr Sutherland from the Auckland branch.

At this time there was a shortage of eligible males for girls like my mother. Nonetheless, the thought of a romantic attachment with my father did not cross her mind until he made the first move. My mother was tall, buxom, with bright auburn hair whilst he was short, near bald and fourteen years older. Walter was quite nervous on their first date and the large box of chocolates he had bought was accidently left under the seat where he had put his hat.

After meeting Walter, Grandma Hunt summed up her impressions for the benefit of her daughter as: 'A nice little man, if he was better looking he would have been snapped up long ago.'

Whereas we don't know my father's views on this lady, he and Freddie got on like a house on fire.

In contemporary cartoons Freddie Hunt at his peak bore a striking resemblance to Horace Rumpole. Imagine Horace being unwillingly dragged to Church by Hilda to hear a very vocal preacher then popular amongst her social set. Mother wrote, 'Canon Burton arrived from England ... he was a wonderful preacher who breathed fire and brimstone and would get so worked up that perspiration would run down his face. One Sunday night we took father along to hear Canon Burton preach, but father was not a success in Church. He would count the verses in hymns out loud, become restless, and ask what we took out of the plate.' His comment about Canon Burton was 'the man's an actor'. (*How I wish I could have met my grandfather!*)

Freddie Hunt loved trout fishing. This allowed him to escape on the weekends to his cabin accompanied by a like-minded female companion. In his fishing box he took a very sharp knife which had been used by a murderer to dispatch his wife. He always referred to it as Mr Guppy's knife. I have often wondered what happened to forensic evidence after a trial.

Anyway the combination of my grandfather Freddie and my mother more than balanced out Grandma Hunt, and mother later reflected on Dad's proposal as follows:

'I began to think about Mr S. Home life was not too bright, mother was constantly nagging me, always holding up Molly's beauty, etc. and I knew Mr S. was well bred and everyone liked him in the bank, and I also thought that he would make a good husband as he was so thoughtful. But the bald head was hard to take. So I wrote saying I would marry him and he wrote father a very nice letter asking for "my hand" in the true Victorian style.'

'My wedding was arranged for 8 a.m. as Mother did not want a lot of bother as she pointed out to me I was marrying an old man so the quieter the better.' *(The old man was 42, bride-to-be 26.)*

Later when she was near seventy Grandma Hunt visited my parents in Sydney: 'She did not like Sydney, nothing was as good as Auckland and she told me I led a dog's life, actually I was very happy and very proud of the children. Mother had taken to smoking—all her life anyone that smoked was "bad" (women I mean) and there she was smoking like a chimney. Even had a supply of different holders and was so proud of her achievement. Suthy saw her off at the boat when she returned to Auckland. It was the last time I saw mother.' *(I'll bet Dad was whistling on the way home!)*

In 1935 my father was sent back to Sydney and my parents established themselves there with two little daughters, Diana May aged five and Barbara Rose aged three.

A visit in the spring of 1935 to Melbourne, where they stayed at my Uncle Charles' rather impressive house (page 55), allows me to pinpoint the site of my conception. I emerged dissatisfied as ever in my parents' bed in Sydney the following June.

A few weeks later my parents' pride in their red-faced puking son and heir was replaced with the prospect of his likely departure due to pneumonia. Margaret, one of Dad's spinster sisters, came up from Melbourne to help. Margaret was one of the first infant welfare sisters and was used to being obeyed on matters outside her experience, a situation akin to a celibate priest's advice on the intimacies of the marriage bed.

My mother's difficult situation is best summed up by a comment she made at the time: 'Living with one Sutherland is bad enough, but two is intolerable.' Just as she was consulting shipping movements to New Zealand, gravity solved the problem when Margaret fell over and broke her arm. She went back to Melbourne and her departure was described as: 'In plaster, spitting chips through clenched teeth.' It was at least ten years before we saw her again.

Dad's position in the Sydney bank was an important one and, with his usual optimism, he was awaiting the 'big appointment'. It was a blow to all concerned in 1939 when the expected elevation became a sideways shunt. Dad was appointed to manage the branch in Bendigo, Victoria.

My paternal grandfather, Hector
McKenzie Sutherland, looking benign

Minnie, Hector's wife, who was greatly
loved by her children and was always neat

My dad as Australia's youngest bank manager in Darwin, 1910 (middle front row)

My mother, Dorothy Hunt, facing life expectantly

My dad as a troop, 1917

My dad after becoming a father in 1930

This proved to be his final appointment until retirement in 1948.

The move to Bendigo was an especial blow to my mother as it took her even further from her friends and relatives in New Zealand. After New Zealand and the harbour views of Sydney she found the scenery around Bendigo rather dismal. On the positive side she soon built up a large circle of friends which she subconsciously divided into two groups: those with breeding and those without. I noticed that some of my friends whom I thought she would classify as 'without' were strangely 'reclassified' because they had good teeth.

World War II commenced, followed by the arrival of my younger sister, Rosemary, on 17 August 1940. The first memory I have at the age of four is of seeing Dad taking mother off to hospital. He did this at dusk in the big Morris 6 which had a splendid chrome radiator. He had bought it from his brother Charles and it regularly broke down. This time it didn't, and the next thing I remember there was a bassinette in the dining room with hoards of people milling around admiring its contents.

From now on, I was not only filled with jealously, but I was hemmed in by sisters. My new sister Rosemary had huge blue eyes and could do no wrong. I became a preschool resistance fighter and sought out other four-year-old males suffering from an excess of sisters.

One bright moment occurred during Rosemary's baptism. I was sitting next to Dad. This would have been one of the few occasions, apart from funerals and including his own, that I knew him to go to church. He was dragged along once to see me as a choir boy (more of that later) and to the odd confirmation. Anyway, I was sitting next to Dad, who was obviously feeling the need to smoke his pipe, when there was a sudden bout of wailing, presumably from Rosemary, off centre stage. It really was a terrible sound so I asked what they were doing to her. 'Cutting her throat', came the terse reply. I sat there quite numb cogitating on this unexpected turn of events. I glanced at Dad but he was looking fixedly forward. Later I found out from mother that he was not too thrilled about becoming a father again at the age of fifty-six.

Growing Up 2

'Wait till your father comes home!'

Often said by Mrs Sutherland of Bendigo
in the immediate post-war years.

Prologue	21
Camp Hill State School: Bendigo's Colditz	22
Recreation	26
Drownings and Violent Deaths	28
Tradesmen and Hawkers	30
Bendigo at War	31
A Burst of Religion	32
Sex and Baldness	34
Choosing a Career and Observing Local Doctors	35
Chemists and a Full-time Job	38
Bendigo High School	40
The school assembly	40
Teachers	43
'Sloid' and other subjects	45
Bombs and Electric Shocks	46
Creative Activities	48
Tennis	49
The Escape	49

Prologue

From my first memories at the age of four until I was eighteen, I was basically restricted to the City of Bendigo and its environs. By 1954 I was becoming dangerously bored, having explored all obvious avenues of interest. Fortunately, it was possible to escape to Melbourne before getting into real trouble.

Situated in central Victoria, Bendigo was a good spot for a leisurely childhood. Gold, the foundation of this very Victorian city, was still being mined. Rows of poppet heads with their mullock crushing sheds followed each deep and still-rich reef. Nevertheless Bendigo was a pretty city and tourists admired its leafy streets and many classic Victorian buildings. It would be hard to beat the famous Shamrock Hotel with its beautiful wrought iron verandahs. Fortunately extremes of weather are uncommon and the air is clean. Indeed on a crisp spring day, provided one has the option of transport elsewhere, Bendigo can be wondrous.

The 30 000 inhabitants mostly knew their place in a society that had few secrets. Second or third generation Bendigonians lived in mansions high above the city centre. Electric trams served the population well and some endure to this day.

What follows are the memories of people and events in my childhood. Some had a lasting effect upon me, others are just stories I enjoy telling!

Camp Hill State School: Bendigo's Colditz

Whoops! I messed up my first day which proved very tiring. I ate my lunch at playtime and later assumed lunchtime signalled the end of that day's instruction. On my triumphant return home Mother was distinctly unwelcoming. She marched her four-year-old son back to school (see map) with no additional lunch. No-one had missed me. 'Wait 'til your father gets home,' had been Mother's warm greeting.

Camp Hill State School still stands as a classic example of monstrous school architecture. Built last century, parts of it resemble a prison. It has narrow windows and cold stone stairways which spread through its rabbit-warren like interior. The whole structure is topped with a cruel-looking tower. It used to sit on an asphalt-coated playground which sloped at 30 degrees. As a result, kicking footballs was either done steeply uphill or recklessly downhill. The whole building and its surrounds seemed designed to cause injury to a small boy's kneecaps.

Something about the slate roof and the tower used to give me nightmares. I dreamt of being stuck on the roof three storeys up, hanging on by my fingernails to avoid crashing to the asphalt playground below (see picture).

The amenities were not impressive either. The infamous shelter shed was also made of bluestone and was unlit. The toilet blocks squatted at the bottom of the asphalt slope and in wet weather were constantly awash.

During the 1940s there was plenty of poverty in Bendigo. Some families with six kids or more lived in tiny miners' huts which might have a primitive extension built out of corrugated iron and bits and pieces removed from old mining buildings. State school photographs showed the kids often dressed in obvious hand-me-downs, either too big or too small for them.

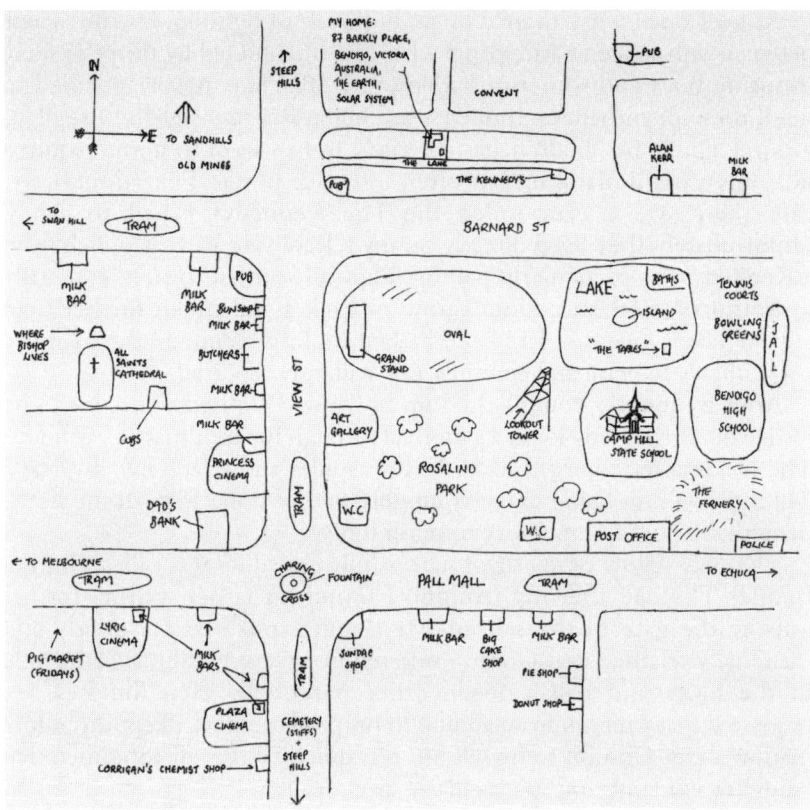

Patches on trousers were quite acceptable and workman-like boots were the usual footwear, often even being worn by the girls. Jam sandwiches seemed to be a routine lunchtime fare. The most popular type of jam was an indeterminate dark purple which had considerable staining properties. Bread and dripping sandwiches were also in vogue. The more fortunate pupils had some of the flavoursome lower parts of the dripping jar mixed with the pure fat.

The fortunes of some families showed marked fluctuations. A bookie's son was a classic example. On one occasion he arrived at school with a new bicycle which was an extremely rare item at that time. However, three weeks later he was not only minus the bicycle but absent from school for a week. The reason he gave was that he had no trousers to wear. This excuse pleased the teacher no end, since apparently he was an unsuccessful punter.

In those pre-antibiotic days, a number of primary children suffered what was probably chronic sinusitis. They would be completely mouth breathers with both nostrils clogged up with green material. I observed this gunk oscillate up and down a fraction of a millimetre or so when they were earnestly making a point.

At least once a day there'd be an outbreak of fighting. Usually it was between unbalanced participants who'd be surrounded by thirty or forty shouting boys and sometimes a few girls. Teachers never appeared to quell these disturbances. Indeed a teacher was a rare sight in the playground. For some children, getting safely from school to home required not only careful planning but often intensive physical exertion. In my case there was a clan called the The Kennedys I had to avoid. Unfortunately, they lived directly on my schoolward path. A punch from a Kennedy was particularly painful. They all seemed to have especially powerful muscles and could throw or kick a ball much further than anyone else. Once one had been bashed up by a Kennedy, the memory was difficult to erase and certainly not willingly repeated.

After frequently being belted up by Trevor Kennedy, one day I saw justice done. Kennedy kicked a football through the headmaster's window. The headmaster emerged like a furious spider and seized my adversary. The sight of him being dragged roughly up the stone steps to the headmaster's sanctum is a memory to relish forever.

The only fellow of my size I successfully belted up was called Gordon Bennet. The day after my triumph I found his father waiting for me outside the gate of the school. Mr Bennet spoke to me gently and earnestly explained that Gordon suffered from severe asthma, had a hole in the heart and half-a-dozen other potentially fatal illnesses. He requested that I refrain from adding to his predicaments. I kept this information about Gordon to myself but was quietly rather disappointed that he didn't suddenly and spectacularly drop dead.

I was seven years at primary school because I failed Grade 3. This came as a surprise to me even though I'd spent most of my time gazing out of the little window by my desk. My daydreaming had gone largely unnoticed, there being some forty or fifty other children in the barrack-like classroom.

It's interesting in retrospect to contrast the behaviour of the female teachers with those of the males. The men were all fairly subdued, grey sorts of chaps. They were all married. The female teachers were all unmarried and prone to sudden and unexpected outbursts of violence.

Take, for example, Miss Tilley my Grade 3 teacher. She could be triggered off by something as minor as the presence of a pencil case on one's desk. These had to be tucked away out of sight. On one occasion there was a sudden hiss of breath in my ear and my pencil case took off in a spectacular fashion. Miss Tilley projected it at almost zero altitude across the schoolroom neatly into the fireplace in the far corner. I was still sitting amazed at this aerodynamic and athletic feat, when a large ruler crashed down on my hands.

This was one of the few occasions when the fire was operative. Its purpose was to warm up a container full of water in which sat small bottles of milk. They were of little interest to me at that time as my milk money had been suspended because I had been caught spending it on licorice blocks at the little shop on the way to school.

If Miss Tilley was formidable she seemed a mere mouse when I moved upwards to the class of Miss Wells. She was a raw-boned poker-faced woman who ruled her class, not with a rod of iron, but a length of cowhide. Her eagle-eye would assess the impact of the strap on the victim's hand with the concentration of a professional golfer. If the hand faded away to absorb the blow another was delivered.

Swearing especially was a trigger point for Miss Wells and the offender would be suddenly grabbed from his or her desk and marched to the neighbouring washroom. Once there yellow soap was shoved into the mouth and rubbed on the tongue. This happened to me once and I had no idea what I'd said, nor did I dare ask. Natural justice and the laws of assault meant nothing to Miss Wells.

In Miss Wells' year I joined the Cubs. This new activity was initially quite enjoyable—I liked the uniform and I admired the collection of badges the veterans wore with great panache. Meetings were held in the local parish hall and supervised by a young and pretty cub-mistress. Cubs at times were extremely noisy so I would go outside now and again for some peace and quiet.

The only cub-lore I absorbed was a variety of calls and yells, some of which went 'dib, dib, dib', etc. Occasionally Mother would ask me to entertain visitors with a performance of these calls. Later I refused point blank to do so.

One night a more senior cub-mistress arrived and pulled the whole class briskly into order. She announced that Grey Wolf would be visiting us next week and we should get our act together. We were given some extra training and ordered to dress immaculately the following week.

The important night arrived and for the first fifteen minutes we were inspected as we nervously awaited the arrival of Grey Wolf. Suddenly, in came this large grey-haired woman covered with insignia. What should I see under the hat, but the steely eyes of Miss Wells? I was aghast. Shortly after, the Cubs and I parted company.

The following summer when my battered psyche had almost recovered from Miss Wells I got another shock. Mother made arrangements for me to water Miss Wells' pot plants while she was on holiday. Mutely I shuffled the two blocks to her place to receive instructions. Miss Wells had a forest of plants which had to be watered twice a day. There was also a large evil-faced ginger cat to be fed each evening. Giving me her usual steely look Miss Wells said, 'If you do the job in a business-like and reliable fashion you will be rewarded.'

The next day when I attended to the cat, it bit and scratched me so I hosed it. I didn't see the cat again for three weeks until I came to collect my reward. The cat appeared quite comfortable and the pot plants, many of which had gone yellow and limp, were basically presentable. The ones which had refused to do a Lazarus-act I'd removed some time previously and I'd taken pains to hose away the large quantities of soil which somehow or other had been washed out of the pots.

Anyway, Miss Wells seemed satisfied and presented me with a parcel which she told me to learn from and treasure forever. I opened it on the way home and it was a book about some ancient Indian scout who had a hatchet face and eyes identical to Miss Wells. I made a point of never reading it. Perhaps because of fear of the ghost of Miss Wells I still have this book in my study. It is called *A Book of Grey Owl*.

The funny thing is that if Miss Wells had been a patient of mine years later, I probably would have found her a dear old lady, quite possibly with a wicked sense of humour.

Recreation

Family picnics were great fun and were almost invariably marked by some memorable accident. Mother and Father sat in the front seat separated by my younger sister, while three kids, a dog, plus or minus several friends were crammed into the back of the old Morris. Our destination would be either a property that was heavily mortgaged to Dad's bank or one of the local creeks.

The moment the car reached its destination the doors would fly open and children would fan out in all directions. Disaster would often strike immediately and it was a race to see which of us would fall in the creek first. I was usually handicapped by nesting magpies who would divebomb me but no-one else. When I see nature films of an eagle swooping down to carry off some small rodent I always recall the alarm of a magpie attack. Sudden flapping of wings, clicking of a sharp beak and those now precious hairs plucked out of my head.

The countryside around Bendigo was alive with rabbits. As an eight-year-old I'd listened earnestly when a school friend instructed me on the theoretical aspects of killing them. The method he recommended was grabbing the rabbit and cracking its neck like a whip. A few days later, when trailing along well behind my Dad, I spotted a rabbit curled up asleep in the centre of a hollow stump. I reached in, seized its legs and pulled it upwards with the intention of killing it. It was a very long rabbit—and it suddenly spun like a catherine-wheel with sharp claws. The outraged animal then thundered off into the distance leaving a 3-inch gash on my left wrist. I still bear a linear scar that any surgeon would be proud of.

My next attempt to reduce the rabbit population was more businesslike. Across the back lane lived a bachelor called Colin who possessed a small car and a .22 rifle. He was also a diabetic and sometimes let me see him injecting himself with insulin, a procedure which I found fascinating.

During the war Colin spent many hours making camouflage netting and the back of the house always reeked with its strange soap-like smell. He had a collection of 'girlie' magazines which had names such as 'The Sunlover'. The ladies all wore what appeared to be opaque body-stockings and as a ten-year-old I could see little point in the collection.

Twice I was allowed to go on rabbiting expeditions with Colin. The first found me staggering along in the heat of summer for five or more miles lugging a cluster of his bleeding and fly-attracting rabbits. I was lured by the promise that soon I would be allowed to have a shot. My only shot sent the rabbit leaping feet into the air and then disappearing like Brer Rabbit into a jungle of reeds. 'Bad luck,' said Colin. I returned home empty-handed at night fall. Empty-handed I might have been but I swear my arms were a foot longer having carried the great white hunter's trophies halfway across Australia.

Being a twit I went out shooting with Colin a month or so later. This time he missed rabbits left and right and, as a result, there was an even longer wait before I was allowed my single shot. This time I was lucky because a particularly stupid rabbit popped up at almost point-blank range just as I was trying to exterminate its more distant companion. Feeling quite superior I dogged along after my companion who became increasingly monosyllabic as he missed other rabbits. In the light of my newly gained confidence I offered him advice. When we got home (his score nil, my score one), he grudgingly offered to show me how to skin the beast. I declined this offer and subsequently spent a frustrating two hours at the bottom of the garden using knives and pliers to try to wrench the skin off the beast. Finally I gave up and buried it; the crack shot was not going to admit that he'd never seen a rabbit skinned!

In Bendigo most motor vehicles had a rest during the war years because of petrol rationing. Our family car spent the duration on blocks wrapped up and tucked away in what Dad called 'the motor shed'. The family cars of several of my friends were in a similar situation and many an hour was spent in them pretending to drive and practising gear changes. One neighbour fitted a peculiar charcoal-burning device to the back of his car and happily pottered around the town for a period of months. Unfortunately, the gas produced by the charcoal so damaged the engine that soon his car was also put on blocks.

After the war new cars started appearing but there was still a great shortage, so many of the cars which had been on blocks for years sold at two to three times their purchase price. Doctors seemed to get first bash at the new cars, particularly large American models. The smaller the stature of the doctor, the larger the automobile he purchased. The small doctors, like Drs Goodman and Sandner, could barely be seen as they hustled from one end of town to another in enormous American cars. Tall doctors, like Jim Sharland, settled on moderately sized Humbers.

Quite a number of small English cars also descended upon Bendigo—Austins and Morrises and the like. Harry Marks, a local pharmacist, obtained a nice little black Austin.

The weekend after the Austin arrived Harry took us for a spin. His wife, Deborah, was in the front seat, his two sons and myself crammed in the rear. Whilst Deborah shrieked in alarm, Harry actually hit 60 m.p.h. on a straight stretch on the outskirts of Bendigo.

The next person to shriek was Harry. He had turned down a narrowing path which came to a dead end, and, after a few clunks, managed to engage the reverse gear. There was a sudden crunch and then the bonnet of the car became more elevated than it had been. Harry got out to investigate. The next thing we heard was, 'Oh my God, we're in a mine shaft.' And indeed we were. The back wheels were dangling over a shaft whose depth estimated by dropping stones was some sixty feet. Harry brilliantly recalled his knowledge of physics and opened the back door to pull the three youngsters out first, leaving Deborah on the other side of the fulcrum. A shaken Harry then removed the strangely silent Deborah from the front seat. It took some hours before the car was pulled back onto the road and on the trip home not a word was said by the adults. In the back seat, the three of us concentrated on breathing quietly.

Sometimes the mine shaft covers would rot and disappear. One lady went into her kitchen one morning and noticed there was more light coming through the window than usual. Looking out she discovered her lemon tree had disappeared down an apparently bottomless hole.

The profusion of mine shafts also led to a plethora of jokes. One of my favourites was: A fellow was idling away some time dropping stones down a mine shaft. It must have been very deep because even when large stones reached the bottom the sound of impact was very faint. Spotting a railway sleeper nearby he dragged it to the edge and with a great effort heaved it into the shaft. As it disappeared from sight, out of the corner of his eye a flicker attracted his attention. Looking round he was astonished to see a large billy goat with its head down rocketing towards him. He threw himself to one side and the goat disappeared down the mine shaft. In a state of shock he went and sat on a nearby stump. A few minutes later a farmer wandered up to him and said, 'Mate have you seen my champion goat? I left it tethered to an old railway sleeper and I didn't think it could chew through clothes line.'

Drownings and Violent Deaths

During summer the centre of social activities for most of the young of Bendigo was the local swimming pool in Barnard Street. The baths were built of timber and had a sixty foot tower situated at one corner of a large lake which also sported a small island. In deep water at the other corner of the lake were 'the tables' also made of timber, where water polo was played.

I nearly met watery deaths in both these facilities. The first time was when I was about seven and had been denied the small coin required to get into the swimming pool. I decided first of all to swim to 'the tables' and then proceed across to the baths proper. I left the shore swimming strongly towards 'the tables' but had only gone fifteen feet or so when I proceeded to drown in the deeper water.

All my splashing and yelling failed to attract the attention of an elderly gentleman reading a newspaper on a park bench on the nearby bank. He continued to read as I proceeded to sink. Suddenly a motorcyclist came past, slowed down, turned and drove his bike straight into the water. He pulled me out, tipped me upside down and shook me. These actions did attract the attention of the elderly gentleman. After some difficulty in restarting his motorbike my rescuer took off leaving me high and dry on the bank. I didn't even get a chance to thank him because by the time I could speak again he was out of sight.

This experience made me determined to learn to swim the following summer. I was successful, but nearly drowned again under different circumstances. Whilst chasing a boy called Peter Shoddy along the timber ramp around the pool I slipped, knocked myself unconscious on the edge and disappeared under the water. This time my rescue had more witnesses and I was brought home with a certain amount of style and attention.

For some reason lots of people seemed to commit suicide in the Bendigo baths. It often took several weeks work with grappling irons before the bodies were found. During this period, if one dived deeply and contacted a submerged object, a rapid ascent to the surface would be made. My best mate Alan Kerr, who lived across the road from the baths, once actually saw a body pulled out with yabbies on it, but got chased away before he could get a closer look. On one occasion, a Greek girl drowned herself and her pet rabbit was found floating by the shore. She left a note on the kitchen table with a knife stuck in it. No-one would tell me what the note said.

A few weeks later there was a multiple murder down the street. Alan and I managed to witness all the carpets and furniture having the blood hosed off them on the front verandah. A very satisfactory sight for ten-year-old boys! We both managed to make our sisters go pale at dinner when we individually and dramatically reported our observations.

There was also drama each time we saw a convicted murderer who sometimes passed down the lane at the back of our house. Many years earlier he had come home at night and found that his wife had killed both their children. His immediate reaction led him to serve twenty years imprisonment. He was a strange, tall, stooped man who wore a large floppy black hat and had an expression not unlike Barry Humphries. We would take off over the fence in a flash if we saw him coming.

Tradesmen and Hawkers

I had found that it paid to cultivate tradespeople, as they were good sources of information and the occasional freebie. Each lunchtime, the baker would park his horse and brightly painted cart outside our back gate. While the horse quietly ate the thistles growing around the gutter the baker would eat his way through half a loaf of bread covered with butter.

The iceman, on the other hand, was a fast moving fellow. He moved so fast he tended to leave sound behind which would only catch up to him when he was on the way out! He used his sharp spike to split ice blocks with the speed of a surgeon operating in the pre-anaesthetic days. One of my sisters had the habit of sitting on the toilet with the door open while conversing with my mother across the hall in the kitchen. She stopped this after she found herself eyeball to eyeball with the iceman as he attended to the icebox in the hall.

I used to like going to the butcher. The younger butcher used to tell me jokes which I barely understood but would later try out on my mother. I was fed biological information such as 'a ram can deal with several hundred sheep in a night' and it would probably only be 'a lick and a promise'. This I duly reported back to Mother.

The elder and bigger butcher looked as though he had evolved from several hundred generations who had started carving mammoths and worked systematically through all the warm-blooded creatures in the animal kingdom. Usually surly, he occasionally had a burst of generosity. Once I went in and asked for a pennyworth of meat to catch yabbies down at the lake. He wrapped up some scraps and said, 'There you are son, you can have those.' I looked him straight in the eye and said, 'Right, I'll have another pennyworth.' The words he used were not reported back to my mother.

The big butcher later committed suicide by taking strychnine. In his rather spectacular death throes he crashed through and broke the doors to the shop. I had a good convulsion-to-convulsion report from a friend which gave me a healthy respect for strychnine.

'You can always tell by a man's face whether he's honest or not.' So would say our bottle-oh when he pushed his rather evil face within an inch of mine. He had a good system of counting the bottles. It went along the lines of 'One, two, three, four, five, how's your Mother? Six, she doing well?, Five, six, seven. Eight, nine, ten, how's your Dad? I haven't seen him around lately. Eight, nine, ten, eleven,' etc. All the time he would be rhythmically stacking bottles in his bag with the smoothness of a well-oiled engine.

One of the most interesting itinerants to visit our home was a fellow called Methuselah. He would arrive at regular intervals to chop the wood. Methuselah was an impressive sight as he stood well over six feet tall, with shoulder length white hair and a beard to his navel. He dressed in obviously homemade clothes and his feet were always bare. When the wood was chopped Mother would bring him his two and six pence, and a tray with a mug of boiling water and one slice of bread with no butter, just vegemite. This was all he wanted.

Apart from his appearance, the thing about Methuselah that captured my imagination was that he always called my mother 'sister'. When he went he always left a little printed card with some appropriate biblical text. For years I really believed that he was my mother's brother.

Bendigo at War

During the war the American Army took over the football oval which was five minutes from my home. They erected tents everywhere and radiated out through the whole district. The local girls kindly held their hands so they didn't get lost or blown away in the wind.

The Yanks' arrival was a great boost to the local economy. One small delicatessen near the oval went into the hot dog business in a big way. Their contents was a matter of speculation. Dad always referred to them as 'mystery bags'.

Watching the activities of the American soldiers became a hobby. I remember seeing an American sentry letting off steam by carrying out bayonet practice on the galvanised iron fence near his guard post.

There was a double fronted shop in View Street which was crammed with fake boxes of chocolates and every conceivable type of confectionery. Imagine the yearnings of those with a sweet tooth who had to pass the shop regularly during the war years! By the time I actually ate a Violet Crumble I'd admired its packaging for over five years.

I was sitting next to John Hial having some Religious Instruction when, for me, World War II ended. I was nine. On this occasion we were studying the beliefs of the Salvation Army. A teacher came in and said something to the Salvo who promptly announced that the war in the Pacific was finished and we could have the day off. John and I hugged each other.

On reaching home I was dispatched to the bank to see Dad. He took me into the newsagent next door and bought me a packet of streamers. The centre of Bendigo, ambitiously named Charing Cross, was filled with cars, trucks, soldiers and civilians singing and dancing and throwing streamers everywhere. Never had I seen adults so happy. There were so many streamers flying I decided to keep mine and went back home.

A few days before this celebration I had studied the front page of the Melbourne *Argus* with some fascination. It featured a photograph of a huge mushroom cloud over Hiroshima. Dad told me with some satisfaction it was the result of the biggest bomb ever made.

No bombs had ever fallen on Bendigo but to me the 'war news' had been as much a part of my daily life as was breakfast.

The first tangible advantage of peace was that we were allowed to play in the trenches that had been dug behind the school. One Saturday ages ago when I had watched Dad participating in digging these trenches, I had become concerned when he eventually disappeared completely out of sight. He was in a bad mood all the way home, possibly because he found the Bendigo ground far rockier than that in which he had dug trenches in France, over twenty years earlier.

A Burst of Religion

Life in Bendigo was basically as boring as it probably was in most other country towns at this time. This may be the reason I never wagged a single day from school. At least something was always happening at school and I found the teachers a fascinating study. The tedium of Bendigo was probably also the main reason why I stayed a choirboy from the age of nine until about fourteen. The local Cathedral, All Saints, was very beautiful with lots of wood panelling and a magnificent organ. Sadly it closed its doors in 1989 because of financial problems.

I really liked the little white collars, the smell of the cassocks and the general feeling of being of some importance. I could only sing about three notes, but these were sung lustily like a little thrush. The choir went more for volume than tune, and the combination of me, with Mr Lynch behind, was highly satisfactory. Mr Lynch ran a Lingerie shop and also had only three notes. These were extremely deep. The pair of us always did a very fine amen.

The Cowlings, father and son, ran the choir. Mr Cowling, Jnr, took the choirboy practice at 6.30 on a Thursday night. He was a bank teller and when he turned up on the Thursday night, he was always in high spirits, with a shiny red face. He would chew peppermints. This may have had something to do with the hotels closing at 6 p.m. in those days.

On Sunday, the face of Mr Cowling, Jnr, was always grey and showed none of the good cheer of Thursday night. It wasn't until I was about fourteen and drank my first glass of mood modifying liquid (a sherry with my mate Alan) that I realised what young Mr Cowling did between the closure of the Bank and his arrival to conduct choir practice.

Mr Cowling, Snr, was a deadly serious watchmaker with immaculately polished rimless glasses. He had everyone scared, including our vicar, the Reverend Dean. He played the organ magnificently and I remember, after one fabulous bit of playing at a full choir practice, I slipped out of the choirstall, approached him at the organ and asked, Oliver Twist-like, whether he would play that last bit again. He sucked in his breath and stared at me with the same eyes as Miss Wells, then hissed 'get back'. I slunk back to the stalls but to my surprise the requested passage was played again. It was later explained to me that we were not meant in any way whatsoever to enjoy the music because it was all to the glory of God. This made me feel a bit guilty.

Two church services at least on every Sunday for four or five years set me up for life. I still consider church-going to be as good for the mind as the soul. It is extremely relaxing to sit doing nothing and one comes out singularly refreshed. The three-hour Good Friday service was a bit steep, however.

On my first Sunday as a choirboy I was disappointed that the collection did not disappear when raised at the altar by the vicar. I was astounded later to come across a number of the church elders in the vestry sitting around the table, sorting and counting the day's collection. I had hoped that my proximity to the altar would allow me to see the collected monies miraculously vanish.

Vicars were subjected to as much scrutiny as schoolteachers. My first vicar was a genial bachelor who was the life and soul of weddings. On a cold winter's night his sermons with outbursts on hell and damnation warmed us all up.

When one of the girls next door got married, I happened to be sitting high up in a fig tree late in the proceedings. The wedding guests were gathered in little clumps, drinking in the neighbour's garden. To my fascination, I heard the vicar telling some quite extraordinary stories to several of the male guests. One was about a fellow who bought a life-sized inflatable female. I didn't catch the end of the story because of the outbursts of laughter. I always wondered how the story ended and considered asking him at choir practice.

My cynicism concerning vicars and deans dates from the time when I went past the half-open Dean's Robing Room. The vicar was fully robed and, to my surprise, practising giving the blessing whilst studying his face from different angles in the mirror.

If this weren't bad enough, I had noticed the same vicar would take special care to make sure there was plenty of wine left for himself at the end of the Communion service. If he found he was getting towards the end of the communicants and the level of the wine (diluted McWilliams port) was getting low in the chalice, his grip would tighten. Communicants would be lucky if their lips were wetted. At times I saw 'customers' struggling to get a decent grip of the chalice while the vicar's knuckles whitened as he fought to control the flow of the blood of Christ.

During school term the size of the congregation increased by twenty-five per cent each Sunday morning. Some forty boarders from the Girton Church of England Girls' Grammar School filed in to fill the first six rows on the left-hand side of the church. The right-hand side at the front was held by ancient survivors of early Bendigo families. Some of these would arrive in chauffeur driven cars.

The Girton girls and the choirboys practised mutual scrutiny, particularly when the girls filed past the choirstalls to take communion. Some of my older acne-faced colleagues took a somewhat unholy interest in these girls. The same goings-on occurred when Bible classes started. The boys would sit on one side of a small over-heated room while the girls, from the Parish in general, would line up on the opposite side.

Most of the instruction on the Good Book was wasted on me and my fellows. One boy kept a week by week report on which girl's breasts were growing fastest or 'on the move' as he put it. Another boy laboriously prepared a list of all the dirty and rude bits he could find in the Bible. The dear man who gave Bible instruction was quite oblivious to the total lack of genuine interest in his subject. He was also unaware of the furtive and calculating adolescent glances which were exchanged.

A man called Riley was the Bishop of Bendigo at that time. He was a very tall, distinguished man who had been Chaplain-General of the Army during the war. I thought he was the nearest thing to Jesus Christ possible. On special occasions he wore fantastic costumes and wielded a massive crosier which was so heavy it must have been made of solid gold, probably from the Bendigo goldfields. Sometimes I had the job of taking the crosier from him as he seated himself in his mini throne. As I transferred it to the special stand I felt positively holy, convinced that if it were dark I would be glowing with divine light.

My last contact with the Bishop was as a sixteen-year-old when I needed his permission to apply for a scholarship to Trinity College at Melbourne University. He literally gave me his blessing, but it did little good as far as the scholarship exam went. I travelled to Melbourne to deal with the exam paper which proved as incomprehensible as Bible lessons. Emerging from this humiliating venture, my frame of mind was not helped by hearing fellow applicants exchanging jolly comments such as 'That exam was a pushover'. It was a depressing experience which was to be repeated a little later.

Sex and Baldness

The activities of courting couples were an important aspect of neighbourhood life. In the days before the ready availability of motor cars, drive-ins, etc., courting couples disported themselves on rugs both on public and private lawns. It all seemed pretty pointless to me.

At the age of eleven the mysteries of human procreation were dramatically revealed to me. Wandering through Coles Store with a classmate, Jack Strellon, we passed the First Aid and Household Remedies section. Jack popped a large leather fingerstall on his finger and said, 'This is what a fellow puts on his dick before he puts it between his girlfriend's legs.' Staring at Jack, I said, 'You must be joking. What on earth would he do that for?' After mulling over this extraordinary and dubious revelation for several weeks, I broached the subject with a fellow choirboy. The boy in question was Jim Short, who until 1997 enjoyed the high office of Senator. I asked him point-blank where babies came from. Jim said, 'By the rape, everyone knows that.' This threw me into a further quandary; the local papers had been filled with details of a man who had been sentenced to fifteen years for rape. A very worried little boy watched his father at the dining room table that night.

When a couple of weeks had gone by and Dad had not been arrested, I evolved a comforting theory which involved baldness. All male Sutherlands over forty seemed to be bald and Dad was no exception. At this time all Australian men wore felt hats and would never go out without one. Frequently I would observe my Dad raising his hat with a flourish when he passed a lady of his acquaintance.

Part of my theory was that a man had to raise his hat when he passed a women so she could see whether he was bald or not. Somehow or other this got entwined with the business of babies resulting from the rape. I concluded that bald men were excused the penalties of rape and therefore did not have to go to gaol. I then found other things to worry about.

Choosing a Career and Observing Local Doctors

When young, I observed sudden death at first hand outside our front gate. A passing horse kicked a small Australian Terrier belonging to a frail and elderly lady. Helplessly, I watched her sitting on the tiny bridge over the gutter nursing the dog as it coughed up white froth. She was clutching and crooning to it as if it were her baby that was dying.

From that day I took a keen interest in first aid. The workings of 'Mother Nature' also intrigued me no end. For example, the return of a local hero was of special interest to me. He was a paratrooper who had been machine-gunned while floating down from the plane. Folklore said he'd had his guts shot out and replaced with rubber tubing. This absolutely fascinated me—I could hardly wait to have a closer look at him at the swimming pool. I hoped to see all his insides working through transparent skin but was disappointed; apart from some silvery-pink scars, his belly appeared normal. He offered me a smack on the chops for staring at him.

I was lucky that my answer to what I was going to do when 'I grew up' never varied. Doctors, and to some extent chemists, were my role models, and a doctor I had to be. I am certain had I not been fortunate in achieving this ambition I would have trundled through life with a touch of unfulfillment.

I found doctors the most interesting people in Bendigo. My first doctor memory at the age of five or six is of seeing Dr Neville setting up his needles and bottles in our drawing room. I found the ritual of skin swabs and drawing the liquid up into a syringe quite fascinating. The screams and anguish of my struggling younger sister when she received her shots was just as satisfactory. About this time, I met Dr Keith Kerr's son, Alan. We met in the back lane and commenced a friendship that lasts till this day. (Alan went on to a distinguished career in the public service. After serving as a Commonwealth Ombudsman he became possibly the most popular Administrator of Norfolk Island during a record term which ended in 1997.) He, like me, had two elder sisters and at the time of our first encounter he was lying low after a row with one of them. Alan's parents were fine people and had an enormous influence on me. In my darkest moments, they were always comforting and encouraging. Any small triumphs I had, they classed as major achievements. For example, when I made a Crystal Set which actually worked, you would have thought I had excelled Marconi.

Alan's mother, Gwen, was of Welsh descent and she was as loving and caring a person as the mother in *How Green was my Valley*. A letter I had from her forty-five years later, just before she died, was filled with unselfish and kindly thoughts.

Like my mum, Gwen was a great cook. Alan and I knew to return from our ramblings around 4 o'clock on a Saturday afternoon in case a large chocolate cake had been iced. The Kerrs also had the first refrigerator I ever had access to, which was another benefit of this new relationship. Incidentally, much to my relief, Keith Kerr was bald like my father and therefore unlikely to be charged with rape.

He had succumbed to tuberculosis whilst in general practice in Finlay, NSW. Upon recovery, he joined the Health Department and was in charge of tuberculosis in the Bendigo Area. Over a thirty-year period he saw the high rate of tuberculosis, especially among goldminers, fall to near zero. Keith's hobbies were crossword puzzles and carpentry, and I spent many an hour sitting near his workbench quizzing him while he produced inlaid coffee tables, etc. He was a very patient man!

Some weekends he would take the two of us for a long tramp in the bush and answer our questions. In simple terms he would describe why patients didn't bleed to death when they were being operated on, how vaccines worked, etc.

When a small girl was struck by a car right outside the Kerrs' home and I saw Dr Kerr in action, I knew I had a good teacher. He had earlier explained to me that 'doctor' came from the Latin for teacher. Furthermore, unassuming as ever, he said he wasn't really a doctor because he only had a Bachelor's degree, and being called a doctor was only a courtesy. Keith had broken off his medical studies to serve in World War I but was sent home after two of his brothers had been killed. At University, he was Captain of the University Blacks, which always surprised me because I couldn't envisage him in the rough and tumble of the ruck.

Most of his old medical textbooks were kept in a big cupboard on the back verandah of his house. Alan, his sister, Lesley, and I became quite adept at locating some of the more revolting illustrations of various medical conditions. Although Lesley was a girl, I greatly admired her. She taught me how to wink. She married Richard McGarvie and for some years lived in Government House in Victoria.

At the age of thirteen, I dropped a kettle of boiling water on my foot and spent ten days in the Bendigo Base Hospital. On the first night there, an old chap died in one corner of the ward. When Dr Kerr paid a social visit the next day, I apparently lent over and whispered earnestly in his ear, 'They are dying like flies in here Dr Kerr.' He would often repeat this story much to my embarrassment.

While gaining this first-hand insight into death and hospital practice I was looked after by Dr Maurie Jacobs. He was a great little bloke, plump, jovial and the oldest practitioner in town. He had graduated in 1905 and years later I managed to locate him in a photograph hanging in the Medical Library at Melbourne University.

Maurie later told me tales of outbreaks of puerperal fever at the Women's Hospital when he was a student. On one occasion a dozen newly delivered women all died within a week. He also used the expression 'Dying like flies' and, although at this time I was dead set on becoming a medical student, I gained the impression that hospitals were dangerous places for sick people.

The Bendigo Hospital was free and the services given by the private doctors were honorary. One just went to hospital, gave one's name, address and next of kin, signed a couple of forms, and that was it. At the age of sixteen while making bombs (see page 46), I cut my hand opening .303 cartridges. It was a simple matter to ride my bike one-handed to the local Base Hospital, be stitched, bandaged, get a tetanus shot and go home. No embarrassing questions were asked. At least not until dinner that night.

My lasting impression is that these doctors' primary interest was good patient care. They were not in it just for the money. Certainly they made a comfortable living, but I knew of many instances when they did not charge patients. Some home visits were not charged for and at times those having a run of bad luck would receive mysterious help. A free load of wood would arrive out-of-the-blue for someone recently widowed and at Christmas an anonymous donor arranged for packages of groceries to be delivered to needy families. After swearing me to secrecy, the grocer's assistant informed me the donor was the elderly and semi-retired Maurie Jacobs.

Doctors appeared to have a fair degree of freedom on what they would charge their patients. Once, as a medical student doing house calls with

Jim Sharland, this matter was discussed at length. He explained that the old Bendigo 'aristocrat' we had just visited was going to be charged five guineas. On the other hand, the previous patient would get a total bill of a guinea covering a number of calls done over the month. 'It balances out overall,' he informed me.

Eugene Sandner was a surgeon very involved in medical politics. He was the local authority on the dangers of nationalised medicine, which at that time was becoming firmly established in the United Kingdom. Eugene had a collection of woeful tales about the way doctors were being treated by their previously compliant and obedient patients. For example, one doctor was called out at night to a remote farmhouse, the farmer agreeing to put on all his houselights so it could be identified in the pitch dark. The doctor drove ten miles in driving rain and criss-crossed the area involved, but could find no house. He got home two hours later and rang the farmer who explained, 'Well, after I rang you the wife started improving. Five or so minutes later she seemed a lot better so I turned the lights off.'

One call took less of the doctor's time. A man rang and told the doctor his wife was pregnant and could he come over urgently. When the doctor arrived, the pregnancy did not appear to be very advanced. The doctor inquired how long she had been pregnant, and the husband replied rather sheepishly 'We think about twenty minutes.'

Upon reflection, no doubt Eugene Sandner, like the vast majority of doctors in Australia, found a dramatic, if temporary, boost in income when Medibank was introduced to Australia in 1973. Many docile lumps which had remained undisturbed on a patient's skin for a number of years were whipped off in the early days of Medibank. There were many new swimming pools installed and affectionately called by their proud owners: *The Medipool*.

In the years since the AMA dropped its opposition to Medibank I have been increasingly disturbed by the avaricious tendencies of some of my colleagues. A handful have really milked the system—God knows what they needed all the money for and they were often the most ardent conservatives who complained about government interference!

Chemists and a Full-time Job

A favourite neighbour of mine was Mr Cocking. He was a homoeopathic chemist, very much of the old school. Standing barely 5 feet tall, he was delicate in build, always precisely dressed and extremely polite. I remembered he cured the eczema on my sister Barbara's hand by prescribing a bottle containing some of the smallest tablets I have ever seen. Like ninety-five per cent of Bendigo males Mr Cocking adhered to a meticulous routine, one aspect of which was purchasing cream cakes at lunchtime from a little shop generally known as the 'bun shop'. This shop had gas fittings and from memory was staffed by a series of pretty teenage girls who invariably developed pimples and obesity during the period of their employment.

The nutritional contents of the little offerings Mr Cocking brought home from the bun shop steadily accumulated in the body of Mrs Cocking. She became almost spherical—with tiny little feet at one end and beady little eyes at the other.

For years Mr Cocking had wanted to go overseas. Mrs Cocking remained opposed to such an undertaking and he had to wait until she died from cake poisoning. The dear little bloke set off with high expectations to study the culture of the old world. Unfortunately, like the composer Franck, he was knocked down and killed by a bus in Paris. When visiting Paris years later, I was particularly cautious in regard to the public transport. I also kept my eyes open for a little monument stating 'This is where Mr Cocking from Bendigo, Australia, got hit by a bus,' but saw none.

Insight into the local doctors was further enriched when at the age of twelve I commenced part-time work as a chemist's boy. The rates were 40 cents for Saturday and $1.50 for the whole week. Duties included cleaning the shop windows and polishing their brass surrounds with a solution of oxalic acid which smelt awful and burnt my fingers. A pleasant duty was washing a variety of glassware in an ancient sink while annoying the junior pharmacist by prattling on when he was trying to count pills. The secondhand medicine bottles I processed gleamed like priceless crystal and would not have looked out of place illuminated in a glass case in a museum.

I found most doctors seemed to have a favourite prescription. Fred Corrigan, the boss, informed me frankly that most of these prescriptions were unlikely to do any good. This information made the loathsome duty of delivering large bottles of medicine even worse. Bendigo is very hilly and during heatwaves both the ancient pushbike and my cardiovascular system were severely challenged.

Corrigan's Bronchitis Mixture was a profitable little item and making batches of it was real fun. The elixir came in two forms: normal strength and special strength. The latter included a shot of opium tinctus, a practice presumably now illegal. Some clients had a bottle of this delivered several times a week.

Bottling the brew was an art. As with jam, towards the end of the dispensing process, I would have several containers that obviously were not going to be fully filled. The crisis was overcome by emptying a few bottles back into the holding container, adding a suitable quantity of filtered Bendigo water and, hey presto, all the containers could be filled.

After corking and washing the bottles came the most delicate part of all—affixing the proud labels affirming that this was indeed Corrigan's

Bronchitis Mixture. It was a real precision job and infinitely preferable to riding the old shop bike up steep hills.

Bendigo High School

In 1949, at the age of twelve I crossed the narrow asphalt path which separated the gloomy, brooding Camp Hill State School from the somewhat sunnier Bendigo High School. My older sister Diana had completed matriculation at the High School the previous year and hadn't been particularly thrilled with it. She informed me that BHS stood for a 'Bloody High School'. She had exchanged Girton Church of England Girls' Grammar School for the High School for her last year of schooling because they did not teach Physics and Chemistry.

The High School was fortunately on flatter ground than the State School, hemmed in on three sides by Public Tennis Courts, the large ancient Bendigo Goal and sections of the Botanical Gardens and Fernery.

One of my first ambitions at the High School was to get a closer look at two extraordinarily fat boys. From the State School I had often seen them drinking at the High School fountain. A crowd always gathered around the fence for a closer look. The eldest boy was called Tas and was 5 feet high and 5 feet in diameter. Being at the High School I was able to find Tas sitting underneath a tree and have a chat with him. He was a delightfully gentle fellow and explained he had something wrong with his glands and had to drink huge amounts of water every day. At the end of one term, he and his brother mysteriously disappeared. Nowadays Tas's problems might be amenable to treatment and he could live a near normal life.

Having satisfied my curiosity about Tas, I settled in to study the new environment and its inhabitants more closely. One immediate advantage was a host of new adversaries for the particular Kennedy who had been thumping me. I settled anonymously into Form 1B. Kennedy having been designated of greater intellect, went into Form 1A. It was most satisfactory to see him getting a thorough belting by a 1A colleague on the first day.

Another great advantage lay in the good variety of teachers. Instead of one teacher day in and out, we had a kaleidoscope of teachers arriving to take different periods. If one subject was boring, or the teacher was boring, you knew there was something better coming along shortly. Far better than seven hours straight of the fearsome Miss Wells. God rest her rigid soul.

The school assembly
Another novelty was School Assembly. This occurred, naturally enough, in the Assembly Hall. It was quite interesting to see what the staff, prefects

and the occasional visitor would do to hold the attention of the five hundred or so students. As a routine, these bi-weekly assemblies included a musical interlude. Just as a motorist must be able to produce his driver's licence on demand, so all the students entering the Assembly Hall were required to clutch the songbook. Mr Tyler, who was known as 'Cactus', supervised the entry of students checking that they possessed individual songbooks. Cactus had a swarthy complexion with only a few spiky little bits of hair sticking out of his head. There was, however, something about Cactus which brought fear to the hearts of all but the most senior boys.

At the appropriate time, Cactus conducted the school en masse in rousing renditions of old Scottish, Welsh and English songs. Occasionally there would be an Italian or even an Australian song. In no time at all we had learnt that the words 'On Ilkey Moor Baht At' should be sung as 'We'll milk the old tom cat'. 'Ho-Ro, My Nut Brown Maiden' became 'Let go my nuts, brown maiden'. I remember all these years later the cheerful words of one song which went 'Forty years on, when afar and asunder, parted are those who are singing today'. I thought at that stage, at least Cactus won't be around.

To become a prefect one generally had to be a large uncouth brute who could kick a football at least half a mile. To a lesser extent, the same criteria applied to girl prefects. In each batch of prefects there were one or two exceptions. The male exception would be a small, neat, deadly serious type with a responsible middle-aged and boring outlook on life. The female exception would be very pretty, petite and with glistening eyes. Overall, the large raw-boned type predominated and they generally faded into oblivion when they left school after their brief blaze of glory.

The god-like standing of one male Head Prefect evaporated in the space of minutes during one school assembly. The actor and hypnotist John Calvert was touring Victoria and visited the school in part to promote his evening performance in Bendigo. The Headmaster was seated on the stage with the male and female prefects in a semicircle around him. Mr Calvert gave a lively introduction with suitable plugs regarding the time and place of his evening's performance. He then asked the five hundred plus students and teachers for a volunteer on which to demonstrate his hypnotic skills. Before some of the local clowns and troublemakers could step forward, the Head Prefect shot to his feet, clenched his jaw defiantly and marched his broad 6 foot frame up to the hypnotist.

The hypnotist waved his hands around in front of our hero. He then turned to the microphone and informed us that the subject would turn into a distressed duck every time he clicked his fingers. He instructed the Head Prefect to resume his seat and prattled on again about starting times for the evening show. We all waited in anticipation. When he suddenly clicked his fingers we saw a remarkable sight. The Head Prefect uttered a terrible quack and threw himself out of his chair. He then ran around on his knees, with his elbows flapping, while he quacked frantically. The distressed duck cut a swathe through the legs of the other prefects and for some time appeared completely out of control. Because he was a particularly noisy duck he could not hear the cut-off double claps the hypnotist was making. When he metamorphosed back to human form, he was hit by a massive chorus of catcalls and shrieks of appreciation which almost brought the ceiling down.

To our delight the hypnotist reactivated his distressed duck one more time but wisely cut the performance time down. After a mere five quacks the duck fell to the floor with a brief flutter of the elbows.

This performance left me with a profound impression and I have often wondered why rival political parties don't attempt to hypnotise the opposing leaders prior to the delivery of an important policy speech. It should be quite simple to do. A politician could be triggered off by a simple stimulus such as a double sneeze. It would make the speech thoroughly entertaining as the speaker changed dramatically from one political extreme to another.

After this performance some of us tried to hypnotise each other without any effect. We may have successfully hypnotised a chook by laying it on the ground and drawing a white line with chalk running from its head to the middle distance. However, the chook was such a stupid creature it was difficult to tell whether it was in a state of trance or just plain sleepy.

Bill Galvin, our local Member of State Parliament, found his visit to the High School a dampening experience. He was an old-style Labor man and at the peak of his career he resembled a large bespeckled toad. To me, Bill has always represented the fluidity of Australian society. Starting out as a labourer, he worked his way up through the ranks and ended as Chief Secretary in the first Cain Labor Government. He had sought and won the hand of the Sharlands' housemaid and as Mrs Galvin she accompanied Bill as he escorted the Queen around Victoria on the Royal Tour of 1954. This put the noses of the Liberal supporters in Bendigo well and truly out of joint.

One day Bill ascended onto the stage at the High School Assembly. An impressive deep-seated carved wooden chair awaited him. Unbeknown to him, and also awaiting him, was half a vase of stale water which had been poured into the chair. Its cool depth awaited the receipt of the parliamentary buttocks while the eyes of a few students were watching Bill with concentrated intensity. There was an almost cruel delay while the prefects and various hangers-on shuffled to their respective positions on the stage. At last the moment came and the great man settled into the chair of honour.

For several seconds nothing happened, then a look of concern or even anxiety spread across Bill's face. With a smooth and almost imperceptible movement, he slid his left hand cautiously down around his left buttock. Next thing he was standing up as were those next to him. The chair, and to some extent, the seat of his pants were dried by someone lifting up part of the stage curtain like a skirt. While this was going on it was noticed that others on the stage were surreptitiously checking the dryness or otherwise of their seats. This was done either by adopting a rocking motion, sliding a hand into the region, or direct visual inspection.

It is doubtful if anyone remembered Bill's speech that day. The prefects and staff directly facing a vast damp patch could hardly keep a straight face. Everyone else was eagerly awaiting a glimpse of the said damp patch as the honoured guest disappeared.

Teachers
My first Headmaster at High School was Mr Smith. He was a tall, stooped, kindly man known as 'Drip Smith'. He always tried to find a personal way of teaching students who faced specific problems. For example, in Form 1B I remember him explaining square roots to the butcher's son along the lines of, 'Well Ken, let's take the square root of 25 chops. The square root of 25 chops will be 5 chops.' Fortunately the undertaker's son was fairly bright!

Our next Headmaster was Mr Crocker who vaguely resembled Harry Truman, the then American President. Mr Crocker was terribly enthusiastic about everything, but in my opinion too easily conned for a man of his experience. Shortly after his arrival, four of us were playing the marbles game called 'Poison' at the very far corner of the schoolyard during a supposedly free study period. To our alarm we spotted the new Headmaster with his black gown flapping in the breeze approaching us by the shortest possible route. By the time he reached us, the marble holes were filled in and a reason for our activities formulated and rehearsed. Mr Crocker was informed that Jack Strellen had lost his grandfather's fountain pen which was of great sentimental value. When we returned to our classes he announced through the school P.A. system that a valuable antique fountain pen had been lost in the school grounds. We all fell about on our desks when he said, 'All pupils please keep a sharp eye out for it!'

Miss Blank was a young Science graduate who would have had difficulty controlling a heavily tranquillised kindergarten class. She was quite attractive and some of the older boys would leer at her for the whole of the class. Miss Blank often wore skirts with a chequered pattern and Roger Nelly would offer the class diagrams of the skirt with indications of the precise position of her private parts.

The rest of the school usually knew where Miss Blank was endeavouring to instruct and guide the minds of her young scholars; the uproar was quite evident and usually only ceased when the Headmaster made one of his routine visits.

Once the class fell strangely silent. This was when the somewhat accident-prone Miss Blank carried out one of the most impressive experiments I have been privileged to observe. She was attempting to demonstrate that, if a small piece of metallic sodium is dropped into water, there will be a brisk and vigorous reaction. She had to take a small piece of the sodium from the safety of an oil-filled container and gently lower it into a large dish of water.

Using a pair of tongs, Miss Blank removed a length of sodium metal from its container. The extracted piece resembled some 3 inches of cabana sausage. She then took a large knife and carefully cut a rather miserly portion from one end of the said sausage. We watched fascinated as the knife was held over a large dish of water. For, on one side of the knife was the minute portion of sodium metal she had cut, but adhering to the other side was the remainder of the 3 inches of sodium metal—it was raring to react with water or anything else available.

In a lady-like fashion, Miss Blank pushed the tiny blob of sodium off the knife then, to her horror, saw for the first time the large rod of sodium adhering to the other side. It lazily rolled off the knife blade and tumbled through the air to disappear into the bowl of water with the grace of an Olympic diver.

For a second or so nothing happened, then a fantastic display took off. Ominous rumblings came from the bowl and puffs of white smoke appeared. There was a sharp explosion and a rod of sodium was seen to leap out of the bowl and back in again like some angry Marlin. The tempo then became frenetic. Benches and laboratory stools collapsed as the class retreated. The bowl was lost from sight behind swirling white acrid clouds. Escape was impossible because the only exit was beside the front bench. This chemical reaction turned out to have quite a sense of humour because each time someone felt it was safe to sneak past and out of the door it would erupt again. Although it was seven minutes before chemical equilibrium was reached and the class could escape the smoke-filled room, it seemed a lot longer. Interestingly, there were no questions in the later exam on the reaction between sodium metal and water.

Miss Gwen Bowles was a local girl who had studied music at Melbourne University. As a reward for her diligence, the Education Department returned her to her home city. Here she attempted to instil into the locals a love of classical music and an interest in scales. Miss Bowles was never boring and adopted a full frontal attack on any unruly behaviour. She was big bosomed and looked like a letter 'P' in profile.

She was quite fond of Handel's Water Music and this would be put on the old gramophone. As she lowered the lid, the whole class would be systematically raked up and down with her scanning glare.

As a pianist, her Chopin's Grand Polonaise was fascinating. She would pound it out with her tiny hands and, as she did so, her chair would move steadily nearer to the edge of the platform. When the rear legs of the chair were almost off the edge, she would give an almost imperceptible wriggle and bring the chair back an inch or so towards the piano. Gwen

Bowles later became a highly respected Head Mistress of MacRobertson's Girls' High School in Melbourne.

Doc Robins, a science teacher, was a wizened little bloke who gave the impression that he considered the vast majority of his students were absolute dolts. He used to mutter asides to himself which seemed to give him a little comfort. He mellowed a little when he found that I was quite keen on science. One vacation I spent a few days helping him clean up the laboratory storeroom and was rewarded with all sorts of junk which I proudly carted home.

At a school reunion twenty or so years later I was surprised to find old Doc Robins still alive. I was even more surprised when he took me by the arm and said 'You must come and meet my Dad'. I though he'd gone mad, but in fact his father was there. Father and son were like two dried peas in a pod.

Miss Styles was an Art teacher who was said to have been born at sea and have no allegiance to any country. Miss Styles taught in an isolated school building which had been built during the Gold Rush and was supposedly haunted—which further added to her mystery.

'Sloid' and other subjects

In those days education at the Bendigo High School could have been described as pyramidal rather than progressive. The combined Forms 1 (or Year 7s) might have some 120 or so pupils, but by the time Form 6 was reached there were barely 20 students. Some of the brightest girls left school at about fourteen basically to become home helps for their mothers. Each form in the first three years was divided into three, namely, A, B and C. The clever kids, i.e. those who could read and write would be in Form A, Form B was the next classification, and kids that could read and write were sometimes in C, i.e. with the dolts and drones. Some of these dolts and drones were 'bus kids' who often travelled up to four hours per day from remote areas to attend school. Sometimes these kids proved to be brilliant when they later boarded in Bendigo and obtained adequate sleep.

On a Tuesday morning in Form 2, boys were given the opportunity not only to travel but to develop expertise in what was described as 'sloid'. I'd always assumed sloid was a parochial term but much to my surprise it is in my Oxford dictionary as a system originating from Sweden 'of manual training especially by means of wood-carving especially in school'.

It was great fun travelling to the old school where sloid was taught. A tram was taken to a spot halfway between the centre of Bendigo and Eaglehawk. One then had to beat a path either around or through some of the best cowboy and indian country in the district. There were huge heaps of mullock, old poppet heads and abandoned mine buildings. By the time we arrived for instruction, we were not only filthy, but utterly exhausted. This didn't matter since sloid seemed to consist of spending ninety per cent of one's time learning how to sharpen chisels or waiting for a particularly foul smelling glue to melt.

On Tuesday mornings while the Form 2 boys were seeing a bit of the world, the girls were engaged in Domestic Science. Security was tight regarding the girls' activities at these classes, although occasionally a batch of scones would be produced. Whatever wood we had been labouring over earlier in the day was tenderer and probably more flavoursome.

Art classes at High School were a singular waste of time. They seemed designed not only to prevent any appreciation of art developing but to eliminate the possibility of discovering any latent talent. One teacher seemed to believe that thirty adolescents would be completely absorbed for an hour if they were given the opportunity to draw a large and extremely uninteresting vase. This teacher also seemed to have a fixation with the floor plans of the Parthenon which had to be reproduced with mathematical precision.

On the other hand, outdoor art classes were a lot more interesting. The fernery in the middle of the main Bendigo Gardens was a popular site. Apart from drawing ferns, there were two popular occupations. One was spying on Gwenda X and her boyfriend who seized the slightest opportunity to display their ardour for each other. The other was tormenting the little bespectacled gardener called 'Jappo', although he wasn't the least bit Asian. We used to annoy 'Jappo' and then split up and escape through the three exits. He was a guardian of public virtue and was known to have turned hoses on courting couples. 'Jappo' probably lived in some sort of burrow deep in his fernery.

Other outdoor art activities were somewhat related. One was hearing Jack Strellon, the Lutheran Pastor's son, reading the latest adventures of Carolyn. These appeared in the central pages of the Melbourne *Truth* and were pretty hot stuff. The other was a tramp through the Bendigo Art Gallery where various displays quite relevant to the adventures of Carolyn would be pointed out by my mates.

Art classes also led to visits to the major churches of Bendigo. The magnificent Sacred Heart Cathedral was slightly desecrated by the largely Protestant class. A number of boys took confessions in the confession boxes and Jack Strellon was caught emptying his fountain pen into the holy water.

Bombs and Electric Shocks

At the age of fifteen I had two highly dangerous hobbies. One was making bombs, the other was dabbling with electricity. Bomb making started off innocently enough with collecting unexploded crackers after the annual bonfire night and covering them with a tin to maximise the bang. Home manufacture of a basic gunpowder followed. In no time at all, cans of homemade gunpowder were being set in cement blocks. These were ignited via fuses which surprisingly enough were available from the local hardware store.

When the cement had fully set around a nice little batch of bombs, I would set off with selected schoolmates to the sandhills on the outskirts of Bendigo. The girls would carry the bombs in their bike baskets. When the bombs went off they were pretty spectacular. However, we didn't know what to do with those that failed to explode. Sometimes the fuse would hiss its way into the concrete canister and then all would be silent. The first time we had this problem we placed the unexploded bomb on top of the next one. It went off splendidly and we thought that was the end of the unexploded one. Then someone suddenly said, 'Hey, have a look at that.' A little black dot high in the sky was starting to descend. The unexploded bomb was returning to earth. The exact spot of its landing is still unknown. Later we dropped unexploded bombs down deep mineshafts.

Afterwards we went up-market, using cordite emptied from .303 shells and detonators. The less said about this the better, other than Mother once reported hearing an explosion during an afternoon game of cards. The test site was on the other side of town! Today I am amazed that there were no serious accidents and that our activities were vaguely tolerated. Once Dad muttered, 'Don't blow yourself up' as he walked past a set of concrete blocks with a black fuse snaking out from each of them. When gunpowder production was delayed because of a shortage of decent charcoal our chemistry teacher helped out with some from the school store. He had a rather optimistic look in his eye as he passed it over.

Another potential cause of sudden death was electricity. At the age of eighteen months I stood up in my cot and reached out to a brass light switch on the wall. Sometime later mother came in and found her son happily humming to himself as he played with exposed live wires. Apparently I gurgled with joy when the sparks flew as the live wires touched. Bits and pieces of the switch had been happily scattered in and around the cot. Dad would often tell me this story with a final summing up: 'My word you were lucky.' Then at the age of twelve I built a simple crystal set, which to my surprise worked perfectly. It seemed quite logical to accumulate bits of old radio sets, transformers, odd electrical instruments, etc.

One summer's evening Mother and Father were having a stroll around the garden. In a nearby shed I was applying the leads of a cheap voltameter to a large transformer. I was trying to locate the 1,200 volt output terminals. The 1,200 volts made quick work of the voltameter and almost made quick work of me. I staggered out and sat on a block of wood outside the shed. Dad said, 'What was that zapping sound?' I have forgotten my reply but I remember watching the muscles in my right arm twitching. This shed at the bottom of the garden was formerly a stable and had ancient power lines which stretched from the house far away. Thanks to my efforts the relevant fuse box was black with use. As it burnt out yet again, the fuse would leap out of the power board and nosedive into the laundry basket. Had he been asked, young Edison could have explained the appearance of mysterious scorch marks on shirts or sheets.

Static electricity also fascinated me and I became interested in trying to accumulate this form of electricity. Eventually I devised a system of belts which brought static electricity up to an isolated chamber where it could be collected. The Victorian Science Teachers' Association had recently started a Science Talent Quest and I decided to smarten up my device and enter this competition. I was quite shattered when Mr Hawkins, the science teacher, said, 'Yes, that looks a bit like a van de Graaf generator. They've just built one at Melbourne University.' However, he encouraged me to go ahead with the monstrosity and surprisingly it won a minor prize. Any success in the city by a country child is amplified in direct proportion to the distance from Melbourne. I thus enjoyed brief but quite undeserved fame, which included a eulogy by the Headmaster at assembly.

Creative Activities

My mother actively encouraged her children to draw and write. The now defunct newspaper *The Argus* had a special Children's Weekend Section, and it was rather unusual for a Bendigo Sutherland not to be represented each week. Many a five shilling postal note duly arrived from *The Argus* for a poem or a drawing. My first proper publication occurred there on 2 February 1951. This was a 460 word piece on Madame Curie for which I received the excellent payment of one guinea. This was followed by a similar sized and equally rewarded article on the discovery of Uranus. I then enthusiastically dispatched a batch of articles with an offer to forward further contributions on a weekly basis. The Editor said that he liked the first two articles but he found my writing almost impossible to read and declined the offer.

Watercolours and pen and ink sketches were a regular outdoor activity. I must have painted dozens of decrepit mining buildings around Bendigo. My specialties were rusty tin roofs and chimney stacks. They were fairly low in artistic merit but high on detail. One of my more bland efforts won Third prize in the local Art Show. Later I had a painting included in an Exhibition in Melbourne and the *Herald* reporter thrilled the family to bits by his enthusiasm for the warm brick colouring. I didn't feel too impressed, however, as the reporter described me as 'a young Bendigo lass'.

Tennis

At the age of sixteen I joined the Ironbark Tennis Club in Bendigo. Injuries were relatively uncommon, embarrassing situations were not. After introducing a friend to the Club he was invited to engage in a doubles match. There was a large crowd of spectators. Charlie the Club President was opposing him close to the net. The very first ball my friend struck was a low hard return which just cleared the net and landed fair and square in the President's crutch. For a dreadful second the ball stayed at the site of impact. President Charlie endeavoured to hold back his tears. He then adopted a strange crab-like hover until the pain eased and was never particularly friendly towards the new member.

My tennis career peaked at the Bendigo Easter Lawn Tournament when I was seventeen. The entry fee was five shillings and the athlete's name and the time of match appeared in a local newspaper. I drew the 8 o'clock time on the centre court against a local champion. It was a great humiliation as the balls arrived at such speed the racket was practically blasted out of my hands. Within thirty seconds of the warm up, I knew I was a goner. (If you can't return a ball when the opponent is hitting towards you, what hope do you have when he is trying to hit it away from you?) Next day, the local paper carried the results on the top left-hand side of the first column. There was no charge for this further humiliation.

The Escape

In 1953 I finished High School. It was necessary to swot like mad for the examinations as if they were a matter of life or death. Medicine was the only career that appealed to me and some dreadful alternatives were being mooted if I didn't matriculate well. Dad thought a job in a bank would be beaut. Mother opted for a career in law, having steered me into five years Latin at school as a useful preliminary. Fortunately my exam results, though average, were sufficient for the purpose. I thus had a socially acceptable reason to flee the predictable life in Bendigo.

I was very lucky in 1954 to embark on the first step of a profession that I had unwaveringly chosen to follow when a youngster.

But first my uncles await us before we explore the worrisome life of a medical student. One uncle in particular still strongly influences my attitudes and work habits. So does his wife. Bless them both.

My maternal grandfather, Frederick Knight Hunt, with Grandma Hunt

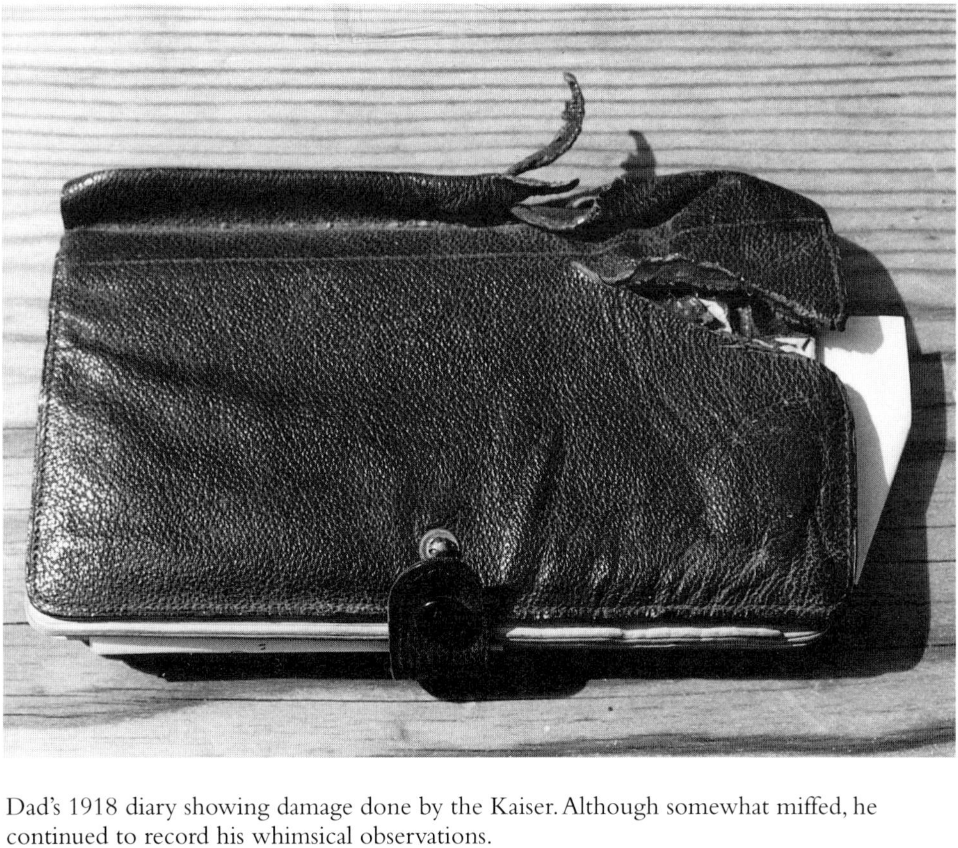

Dad's 1918 diary showing damage done by the Kaiser. Although somewhat miffed, he continued to record his whimsical observations.

Etching from *Romancing of Ballooning*. My great-great grandfather (James Glaisher) in big trouble at 26 000 feet in 1862. His companion, Henry Coxwell, unable to move his fingers because of the cold, pulled the valve-cord with his teeth. They safely landed, having reached the greatest height yet attained by man (29 000) feet).

Uncles 3

'Oh no, thank you, I only smoke on special occasions.'

Labour minister when offered a cigar
by King George VI

Dead for a long time have been my Uncles Angus, Charles and Ian

Uncle Ian	51
Uncle Angus	53
Uncle Charles	55
Life on the farm	63
Charles's influence	66

Two of my uncles had had significant influence on how I viewed the non-Bendigo world. They were both very idiosyncratic, determined men and I drank in every word they uttered. In comparison the eldest uncle, Ian seemed a shadowy recluse.

Uncle Ian

Uncle Ian was Dad's eldest brother and a completely different kettle of fish being a sombre and confirmed bachelor. He'd obtained one of the first Master of Civil Engineering Degrees at the University of Melbourne in 1908.

As an engineer with special interest in hydraulics he had a high reputation. His memorial is the Maroondah Dam which he designed; he was also resident engineer during the many years of its construction. This project suited his bachelor status and he certainly got plenty of fresh, crisp air. The filling of the dam started in 1926.

As the eldest son, Ian did most of the work in winding up his father's estate and problems over the 'Novar' subdivision probably triggered off his interest in financial matters. Instead of engineering he wrote several books on economics which he published privately. *The Monetary Puzzle and a Possible Solution* was published in 1933. He proposed amongst other things that upon the death of an owner of any income-producing property, after payment of debts, etc., the property would revert to the Crown. Twenty years after Ian's death I found in his bedroom boxes of this book in mint condition. I tried unsuccessfully to read it in its entirety.

The letters this book invoked from some of the economists of the day were at times quite fiery and far more readable than the book itself. His correspondence with the famous J. M. Keynes got off to a grand start in January 1935. Keynes thought *The Monetary Puzzle* was a brilliant little book and told Ian that he was 'substantially on the right track'. Furthermore 'such quantities of rubbish looking not unlike your book reach me that it is an exceptional pleasure to read something with so much insight and genius'. By May 1935 Keynes had cooled off and doubted that the book could be published in England. By October Keynes was pointing out how difficult it was for someone like Ian who worked in isolation. He considered Ian's 1942 book *Property and War* not equal to the earlier book and closed with the advice 'you do not really know what it is you want to say, and, therefore, probably you ought not to have been writing a book'. Sadly, *Property and War* is respectfully dedicated to Mr John Maynard Keynes.

The most vicious criticism of Property and War came from a J. H. Wood who considered that the book had no justification for its existence at all. He got quite personal and wrote, 'You are young evidently and immature but your worse fault like most Australians is you don't think for yourself.' Another gem is, 'One gets sick and weary tracking down the errors, falsehoods, etc., so rampant. I have met naught like this since I read *Mein Kampf*.' He concludes: 'I am damn sorry that the reading of your book and writing this have been for me so painful.' This letter must have really made Ian's day. I would have torn it up and burnt it but he neatly kept it under 'correspondence', a file which already held a fine collection of signatures of the world's leading economists.

I can only recall meeting Uncle Ian once. At the age of about ten, Dad and I went round to visit him. I recall a roar of a large motorbike starting up in the garage as we came in at the gate. Playing with this machine was a greyfaced man with a very dark and rather evil moustache.

He died several years later of a heart attack while striding up one of the steeper hills at the Melbourne Botanical Gardens. According to Dad, a month earlier he had been told to limit his exertions. After his death, Dad told me, 'You may draw your own conclusions.' I'm still trying to draw my own conclusions some fifty years later.

I never heard any jokes attributed to or made about Uncle Ian. But apparently his life was not all gloom. After his death, some underclothing supposedly belonging to a distant female cousin was found in his chest of drawers. He left half his estate to this relative. As a consequence, one of my aunts refused to sit in the solicitor's office in her presence. I like to think that on the one occasion I met him he was starting up his motorbike prior to setting off for a romantic picnic or some other pleasant activity with this cousin. A recreational bonk may not have been out of the question.

Uncle Angus

Uncle Angus was my Dad's favourite brother and his imminent arrival in Bendigo always caused a flurry of activity. At dawn Dad would rake the front path, which was his equivalent of putting out a welcome mat. Angus was a land valuer for the State Savings Bank and drove intriguing cars. My favourite was a Ford Mercury with multiple little glass panels in its front end. It would be carefully parked in the backyard overnight, giving me plenty of opportunities to inspect it thoroughly.

Angus was short and stocky with a deep mellifluous voice. He had a very Scottish face and years later when I was in Scotland I could spot Anguses all over the place. In the film *Whisky Galore* there are at least five facsimiles of Angus in the cast.

After the formalities of his arrival were over, Angus and Dad liked to withdraw out of earshot. They'd generally settle on a heap of bricks down at the bottom of the garden. When this conference was over we could all enjoy Angus's warm and entertaining company. Angus had two other attributes which made him special. He was an inventor and a serious practical joker so he often had some tricks up his sleeves to amuse a small boy.

Angus had been born in 1892 and christened Hector Anaeus. He served in World War I and in the old tin trunk I found a letter he had written to his father who was opposed to his enlistment. These extracts give insight into his tenacity.

Nhill
14th January, 1914

Dear Father,

You will have seen by my letter to Mother last week that I have not given up the idea of getting into the Exped Force.

Three months ago I had the chance of getting into the light horse here but you did not want me to so I just gave up the idea, and when I was on holidays I was thinking of making enquiries in Melbourne but you put me off then too, since then I have had another good chance of going, the Manager of the State Savings Bank here is a Captain and is going down to camp to train men and would help me ...

Niman (the accountant here) and I have both been thinking about this for some time and last night, without binding ourselves in any way, we got the form for volunteering and were examined by the local doctor, he was doubtful about Niman being accepted but said he had no doubt about me and signed my papers, so if I sign the form enclosed, with his signature on it, I will be accepted ...

> *I am enclosing the form in which you can see the doctor has passed me, and am enclosing an envelope to return it in, I would like to hear from you, but I do not see any reason why I should not go as it will not affect anyone. In any case I have about two weeks to make up my mind. With love to all at home.*
>
> *Angus.*

My cousin Dr Rod Sutherland has given me the following information about his father. Clearly Angus's prediction that 'it will not affect anyone' turned out to be unduly optimistic!

According to Rod:

> Angus served in the 7th Battalion in France and was wounded at Pozières. Several men were buried by a shell and their bodies were dug out from the mud. Most were dead but Dad was still alive. He was sent off to England and woke up two days later in Guys Hospital where he lay paralysed from the waist down. After six weeks he began to recover the use of his toes and returned home on a Hospital ship. On arrival he was able to walk on crutches and eventually recovered enough to join the Royal Australian Flying Corps. The war finished before he'd finished training and he became a soldier settler at Yatpool. He had several good seasons on the land but his injuries had left many weaknesses in his back and leg muscles and he moved to a job with the Lands Department in Mildura. Later on he applied for a job with the State Savings Bank as their Land Valuer in the Mallee and Wimmera area. Two hundred people applied, the Depression had begun, but he managed to obtain the job which he held until he retired.

It is strange that I should first have heard of this miraculous escape at the age of sixty.

On a lighter note, Rod related a fine example of quick thinking by his father when he joined the Royal Australian Flying Corps.

> At his initiation he had to sit stark naked on a block of ice and sing a song. He chose the National Anthem and was able to stand up!

Angus was a part-time inventor and some of his devices were manufactured and marketed. One that I remember was a metal tea caddy which had a little knob near its base. If the knob was pulled when the caddy was held over a teapot, exactly one teaspoon of fresh tea fell out. Rod Sutherland told me about two other inventions. One was a fire detector which used a type of firecracker and the other was a device to disguise a machine-gun. Testing prototypes must have been fun.

Angus handmade some of the paraphernalia for his practical jokes.

During the sixties, trick shops and novelty stores across Australia were selling life-sized model snails. These little artistic delights were used to garnish the salads of the unsuspecting. They were equipped with genuine lacquered shells, realistically slimy looking bodies and eyes on stalks in the 'go' position. This activity dramatically reduced the snail population in the region as snails were collected from the gardens of all and sundry. They were mass produced by Angus in his garage in suburban Camberwell, Melbourne, and although somewhat of an embarrassment to the family were apparently fairly lucrative.

Another of Angus's popular creations was a set of lifelike fingers which could be wrapped around a door. They were particularly effective on a sliding door. It can be quite heart-stopping to be alone in a house on a dark and windy night and spot a set of fingers clutching the edge of a door!

Angus liked testing prototypes on his nephew. Once, when bidding good night at his front door, he excused himself for a minute, then returned and solemnly proffered me a hand to shake. A cold clammy hand remained in mine as he turned away. It was a rubber glove filled with sawdust and precooled in the refrigerator.

At an aunt's house years later I came across a set of tiny metal type of the kind used last century by children for making name stamps, etc. On the underneath of the little wooden box holding the type, someone had neatly stamped several times: 'Angus is a pig. Angus is a pig'. Perhaps one of his siblings had been moved to immortalise their attitude to him after a practical joke. Angus would have made a wizard of a backroom boy designing special devices in World War II. Imagine Germany being invaded by exploding snails!

At the age of eight I once got the better of Angus with a spur of the moment comment. The day before his visit, Dad and I had raked the front footpath and then proceeded to clean out the chook house. The chooks had become infested with lice, insects I had never seen before. They were fascinating, the lice that is, and it was necessary to paint all the perches with disinfectant and change the straw. This must have made a great impression on me because next day, when standing beside Angus, I looked up at him as he was patting my head and said, 'I wouldn't do that, Uncle.' And when he said, 'Why not?' I said, 'Because I've got lice.' His hand froze on my head as he looked inquiringly at my father.

Uncle Charles

For many years at Melbourne Zoo there stood an elderly adjutant stork in solitary confinement. No-one knew how long this animal had been in residence. The fine old creature observed the passing public with a pair of extraordinarily bright blue eyes. Its eyes were identical to those of my Uncle Charles. The other similarity between the stork and my uncle was that both were very bald. The rest of this ancient bird was not in the best of condition. Its long bill was discoloured and gave the appearance of having been attacked by termites. Many of its feathers were missing and its legs, quite frankly, were an embarrassment to look at. Often when the stork and I were contemplating each other, passing children would pause and call their colleagues over. They would make comments like 'Yuck, isn't it horrid.' After they had rushed away, I sometimes imagined a hint of a tear in the old bird's eye.

The stork refused to share its allotted area. Attempts to introduce other fauna into its territory confirmed the fact that, although its beak looked fragile, it was still a formidable weapon. It was a sad day when I came up for some mutual meditation with this old friend and found him gone and his enclosure filled with a collection of rather insipid marsupials. I felt I had lost a tenuous but precious link with my Uncle Charles who had an enormous influence on me.

My first recollection of Charles was near the end of the war when I was seven. Dad was in a double flap and must have raked the front path at least three times. Clearly this was going to be An Event. I can remember the dapper fellow in the uniform of a Major who wore a watch which had both a leather cover and a protective grill. Much to my delight I was allowed to open and close this particular adornment a number of times.

Charles was undoubtedly the cleverest of my uncles. Very thin, about five foot, six inches in height, he always moved about at high speed. His unobtrusive English accent was the result of years spent overseas. Charles was one of the first Allergists in Australia and, at his peak, was probably the best known one. He founded the Australian College of Allergists and was its first President.

His medical course at Melbourne University was particularly arduous because being the youngest he was the last son left at the Dandenong homestead. Apart from lengthy train trips each day he had to milk cows and do other jobs around the homestead, prior to leaving and upon his return. He graduated in 1920 and initially worked at the hospital for sick children in Brisbane. He then went into general practice at Alpha, Queensland, where he had some hair-raising experiences. These are meticulously recorded in his diaries which he kept from 1917 to 1955.

In 1922 Charles went to London to study allergy. He returned to Melbourne two years later having gained membership of the Royal College of Physicians and been trained in the latest techniques in the treatment and investigation of allergies.

He set up in Collins Street and soon had a large and busy practice. In those days, before antihistamines, antibiotics and cortisone, severe allergies were usually treated with desensitisation. It was often the asthmatics' only hope of regaining their health.

Apart from being an enthusiastic clinician, he was very keen on laboratory work and set up a small laboratory in the grounds of his home in

Melbourne. From his birth in 1896, he led a frenetic life, which was sadly cut short when he developed Parkinson's Disease. He died in 1962 before the release of new treatments which might have both ameliorated his symptoms and prolonged his life.

Charles was undoubtedly the pride and joy of the family. At thirty-five-years-old he was considered a confirmed bachelor and lived in a huge house called Duncraig which still stands in Armadale, Melbourne. Comfortably ensconced in this seventeen room house were his elderly mother, two unmarried sisters and the brooding bachelor, Ian. About 1935 this cosy arrangement was suddenly shattered when Charles suddenly became more sociable and started attending society dances in Melbourne, where he met Marjory Minifie who was the only daughter of the founder of O-So-Lite flour mills.

Before he married Marjory in 1936, all his kith and kin were moved en masse to a house several blocks away. In 1937 Marjory nearly died from an obstructed labour and spent many months recovering. She was saved by some of the first multiple blood transfusions carried out in Melbourne. These were performed by Dr Ian Wood who was to become my boss, and was later knighted Sir Ian Wood.

According to Sir Ian, Charles practically abandoned his practice for the pursuit of suitable blood donors. Day after day, he would hold what he called 'coffee parties' while the blood groups of prospective blood donors were screened. Although unable subsequently to have children, Aunt Marjory regained her extraordinary vigour and played tennis and gardened well into her eighties.

Charles and Marjory were in England at the outbreak of World War II but were soon separated. Marjory spent much of the time in voluntary work in England, whilst Charles became a Major in the British Army and ended up in charge of a hospital in Colombo. He described operating on patients who were injured during the Blitz, sometimes with only the light from kerosene lamps. I asked him once what he had been most frightened of. He replied without hesitation 'Flying glass'.

Back in Melbourne after the war, his practice flourished. He was at the forefront of a number of advances in his field and was the first doctor in Australia to use ACTH (Adreno Cortico Trophic Hormone). He and Marjory were a very dynamic couple and nephews and nieces benefited from the special interest this childless couple took in their young relatives.

Like us all, Uncle Charles recognised that my eldest sister, Diana, was a very bright cookie. Like me, from the word go she wanted to do medicine and Charles gave her every encouragement. She had a fantastic memory. At the age of fifteen, she did very well in a national radio quiz programme. We would gather around the radio and with wonder hear her spout correct answers to questions such as 'What is the population of Australia per square mile?' and were stunned when she was knocked out of the finals by a memory block. She was asked the name of a famous

British comedian who was visiting Australia. It was Tommy Trinder, which even I knew. Despite this disappointment, she became famous in Bendigo and I took pride in being known as Diana Sutherland's brother.

Diana had a laboratory in the cellar of our house which she used for all sorts of fascinating pursuits. She once managed to create a chewing gum which she tested out on me. It was a gritty, grey material with a distinctly metallic taste.

Right from the start I took a great interest in her activities. Anything she had finished with, I seized. Grade 6 saw me giving a demonstration of indicators to my rapt classmates and a surprisingly interested teacher. A container of colourless liquid would be miraculously rendered deep purple by the addition of another uncoloured liquid. Then all the colours of the rainbow would appear by adding further indicators. For a finale, the whole vessel erupted like a giant alka-seltzer when I lowered in marble chips. This was great fun, but subsequent experiments became more dangerous. About the same time, I made a batch of soap which turned out to be amazingly long lasting. The end product would not lather, smelt foul, and blocks of it can probably still be dug up from the garden where it was buried many years ago.

There were extraordinary tensions in the house when she was doing Matriculation, including regular and spectacular flare-ups between her and my father. Mother tried to prevent these, but there seemed to be an almost physiological need for both of them to let off steam regularly. The rest of us just took cover and it was pleasant to have the flak drawn away from me. It was all worthwhile when we heard that she had matriculated very well, obtaining as a sideline First Class Honours in British History.

This was good enough for Charles and he proposed to support her during her medical training. This was grand since there was no way Dad could have financed her completely. Dad had supported Charles on and off during his medical course so his help was accepted gratefully, although I recently found a letter written to Charles by my mother about this time pleading with him not to support Diana as she did not believe women should do medicine.

In 1949 large numbers of ex-servicemen wanted to start university courses and because of this overcrowding a branch of the university was set up in Mildura at a former army base. This campus housed first-year medical and dental students and heaven knows who else. The leap from being a schoolgirl in Bendigo to a resident at university produced a few startling changes in my eldest sister. Her language became more positive, particularly when directed at me. On her first day at home, I followed her around asking annoying questions and I still remember the shock when she suddenly turned angrily upon me and sharply told me to 'piss off'.

During her first summer vacation Diana obtained a job at the Bendigo Jam Factory. Before setting off, she proudly told me of the large financial rewards that were close at hand and I was promised a new tie. However, the job turned out to be awful and she had to quit on day three after cutting her finger. Work at the Jam Factory was never mentioned again. Nor was the tie.

After her first year in Mildura she became a Resident at Janet Clarke Hall. She worked like a demon and the only major hiccup occurred in the final year. Charles had an obsession about tax deductions and he insisted she do a certain amount of library work researching aspects of allergy for him. Final year is tough enough without any overloading and after trying it for a while, Diana refused to do any research. Her stand was vindicated when she came top of the medical course. Sadly she got none of the main prizes because they had all been won by separate individuals in each subject. I was now even prouder to be known as Diana Sutherland's brother, especially when she became a Resident at the Royal Melbourne Hospital.

From the age of about eleven, I stayed several times a year with Charles and Marjory at Duncraig. It remains the largest private house I have ever stayed in in my life. It was built in 1857 and was added on to from time to time and still exists having been classified by the National Trust. However, the large garden has been subdivided and the tennis court and an assortment of interesting outbuildings have disappeared.

This house, its inhabitants and their visitors had a dramatic effect on an eleven-year-old from Bendigo. The style of living was both gracious and austere. My lasting memory is of breakfast served on the patio in the beautiful garden with freshly ground and brewed coffee. Marjory was a frugal shopper but produced large and unusual meals which more than satisfied a growing youngster's appetite. At first I was appalled at the thought of tucking into a sheep's head but soon fell in line with everyone else. She never stopped, and thought nothing of packing in a game of tennis between gardening and other activities. Before Charles went off to his rooms she would pick flowers for him to take into the city and then drive the car out of the garage and dust it down. We would then line up and wave him good-bye.

In those days, Charles was almost guaranteed a parking spot in Exhibition Street, near his Collins Street rooms. Post-war, like the Bendigo doctors, he had opted for a new American car. Being a small chap, he had a large Oldsmobile. It had a massive front grill which he described as being like the grinning lips of a barmaid with no teeth.

Although Charles belonged to the Melbourne Club, he usually ate alone in his rooms. His lunch would consist of a couple of slices of bread with an extremely mundane tasty cheese. Like a lot of my family, he liked eating alone and thoroughly enjoyed his own company. I think one reason he bought his little farm at Macclesfield (of which there is more later) was so he could get away from humanity.

For thirty years Charles treated patients free of charge at his Allergy

Clinic at the Alfred Hospital. He would be particularly annoyed if the patients or their parents proved that they were not candidates for his free treatment. One pet hate was for them to be caught smoking tailor-made cigarettes. This would put him in a bad mood for the rest of the day. As a result his return at nightfall was something to be faced with trepidation by his family.

Once I temporarily annoyed Charles. We were discussing the different types of doctors and I brightly told him that our local chemist (Mr Corrigan) had said that most of the Collins Street specialists 'Charged you £10 just to hang up your hat'. He had had a tiring day and this did not go down very well.

Charles and I were always pleased to see each other when I came to stay. I remember the surprised look on his face when he found I had grown 2 inches taller than him since our last meeting. Staying at Duncraig, however, was no rest cure because my Aunt and Uncle thoroughly organised me. Certain routine jobs were mandatory: washing-up, replenishing the stack of logs in the bins near the open fireplaces and lumping coke to the proximity of the hot water system were all performed semi-automatically. Like magic, jobs would appear in the garden to be supervised by Aunt Marjory. A nephew was not allowed to lounge around the place while his uncle was at work.

When I'd done two tough days I was exhausted and somewhat disheartened, but after dinner on the second night Charles announced I'd done a good job and had already earned £3. He said his going rates for students were five shillings an hour. This put a different slant on the whole situation because, not only was I pleasing my Uncle, but I was also well on the way to becoming a capitalist. My cousin Rod, Angus's son, a very cheerful fellow who was at that stage halfway through his medical course, was my co-worker in these activities.

When work at Duncraig petered out, Marjory let her friends know she had an enthusiastic gardener for hire and I went forth to mutilate plants and lawns over a wider area.

When these jobs also faded out, Charles introduced me to the world of free enterprise. He placed an advertisement in *The Age* stating that a student gardener was available at six shillings an hour. These new gardening activities were a different kettle of fish. With Aunt Marjory's friends, I would just be getting up a sweat after some thirty minutes of work when they would come out and offer me cups of tea and cakes. They would also warn me against 'bursting my boiler'. The respondents to the advertisements in *The Age* were all for student exploitation and the completion of Herculean tasks. They made Duncraig look like a rest camp.

With some of my new-found wealth I bought a little white statue of a horse at Myers for my mother. I showed it proudly to Charles who cocked his head to one side and muttered something about 'Beauty, I suppose, is in the eye of the beholder'. Unfortunately I dropped it when unpacking at Bendigo and its left ear broke.

I brought home assorted bits and pieces of radio equipment purchased at the Army Disposal Store and I also did some shopping at Selby's for test tubes and various chemicals. At that time the store was at the top end of Swanston Street and had a fine little device in the window. This consisted of a sealed glass container with a small metal fan which spun around enthusiastically when the sun shone upon it. They are now back in fashion and we have a cute little one in our living room.

Weekends with Charles were spent either at Duncraig or at his little farm. Both were completely different but similar, in that I followed him everywhere like a shadow. He was always doing interesting things. His workshop was filled with all sorts of modern gadgets and, like most of the family, labels abounded everywhere. Charles always wore rubber or leather gloves when doing any manual work and this impressed me even further. During his weekend activities he wore a white floppy hat to protect his dome. Even now I can see his beady blue eyes sticking out from underneath it.

Some parts of Duncraig were always in need of repair and he would canvas opinions from suitable tradesmen amongst his patients as to the best way to tackle a problem.

Odd jobs aside, our greatest mutual interest was in the laboratory he had built beside the garage. Most allergen extracts used to treat patients were prepared in his rooms in town by a science graduate, but some of the larger batches were produced in this laboratory. His major interest was in the preparation of allergens from house dust. The contents of vacuum cleaners from different parts of Melbourne would be extracted and compared.

It was great fun putting house dust suspensions through a wine press and obtaining a beautiful clear liquid not dissimilar to a light rosé. He had spent some time working in America learning the latest techniques of allergen extraction, and his laboratory contained the most up-to-date equipment. Sadly, all this work led nowhere and only shortly after his death was the root of the problem discovered. Overseas workers identified a minute house mite in the dust as the primary cause of allergy. Fruitless although his research may have been, we were not to know that at the time and I revelled in tinkering with the equipment.

We made other extracts for treating Jumper ant and Bull ant allergy. I don't remember who provided Charles with the little bags filled with ants for processing. However, it is now known that 'whole body extracts' are not effective in treating this common and often very dangerous allergy. Only the purest venom can be used to desensitise the patients; the treatment starts with minute doses of venom which are slowly increased rendering the patient immune rather than allergic to ant venom.

It's strange to think that some fifty years later I was involved in the dissection of more than 30 000 ants to obtain purified venom for patient therapy. (See page 313.) Charles would have been fascinated both by the extraction process and the new type of equipment. Now there are also laboratory tests, which he could only dream about, to follow the patient's progress.

On Saturday afternoons there would be vigorous games of tennis with all sorts of regulars dropping in. Charles was resplendent in a better quality white hat and long white trousers. This was just as well because his knees were even funnier than mine. At first I wore my Bendigo tennis-playing gear and this apparently stimulated in my absence a discussion on my apparel between Charles and Marjory, resulting in Marjory marching me into a shop in Prahran where I was suitably bedecked in tennis clothing of a uniform whiteness.

My Bendigo-style tennis also needed a little attention. Although moderately effective, it had a certain eye-catching quality reminiscent of *Monsieur Hulot's Holiday*. Marjory took me to classes several times a week, which I suspect were run by an old school chum of hers. She was a large corset-clad lady who tended to wrap her arms around me to show the position to adopt prior to serving, etc. This would make me squirm but she would murmur reassuringly in my ear, 'There is nothing personal in this.' I took my revenge when Marjory wasn't watching, by hitting the ball further and further away from my instructor which tested her heaving corsets to the limit.

Now resplendent in white and fully trained, my singles matches with Charles became more spirited. Once, to our mutual surprise, I actually beat him and it became more difficult to convince him to come and have 'a quick hit'. It was an early lesson in one rule of life: to remain popular, it's best not to be too good at anything. A week later I at last beat Marjory and bragged to him of this fact. He quietly told me that I should not take any pride in my victory because she had been very ill once after falling off a horse. Marjory looked quite confused a little later when I asked her about the accident, mainly because she had never fallen off a horse and her injuries were the result of her prolonged labour.

A revelation is the best description for a Dinner Party at Duncraig. To be seated with ten or more people by candlelight in the huge dining room was a pretty spectacular experience—a bit different from dinner in Bendigo. At Duncraig the youngest fellow guest would usually be at least thirty years my senior and generally I would be seated between two of the more senile guests.

Lack of immediate distraction allowed me to absorb the whole atmosphere and to follow as much of the conversation around the table as possible. Charles would be particularly genial on such occasions and his face would take on a pleasant glow. Way down the other end of the table, Marjory always seemed to catch his jokes and laugh infectiously. I had asked Charles earlier why they dined by candlelight and he had looked at me quite benignly and said, 'Marj thinks it enhances my classic features.'

Candlelight certainly did something to a dining room which was twenty feet or more long.

After dinner the ladies would literally withdraw into the drawing room whilst the men gathered around the fireplace in the dining room to smoke and have a port. At first I missed out on after dinner drinks as Charles informed me a glass of Benedictine would blow the top of my head off. I'd sit on a stool in the background and listen to such luminaries as Sir Macfarlane Burnet and Professor Trikojus exchanging jokes that were described as 'old chestnuts'. I felt as though I was 'one of the boys' when I was allowed to join in the laughter which followed Professor Trikojus' joke about a phantom piddler. At other times they would deliberate on ponderous matters which concerned scientists, such as the drift of water molecules away from earth into outer space.

After the small dinner parties we would move into the drawing room and listen to music. Sometimes it was recorded and came from a speaker fitted high up on the wall at the far end of the room. It had been set deep into the wall and was not obvious. Usually the lights would be dimmed and the music chosen was often Chopin. A large open fire illuminated the meditating faces of the guests.

One old chap amazed me because he would fall asleep in the semi-darkness and start snoring within thirty seconds of the music starting. He would awake with a jolt when the old 78 record got to the end of the track. A few seconds into the new record and he would be snoring rhythmically again. Like Harvey's demonstration of the flow of blood up the vein upon the release of pressure, this exercise could be repeated innumerable times.

After-dinner music was also made by Marjory's mother, Mrs Minifie. Then in her eighties she was tiny with silver hair, a tight bun and a traditional brooch and black ribbon around her neck. She had brought her own Grand piano to Duncraig when she moved in many years earlier. Mrs Minifie had an enormous musical repertoire. Before I went to the first film of an Opera I had ever seen, she ran through the main arias and included a fairly vigorous rendition of the 'Anvil Chorus'.

Mrs Minifie was from a bygone age. She ate like a sparrow and pronounced 'room' as though it was spelt 'vroom'. I was told she always had a cold shower, never a hot one. She had a regular daily routine, part of which included half-an-hour raking up leaves from the garden. She and Marjory now rest together at the Brighton General Cemetery. I always visit their graves each New Year's Eve on the anniversary of Marjory's death.

Life on the farm
Charles owned a small farm at Macclesfield in the Dandenongs. It consisted of a few cleared acres, a little farmhouse, a tenant and a number of cows. The rest of it was natural bush with a small creek and, in the

middle of this wilderness, Charles set up a semi-permanent camp. He liked to spend as many weekends there as possible, preferably alone.

Apart from functioning as a retreat from humanity and giving the normal Primary Producer Tax deductions, parts of the farm were used to grow various grasses for preparing allergen extracts. The farm also provided firewood for Duncraig and functioned as a mini rustic penal colony for nephews.

Two types of weather were available at Macclesfield. It was either wet and cold or it was extremely hot with blustering north winds. If it did stop raining temporarily, there was always enough moisture dripping from the foliage to maintain one's state of wetness prior to the arrival of the next shower. Mud was everywhere and this strengthened the leg muscles, as one would clump around in gumboots with half a ton of mud adhering to each boot.

In the summer, the air was filled with the most virulent pollens possible, especially from Charles' rows of allergenic weeds. I have never suffered such extraordinary hay-fever since. I would stagger along behind my oblivious allergist of an uncle, with swollen red eyes, blocked nose and a red-raw throat. Temporary relief could be obtained by submerging my head and neck into the creek and wiggling them around which allowed the water to flush through the various orifices and crevices involved. I would then have to catch up with Charles, who had briskly gone ahead clanking away with the spades, axes, etc., he carried.

If Duncraig was busy, life at the farm was at an even more hectic pace. Charles wanted me exhausted and preferably out of his hair.

Talking of hair, or rather lack of it, I made a blunder shortly after first arriving on the farm. Charles had cracked his dome on a low hanging beam and uttered the moan which I was accustomed to hearing my father produce when striking this particular region. The sound is only heard coming from bald men and is one I can now faithfully and almost unconsciously reproduce. When Charles had regained his composure I brightly asked him why bald heads are so funny. He muttered he had had no reason to understand that was the case.

I explained that recently at school we had all had a good laugh about a bald and elderly Greek poet. He had been walking around a lake when he had been struck on the balders by a turtle. The turtle had been born aloft by some bird of prey which had jettisoned it with a precision and timing that would have done justice to an aircraft's bomb aimer. I roared with laughter again at the idea of this fluke accident involving two convex surfaces hitting each other with such a resounding thump. The icy silence from Charles suggested I should not push the funny side of bald heads any further.

Anyway, back to the farm and jobs. When I was about twelve Charles purchased a new axe suitable for my small and relatively ineffectual physique. Initially I quite enjoyed chopping down trees that were designed to go up the chimneys at Duncraig. The whole of the Dandenong Ranges would ring with my screams of 'timber' as I cut down saplings that Charles described as 'probably no thicker than his wrist', but this soon wore a bit thin when the felled trees had to be chopped into four foot lengths. Pulling down bigger trees with a winch was far more exciting because it had elements of great danger, involving steel cables which might snap and so on. However, using the hand cross-cut saw became boring after a while and every time we paused I fancy Charles was just as ready to have a rest as I was.

The machinery on the farm was its greatest attraction. In letters, Charles would inform me of the latest acquisition, often with stick figure cartoons showing the new equipment chasing his nephew. An early favourite of mine was a motorised grass-cutter which moved at reasonable speeds while one ran along behind it endeavouring to control its throttle and brake. After a bit of practice, I was sent to cut the grass in the paddock around the tenant's home with the instruction not to run down old Andrews (the tenant) because he would probably clog up the cutting blades! Charles was very pleased with this motorised scythe because, for the price of a gallon or so of petrol, it would keep me occupied and out of the way for hours on end.

Later when he got a little Ferguson tractor I was in heaven. That year, the most thoroughly ploughed patch of land in the universe could be found on a small field at Macclesfield.

We seemed to spend hours and hours repairing fences and cattle walks, etc., but I doubt if any evidence exists today of all this activity. However, Charles certainly enjoyed it and was always in a much better frame of mind when he returned to Duncraig on Sunday night than when he had left the day before.

The first dawn that I experienced on the farm with Charles got off to a bad start. I was lying awake wondering whether I would need a blood transfusion after the efforts of the mosquitoes during the previous two hours. Suddenly the rays of the sun struck a glistening object on the bush table where the water drums were kept. Curiosity got the better of me and so I eased myself out of the creaking camp bed and waddled over to explore. It was a glass teacup, rarely seen nowadays, and in it were three pristine white blocks. Little crevasses in the blocks were filled with gold. Unexpectedly Charles approached me from behind after having had a drover's breakfast (a quiet wee and a look around) in the distant scrub. He angrily claimed his false teeth and ignored my questions a little bit more than usual for the next few hours.

Once, I was left up on the farm by myself for a week when Charles went back to town. I was issued with rations and a list of jobs to do. However, I had no watch.

When Charles arrived at the camp the following Saturday at 9.30 a.m., he found me tucking into lunch because I'd lost all sense of time. He was a bit annoyed at first because he thought I was merely having a leisurely breakfast. Upon reflection, he then decided it was all rather funny. He said that if he'd left me there for another week, I would have been completely out of time scale and be having breakfast at sunset. His cheerful banter waned somewhat when he found that I had eaten most of the emergency rations, some of which he intended to use that weekend.

I pointed out that I had cleared about 5000 acres of bushland during the week, an activity which gave me an extraordinarily healthy appetite, but he muttered something about 'how did I manage to find time to do any work whilst devouring so much food'. By lunchtime, when I was ready for my main meal, he was not in a particularly good mood. Apparently I had downed several trees on the adjoining property, having not noticed that the half buried strand of wire was the remnant of a dividing fence.

Since all the tinned meat and sardines, etc., had been disposed of during the week, lunch consisted of a tin of tomatoes on toast. However, this was not your ordinary tin of tomatoes. After being submerged in a billy of boiling water for twenty minutes, it was removed and placed on the primitive camp table. When he'd allowed it to cool for about five minutes, Charles approached it confidently with a can opener.

Within a millionth of a second of the can opener penetrating the lid, Charles was hit in the eye by a jet of hot tomato fluid. I thought it served him right, but said nothing. I watched him cautiously approaching the can again after a five minute delay. When his hand was only six inches from the can it erupted again, this time firing a jet of hot liquid straight up his sleeve. Charles must have known that if a slapstick act had a good response from the audience, it should be repeated as soon as possible. It is better the second time when you forsee the inevitable about to happen. Charles was quite subdued as he ate his lukewarm tomatoes. He was probably contemplating the blessings of being childless.

I found it hard to understand why Charles found pleasure in the camping life away from the comforts of Duncraig. To be woken before dawn by mosquitoes, cold and stiff in a very uncomfortable camp bed, is not an ideal way to start the day. Without a hot shower before breakfast, I did not make a contented pioneer. On the other hand, the smell and sounds of the bush could be magical and the companionship around a camp fire quite marvellous. Nowadays I prefer to grill a Sunday chop in the bush and watch the smoke from the camp fire drifting up and around the trees, but get home again in the evening leaving the bush to itself.

Charles's influence

In my pre-adolescent days when Charles and I were getting on very well, he proposed to my parents that I become a boarder at Melbourne Grammar School. I wasn't very keen on this idea as I had heard all about boarding schools from some of the locals and I was quite happy at my Bendigo school. Charles was a little annoyed at my lame excuse that I had to stay at home to chop the wood. Another factor was that, although Charles offered to pay the fees, my parents would have had to provide all the school uniforms, which was an outlay they could hardly afford.

Of the ten boys that I knew who had been boarders in Melbourne, each and every one of them absolutely loathed the experience. In the long run, most did poorly academically. Alan even caught ringworm and had to have his hair shaved off and be painted with blue dye. He was a remarkable sight. His appearance led to the only time I can recall when we actually engaged in a punch-up. The cause I believe was that I called him baldy once too often during this period of extreme self-consciousness. As his hair began to sprout again, he was given the nickname of Fritz which lasted for a period of many months. The risk of being made bald and painted with blue dye seemed another good reason not to become a boarder.

In spite of this Charles had an increasingly sound influence on me. He was the first person I ever met whose schoolboy interest in science had continued strongly into his middle life. He was very well read and a mine of information on all sorts of things ranging from how atomic bombs worked to how plant hormones were used to kill blackberries on the farm. He was often astounded by the complete ignorance of an eleven or twelve-year-old boy. For example, there is a street called Valency Street not far from Duncraig. Driving past it one day, he wondered aloud why it should be called Valency. I said 'What is Valency?' and he went into a broody silence for some time until I repeated the question. Charles then gave me a precise run-down on how atoms had either one, two or three hooks, and so could link together. The number of hooks indicated the extent of the valency. The following lunchtime he popped out of his rooms in the city and bought me a book on basic chemistry. Over the next few months I accumulated a mass of useless information on the melting points of elements and their various other properties. All this had to be re-learnt several years later with more complicated explanations than my uncle's little hooks.

Once, after discussing the difference between germs and viruses, Charles presented me with Paul de Kruif's *Microbe Hunters*. I received this in 1949 at the age of thirteen and still have the same well-thumbed copy. It was the 22nd printing of the book which first appeared in 1926. It is a highly readable saga describing the work of Pasteur, Koch and many others. Mother said that for months on end she would always find it underneath my pillow.

Here is a sample of the galloping style of de Kruif as he describes Robert Koch in pursuit of the bacteria which causes Tuberculosis:

> Now he went to hospitals everywhere in Berlin, and begged the bodies of men and women who had died of consumption. He spent dreary days in dead houses, and every evening before his microscope in the

laboratory where the stillness was broken only by the eerie purrings and scurryings of guinea pigs.

He injected the sick tissue from the wasted bodies of consumptives who had died, into hundreds of guinea pigs, into rabbits, and three dogs, thirteen scratching cats, ten flopping chickens and twelve pigeons. He didn't stop with these wholesale insane inoculations, but shot the same kind of deadly cheesy stuff into white mice and rats, and field mice and into two marmots. Never in microbe hunting has there been such appalling thoroughness. 'Ach! This is a little hard on the nerves, this work' he muttered. (Thinking perhaps of the lightning move of the paw of one of the cats jabbing the germ-filled syringe needle into his own hand.) For Koch, hunting his invisible foes alone, there were so many disagreeable and always imminent possibilities of excitement—of something tragically more than excitement. 'But the hand of this completely unheroic looking little micro hunter never slipped, it just grew drier and more wrinkled and blacker from its incessant bathes in the bichloride of mercury—that good bichloride, with which in those days the groping micro hunters used to swab everything down, including their own persons.

This book was almost as exciting as Biggles ('with his jaw set grimly, Biggles kicked the rudder bar as he swooped down on the Messerschmitt'). But the tales of the fight against the various diseases and the dangers the researchers faced, were far more absorbing and were remembered longer.

At times I would have trouble with Charles because of my limited vocabulary. Once he embarked on a mini discussion concerning the importance of avoiding a mediocre career. From my painting I knew ochre was an off-yellow, but I couldn't see why such a colour as mediocre should be used to describe a career. It took me some time to work out the drift of the conversation. On another occasion he began to talk about what I thought was the importance of Korea, as it was the height of the Korean conflict. Again, it was some time before I caught up with the general drift of the conversation and realised he was talking about my career. Despite these various hiccups in communication, I always listened avidly to all he said, either directly or indirectly.

My closeness to Charles probably peaked when I was in Year 11 at school and then sadly declined for some years, as I sought greater independence. However, communication with him was continuous with regular letters being exchanged as I progressed through High School in Bendigo.

A Student and Hospital Intern in Melbourne 1954-1961

4

'... a parcel of young cutters and carvers of live people's bodies, that disgraces the lodgings ...'

Mrs Raddle, landlady, describes medical students in *The Pickwick Papers*. Charles Dickens (1812–1870)

Prologue	70
Fruit Picking for Fun and Profit	70
Settling in at University	72
Academic Innocence and an Approaching Storm	74
The da Vinci of the Dissecting Room and Romance Nipped in the Bud	75
Clever Rabbits: Later to be Recalled	76
Joys of Physical Chemistry	76
The Rules of the Game	77
The Sky Darkens	77
Cleaning at the Women's Hospital	78
The Axe Falls	79
Becoming a Troop	81
St Richards' College for Young Gentlemen	83
St Richards' breakfast cuisine	84
Sisters of the Holy Name	85
Things to do in a boarding house bathroom	86
The cleansing of St Richards'	87
People Under the Microscope	89
Oral Examinations	90
Death of a Sister	90
Earning a Crust 1956 to 1958	91
The Myer Emporium	91
Hot Cross Buns—becoming a master pastry cook in eighteen hours	94

A glimpse of the inside workings of the Victorian Railways	95
The neutral ground of a hospital orderly	97
An Intern in Casualty 1961	101
Certifying the dead	101
Poised for disaster	103
My first professional house call	105

Prologue

I found the medical course a long, hard slog. Having followed my sister's and cousin Rod's progress beforehand I knew what I was in for. The length of the course, combined with its sometimes formidable challenges, such as the anatomy of the head and neck, is why camaraderie is particularly strong amongst medical students. By and large we were a serious and highly conservative mob. This grey background only highlighted the activities of the occasional eccentric or non-conformist student.

The life of a medical student and, for that matter, an intern is largely programmed by others. What follows is an idiosyncratic selection of memoirs rather than a comprehensive account of the medical course and my colleagues.

Suffice to say I went in at one end of the medical course a wide-eyed country kid and emerged seven years later an over-confident prig very relieved to have passed. As we shall see some fun was had along the way.

I graduated in December 1960 and was lucky enough to be appointed an intern at the Royal Melbourne Hospital. Except for some of the Casualty section, this chapter is confined to student days. There will be a bit of 'doctoring' in the next chapter!

Fruit Picking for Fun and Profit

In March 1954 I set off for Melbourne to start a medical course. Although low in funds I was full of optimism. Like practically everyone else who passed Matriculation, I had a Commonwealth Scholarship which paid most University fees. I was not eligible for a living allowance, however, because dad's bank pension was marginally greater than the £19 ($38) per week cut-off point. Come to think of it, I never came across anyone who received living expenses from this Commonwealth scheme.

During the previous month I had gone flat out to make money. Armed with a free bus ticket from the CES I set forth to pick fruit at an orchard some forty miles north of Bendigo. The experience was not as diverse as I had expected. It would have been better described not as fruit picking but as pear picking. Probably half a million pears in fact.

Pear picking started at dawn with hands numb with cold and limbs stiff from the previous day's exertion. By 8 o'clock, having eaten a massive country breakfast, we trooped back to the orchard to balance on ladders and get sunburnt and scratched. At the end of a twelve-hour day I would be utterly stuffed and unable to be cheeky to anyone.

This was the first time I had lived away from home, surrounded by strangers. Thirty or so male pickers of all ages, races and backgrounds lived two to a hut. The novelty was the continuous physical labour day in, day out. My physique did not seem specifically designed for continuous hard yakka, although fruit picking turned out to be the highest paid job I was to have during the next three or four years. I preferred short bursts of activity rather than a twelve-hour slog. My belief that I was more of a sprinter than a marathon runner was confirmed some years later when I was employed as a builder's labourer. My job was to lay part of the foundations of the Biochemistry building at the University of Melbourne and at lunchtime, while my co-workers went to the pub, I would bolt my lunch and fall asleep until the whistle blew. My lunchtime milk bottles are deeply embedded in those foundations—a fact I recall daily when I park beside that building.

The worst part of fruit picking was trying to get going in the morning. My first impression on waking was that I had been struck by poliomyelitis during the night. For some reason the local Radio Station played 'Swedish Rhapsody' at around 5.30 a.m. every morning. Thereafter I associated this particular music with aching muscles and groaning fruit pickers.

Even though I hadn't even started my course, I was called 'Doc' and frequently asked to give advice on a variety of medical complaints, most of which I had never heard of. Of particular embarrassment were probing questions on sexual matters, especially in regard to the female anatomy. My hut mate was sex mad and I fancy some of his questions would have caused eminent gynaecologists to pause in their tracks.

More tedious in many ways than the actual fruit picking was listening to what is best described as 'monotonous bad language'. One particularly dull fellow managed to use the 'f' word at least twice in every sentence he uttered. When this bloke was injured he became quite speechless, having exhausted his supply of expletives!

On Saturday nights things could get a bit rough around the huts when the boys got back from the pub. I found the only way to get a decent sleep was to bunk down on a batch of pear boxes in the Packing Shed. One night there were so many fights and altercations that the boss came along to quieten things down. Next morning I was the only one who fronted up for breakfast. This was one of the dividends of being a non-drinking virgin.

Some days when spirits were high the odd pear would be thrown about. Once a pear whistled past my ear and I turned around and saw in a neighbouring tree a German we called back in those politically incorrect days Slimy Siegfreid. He was smiling a little more slyly than usual. Taking a good grip on both my ladder and the tree, I fired off a hard Williams pear. This got Slimy Siegfreid, to my surprise and probably his, right behind his left ear. Slimy and his ten foot ladder disappeared out of sight into the tall grass and, after a moment's silence, he emerged and gave a roar of anger, something metallic glistening in his right hand. I shot off my ladder and onto the trailer load of pears which was speeding past. Fortunately, when we met again at lunchtime he seemed to have cooled down. Thereafter I watched him warily, especially since his favourite entertainment was the late afternoon slaughter of the sheep destined to feed us the next day.

The best part of fruit picking was sitting around the huts in the evening listening to the extraordinary stories. The oldest resident picker was Arthur King and he was a natural story teller. A fine looking silver-haired man, Arthur was a World War I veteran and a philosopher. We got on very well, he being the oldest and I the youngest. Another bond Arthur and I had was dirty jokes. He didn't like them and I usually didn't understand them.

In late February I arrived back in Bendigo, the left side of my face grossly swollen from an infected scratch, with the grand sum of £150 and a life-long aversion to pears in any shape or form. Dad presented me with a smart all-leather attaché case and I set off optimistically to Melbourne to start my six-year medical course.

Settling in at University

Enrolling in Medicine at the University of Melbourne seemed pretty straightforward. Almost anyone who had matriculated could enrol in the pre-medical year. Once I had done this I could wear the skull and crossbones badge of the Melbourne Medical students as well as the University of Melbourne badge. These were proudly displayed on the lapels of my old sports jacket for the benefit of discerning tram and train travellers.

Fully badged-up, my next job was to find somewhere to live. Discussions with the University housing advisers made it clear that, with the funds available, I could either eat regularly and have nowhere to sleep, or vice versa. Clearly I needed some kind of live-in job, and this they found for me. Mr and Mrs Tom Blamey of South Yarra had a position for a male live-in baby-sitter so off I went to investigate the situation. They lived in Rockley Road, South Yarra, in a comfortable old house which Tom had extensively and continually renovated. He had been a Lieutenant Colonel in the Second World War, but his father had done even better and had ended his army career as Field Marshall Sir Thomas Blamey.

Whilst in America during the war Tom had met Georgina and they subsequently married and had two sons: Ted who was eight years old and a little bugger, and Terry who was a cute little four-year-old.

Mrs Blamey was a straight talker who made it clear that I was directly responsible to her and Tom had very little to do with the arrangement. My job was to feed the kids most nights and to be there to baby-sit up to five nights per week. Some gardening was required, and for this I would get full board and my laundry done.

My first day at the University of Melbourne was almost as bewildering as my first day at Primary school. I got there smartly before 9 a.m. only to find the action didn't start until the first lecture at midday. The next three hours I spent alternatively drinking coffee and going to the toilet. I then joined two hundred students crowding into the Botany Lecture Theatre and there wasn't a familiar face amongst them. Most seemed to know each other and the air was filled with noisy greetings. Manoeuvering my way through the throng down to my allotted seat, I found myself sitting next to another Sutherland, a nice quiet fellow, Peter Sutherland, dux of his school with seven first class honours, after a double Matriculation. He is soon calling me 'Sutho' as he does for the next forty years or so.

A sudden disturbance on my other side marks the arrival of a third Sutherland, this one called Neil but nicknamed 'Abdul'. He is greeted with great enthusiasm by some of his colleagues who are also repeating the year. A wild man is Abdul, in his early twenties, bald as a coot and renowned as a double bass player, amongst other special skills. I subsequently never failed any oral exams, probably because I always followed Abdul's path. The examiners appeared too angry, numb or flummoxed by Abdul's efforts. I fancy the frightened, nervous conservatism of my replies allowed me to slip through without any particularly horrid questions while they were in a recovery phase.

Years later, Abdul brought a burns victim up to the hospital where I was working and casually asked me if I was doing anything the following Saturday. I said I was on duty, and he said, 'That's a pity, can't you get off for a while because I'm getting married and I need a Best Man.' I did get two hours off for the brief ceremony. I also did all the right things at the rather posh reception, read telegrams, didn't have a drink, and was back at work at the appointed time. Several months later when I was using my dinner suit again, I found a pocket full of telegrams which had been inadvertently shoved in the left-hand pocket and remained unread at the reception. Unfortunately, by this time the marriage was practically over and so they were furtively disposed of. Marriages at which I have been Best Man have had a 50 per cent survival rate—which is actually good compared with the statistics.

Anyway, Abdul plonked himself down beside me with a brisk 'G'day' just as the lecturer entered. As Miss Margaret Blackwood, later Dr Blackwood, and even later a Dame of the British Empire, was about to open her mouth, Abdul leaned towards me and said, 'She's built like a female commando truck driver isn't she?' I'd never seen a female commando truck driver and his comments rather shattered the awe I was feeling towards my first University lecturer.

Naturally I was unaware of some of the wonderful, almost unbelievable, things that lay ahead. Twenty-four years later, as Vice-Chancellor, Dame Margaret helped me adjust my academic dress prior to us processing into Wilson Hall where I received a doctorate.

When the lecture was over everyone dispersed, the veterans to the local pub or billiard hall. As I left the rear of the lecture theatre I could see the Royal Melbourne Hospital from the window and wondered if I would ever follow in my sister's footsteps to that Hospital, now that I was aware that many students failed first year. Then, since practical classes hadn't started yet, I went back early to the Blamey household and was rewarded with some extra gardening jobs.

Life with the Blameys was pleasant enough and certainly educational. Mrs Blamey was way ahead of her time as far as assertiveness went. She was also direct, fair and consistent. There were rational and clear-cut house rules. I could invite friends around and was even allowed access to the Blamey's bar. I didn't drink in those days but soon learnt how to mix drinks for friends. I hardly ever went out and became a bit of a swot. I had to learn how to cook, and fast. Mrs Blamey was an excellent cook and the kitchen was modern and well-equipped, particularly with American gadgetry. In no time at all I could knock up pancakes, sweet corn, garlic steak, etc. I also learnt to like small children to some extent. Terry was a great little bloke and would often quote his mother's americanisms, sometimes in a fashion hard to interpret. Once, he described someone as a 'cinnamon pinch', which I later decided was 'a son of a bitch'.

Young Ted in my eyes was not a dear little bloke at all and we never fully established a harmonious relationship. Part of the difficulty arose because I didn't know what he had authority to do and what he did not. On one occasion he informed me he had had permission to use an old fountain pen from his father's desk. I queried it—but no, he was adamant. Uproar later ensued as this was the pen used by the Field Marshall in Tokyo Bay to sign the Surrender documents with Japan. Fortunately for me, Tom straightened Ted out on that particular occasion.

Academic Innocence and an Approaching Storm

The pre-medical year of 1954 started with some ominous rumblings about the 'quota' for the first year of the actual Medical course. More than 100 ex-servicemen and others had qualified by various means, and there were hardly any spaces for them in the first year. The pass rate of pre-med students was going to be reduced so that batches of these other students could be channelled into the course. Since I was studying every night and taking the whole matter very seriously, failure seemed remote. The 'quota' problem was put aside as being more of a problem for the wild boozing billiard playing students.

The da Vinci of the Dissecting Room and Romance Nipped in the Bud

Slowly a few problems developed. Most of the students had done Matriculation Biology, and since this had not been available in Bendigo, there was quite a bit of catching up to do. Except for the 'classification of animals', I was really into Biology and thoroughly enjoyed dissecting rats, etc. This was evident from my practical book, in which the use of mapping ink and a 2B pencil produced wondrous pictorial representations of rats in various stages of being disassembled. A shy little demonstrator took quite an interest in these drawings and, to a lesser extent, in the artist. Actually her interest was more along the lines of liking the drawings, the rat and then me. I did like her cool fingers guiding my hand as we collectively sought some insignificant structure deep in the rat's pickled innards. Both her hands and fair hair smelt of formalin. She called me 'Mr Sutherland' which impressed me no end, being the first time anyone had addressed me so. My only wish at this time was that together we could progressively dissect our way through the whole animal kingdom.

At regular intervals our practical books were collected and later returned with comments. One weekend my masterpiece was borrowed by a worldly sort of bloke called George. He returned it just in time for me to hand it to 'she of the cool hands'. Imagine my surprise at the subsequent class when she swept past me and without comment slammed my book on the bench. All was explained when I opened it. To my horror, crude and vile labels had appeared over my rats and frogs. Large arrows directed at orifices and reproductive equipment led to lists of common names. There were also frank descriptions of the functions of these bits and pieces. Some of the words were quite new to me. A few of them may have been new to my little demonstrator who subsequently never demonstrated anything further to me. George did not become a doctor—which was probably in the public's best interest.

Sadly, I now ploughed through Biology in relative solitude assiduously learning everything that came up except of course for the 'classification of animals'. This looked more and more uninteresting each time I glanced in its direction.

Clever Rabbits: Later to be Recalled

One of our lecturers, Mr Boardman, was a stocky greyish sort of character, who didn't really attract a great deal of interest but managed to keep the student body pretty quiet. He had a very firm jaw and a tranquillising monotonal delivery. However, there was one moment in a lecture which made me raise my head and stare at him in wonder. He was describing how rabbits could be made to make antisera to the serum of other animals. 'Inject, say, some mouse sera into a rabbit a few times,' he said, 'then later collect some serum from the rabbit and mix it with the mouse sera. Over a few minutes a fine white precipitate will form, but only with mouse sera, not with any other serum from any other species of animal. This is as specific as any chemical reaction.' How wonderful, I thought, little realising that I would later recall this bit of information so many times. Mr Boardman then went back to more boring stuff while I surveyed the two hundred or so heads busily scribbling away below me and meditated on the cleverness of the rabbit.

Physics was a tidy subject thanks to a fine little book written by the lecturer, Dr Rogers. His *Physics for Medical Students* was right up my alley—just nice facts to learn and no complex mathematical concepts to wrestle with late into the night. The practical classes were also good fun, with attractive equipment made of brass and polished wood.

Joys of Physical Chemistry

Chemistry, on the other hand, seemed to have got out of hand in the four months since I had left High School. The peaceful descriptive chemistry of melting points and how a Bessemer converter cleaned up pig iron, etc., was pushed aside. The pre-med student was expected to revel in complex problems of physical chemistry. To solve these one needed not only real intelligence but a complete understanding of the twists and turns of the theories involved. Unfortunately my brain did not find this challenge remotely comfortable.

Just as I was despairing, things took a significant change for the worse. Dr Brown who gave the lectures appeared to be a pretty cold sort of fellow. The faces of the 200+ medical students were not actually radiant with expectation whilst awaiting his arrival. Immediately before his entry into the huge Chemistry Lecture Theatre, two functionaries would arrive and perform two distinct activities. Since lectures were compulsory, one would record the numbers of each empty seat. The other would activate two sixteen-foot high continuous belt blackboards to ensure that nothing was written on the hidden side that would appear in front of Dr Brown when he pulled up the next section to write upon. This inspection had been introduced by Dr Brown after he found himself (and the students) facing a description of Dr Brown in as they say 'less than favourable terms' on an earlier occasion.

Some of the students repeating the year seemed to take great pleasure in upsetting Dr Brown and his reactions made things worse. He developed the rather dramatic habit of tossing his locks and walking out, leaving uproar behind him. It also left me contemplating many unfinished sentences which on each occasion might have been able to unlock all the secrets of physical chemistry. It got to a stage when even minor things would set him off. One day Dr Brown was pouring liquid from a large glass cylinder into a large glass beaker. This, I believed, had something to do with physical chemistry. As the sound of the liquid being poured into the large beaker echoed around the theatre a heavily disguised voice from the rear piped up with 'I'll have a beer'. There were roars of laughter and then yet another walk out and unfinished sentence.

The Rules of the Game

One distinctive thing about a Medical course is that you have to pass every subject, every year. Fail a year and you have to repeat every subject, including all the practical classes, even in the subjects you passed. The lectures are all compulsory and lecture fees must be paid again.

All in all, it's more convenient to pass at the first go and it also makes good economic sense. The average medical student is thus more likely to feel the odd twinge of worry at the mention of examinations than those doing less rigid courses. This stress might be a prelude or conditioner to the incidence of alcoholism, depression, suicide and other delights that doctors are prone to suffer.

The Sky Darkens

As the examinations approached a few things started to go astray. Mrs Blamey and the boys went to America to visit her father and, although Mr Blamey said I could stay on, it seemed best to seek a situation elsewhere. This time I went to an all-American family with a disciplinary bent. The less said about this experience the better, other than to observe that some Americans still practised slavery. Funds were getting low but I approached the examinations optimistically having led a pure life and studied diligently.

A little flash of joy came when it was announced that Peter Sutherland and I had submitted the best essays after the lectures on 'scientific method'. These weekly lectures were not compulsory, and not examinable. Everything else was, as I now found out. Take for example the Biology examination paper. This consisted of four questions, all compulsory.

Question No. 4 being to describe in detail the 'classification of animals'. Jesus wept. Various smug bastards later described the whole paper as a dream.

The Physics paper consisted of a number of quite sensible questions, but Chemistry had a strong leaning towards physical chemistry. As Dr Brown had said as he walked out of his last lecture, 'I hope you get a question on this.' We did. In 1988, out of curiosity, I went to a lecture by Dr Ron Brown of Monash University. He was silver haired and gave an entertaining presentation containing his research into novel compounds which was of the utmost clarity. It made me think.

Cleaning at the Women's Hospital

The day after the exams I began training as a cleaner at the Women's Hospital in Melbourne, and the wait for the results began.

When I reported for work my first instruction was to find old Clarrie and give him a hand in the Nurses' home. After searching for half-an-hour I finally located him ensconced in a tiny room labelled 'Cleaning Equipment'. He offered me an upturned bucket as a seat and after indicating a heap of magazines, he resumed his interrupted reading. After thirty minutes of reading in the dim light I was getting a bit worried, but after sixty minutes, frantic was a better description. Finally I said to Clarrie, 'Isn't it time we got moving?' 'By crikey', he replied. 'You're dead right,' and led me down to morning tea.

A week later I emerged blinking into the daylight, having read some 120 editions of the *Australian Post*, and with an expert knowledge of Clarrie's medical history. The highlight of the latter was when he injured a testicle and the doctors wanted to remove it. Clarrie declined and said he would 'go home and try it out'. He was apparently still satisfactorily 'trying it out' twenty years later.

One day Clarrie stood up in our cubbyroom at a time which was not a scheduled meal or tea break. This was surprising enough, but I was quite astounded when he announced 'Time to do the polishing.' We went up to the top floor where in a remote corner of a corridor reclined a large commercial polisher. Clarrie let out 65 feet of cord, plugged it in and instructed me to grasp the handles, squeeze the control switch and start polishing. The result obviously pleased him. The polisher took off in a viciously tight semicircle and tried to climb up the wall like a motorbike in the 'Globe of Death'. With me hanging on firmly and the On-lever still compressed, the hellish machine then turned its attention to me and tried to run up first one leg and then the other! I finally released the On-lever and the polisher and I slowly regained our normal spatial relationships with the corridor and each other.

Having satisfactorily humiliated me, Clarrie introduced me to the art of using a huge floor polisher. Raise the handles a little and the brute glided to one side. Lower them and it eased to the other.

For the next week I waxed and polished corridors, trying not to lay bare the ankle bones of passing nursing staff. Clarrie was nowhere to be seen, having casually swiped all the magazines from the nurses' visiting room. Later, I heard him moaning to the boss that the job was like painting the Sydney Harbour Bridge and he would need regular overtime to finish it by himself. Next thing I knew I was to be let loose on the floors, walls and windows of the hospital proper.

Hours of floor polishing have bonded me to cleaners everywhere. For forty years or so I have always chatted to cleaners and polishers in all manner of institutions. These men and women have the covert ability accidentally to slide the polisher laterally and crack the ankles of anyone they don't like. They also often know more about what goes on in their building than anyone else.

During my third week, I and some of my fellow cleaners were invited to star in a public relations film being made about the Hospital. After being marshalled we were told to swap our overalls for street clothes. This is how, at the age of 18½ years I found myself in the front row of a mock expectant fathers' information class. The white-coated doctor appeared, pointed a stick at a large diagram and said, 'The egg comes down this tube.' The camera scanned the front row and that was that—back into overalls. I was a bit shirty about this episode, not only because I had been settling down for a bit of interesting instruction, but also to this day I worry about how the hell I could explain being in that particular class.

One cleaner, Frank, was a violin maker but could not make a living practising his craft. Unfortunately I didn't really help his economic situation when I bought a fiddle from him. Funds were a bit short at the time and his repeated assurance: 'I trust you Johnny,' only prolonged payment even longer. (I usually went by the name of John or Johnny in these casual jobs, since Struan produced too many variations—I've seen it spelt at least twenty different ways.)

Frank was an Austrian and a very gentle man. It seemed sad that hands that could make and play a violin so beautifully had to push a bucket and mop during the day. He lived in a tiny little house with a wife who looked thirty years younger than himself. She seemed to spend the whole time keeping their infant away from the workbench tools and instruments. When I went to make my last payment they had disappeared without trace. Sorry Frank.

The Axe Falls

The call 'Surfs Up!' activates the student body with joy and expectation. The call 'Results are Up!' certainly arouses expectation but after a hard day's window cleaning during a heatwave my body was difficult to activate. A vacant hospital bed looked quite welcoming. A strange desire to hang back a little just in case overcame me as I crossed Swanston Street. At the University quadrangle students were swarming around the noticeboard. An ashen-faced friend said, 'They've only passed about a third, and there are hardly any supps.'

My number was there in the passed list for Physics, but not Biology or Chemistry. However, I had scored supplementary exams in both these subjects. Another chance in two weeks' time! I'm lucky. Of the 240 students, only 58 had passed outright and 20 had supplementary exams. Back to the books. My knowledge of the 'classification of animals' was now second only to that of God. Chemistry notes were read as avidly as if they were long awaited letters from a beloved. Twelve of us trooped into the Biology exam which seemed straightforward. Unfortunately there was nothing about the 'classification of animals'. The night before the Chemistry exam I stayed up all night going through my notes again and again.

The results came out when I was at home in Bendigo, just prior to starting National Service. Passed Biology, failed Chemistry. Not an auspicious start to a Medical career and in marked contrast to my sister's achievement in coming top of the Finals the year before. It was not a particularly cheerful Christmas as my self esteem was rock bottom. The only comfort was that many of my friends were in the same boat. Should one repeat the whole year or give up? What if one failed again? In my first week I had asked a student who was repeating for the third time why he did it, and he had replied with vehemence, 'Because I want to be a Doctor!' This too was my long held ambition.

Only 70 students passed pre-med in 1954, which became known as the 'Year of Massacre'. The 64 who successfully repeated the year in 1955 were a pretty hardened band and, almost without exception, sailed through the course in pretty fine style. Looking at this group years later they appeared to have a higher-than-average proportion of professors and outstanding surgeons.

Many became second or third generation doctors and from family experience knew what sort of life lay ahead of them. This contrasts with the present situation where quite a few students don't seriously consider doing Medicine until their Year 12 results come out and they suddenly find that Medical School is a possibility.

The only way to set my life back on the rails was to pass pre-med in 1955 and it seemed a good idea to review the relationship between Chemistry and myself so I sought an interview with Dr Brown. The interview was short, icy and unproductive. I was told I had failed, 'By the skin of my teeth' and that I should, 'Spend more time going over old exam papers with my tutor.' I didn't have a tutor. And I didn't look forward to another year of Dr Brown.

Failure meant the loss, hopefully for just a year, of my Commonwealth Scholarship. So now I'd have to pay all lecture and practical class fees while doing every damn thing again. My mother said that, if necessary, she would loan me the money. And my summer income would be reduced due to involuntary employment in the Army. However, National Service turned out to be a blessing in that it was a change of lifestyle.

Becoming a Troop

In 1955, the Australian government was not planning to have the cream of Australian youth shot at by non-Australians on non-Australian soil. This made the six-month national service much less dangerous and quite good fun. The intake was basically a student one, the long period of training being done in the first quarter of the year. The recruits were sent to Puckapunyal from all over Victoria.

The night before the call up, my mate Alan Kerr told me how to do left and right turns and threw in some advice on saluting techniques. Alan had been a school cadet at Wesley College and was a mine of information on military matters. Dr Kerr, who had served in World War I interrupted the military manoeuvres going on in the living room to advise the two idiots that he would drive them to the station early next morning. This he did, and he pressed a one pound note into my hand and advised us to 'Stick together and toe the line'. He then solemnly shook hands with both of us and, as we boarded the bus, I wondered if I would ever see Bendigo again. My dad had carried out a similar ceremony earlier, but without the pound note.

By midday, we possessed a huge heap of military clothing. Our civilian clobber had been taken away and we wore slouch hats, at a variety of angles, as we staggered from one place to another in a mixture of strange clothing. Two habits were quickly established. First, you eat all of anything offered and second, whenever possible, sit down or better still, lie down.

By nightfall, fifty of us had been grouped into a happy family unit called a platoon. We were mothered by the big-gutted Sergeant Lewis and his two regular army corporals, Dyke and Hammersley. While we 'Natios' occupied two big tin sheds the sergeant and his two corporals lived somewhere on the base in an area off limits to the likes of us. In this haven they could drink large quantities of mood-altering liquid as they ruminated over the events of a long day. Some mornings they were particularly friable and their eyes, if visible, were inflamed. Corporal Dyke had the habit of becoming Lance Corporal Dyke from time to time for reasons which remained a secret, but had something to do with evening festivities.

Sergeant Lewis insisted that we call him 'Sir' because he was in charge of a platoon, but fortunately we did not have to salute him. I didn't like the look of Lewis from the start so merged in with the mob and let others clash with him. I was a wizard at drill, thanks to Alan's private tuition. One day, upon Lewis's command, I marched out of the rank and did a real Sandhurst/Westpoint series of manoeuvres. Towards the end of these, Lewis shouted, 'Sutherland, you march like a Clydesdale!' What a bastard! Lewis maintained he had risen to the rank of Major in the British Army. In wartime anything can happen.

Dyke and Hammersley were Korean War veterans and everyone liked them. They were not too bright and spent a lot of time arguing about what to do next. They and Lewis had the difficult job of keeping fifty youths in a state of exhaustion for weeks on end. They were helped by promoting to NCO some of the troops who had been school cadets. One of these was Peter Pockley who is now well known to ABC listeners. He was a former cadet officer and he almost ran our platoon. His humour and intelligence were more than a match for certain anti-authoritarian elements in the rank and file. This allowed Lewis, et al., to concentrate on the yelling and shouting while Corporal Pockley handled the more intellectual aspects of training the troops.

From the outset, there were a few things about military life which I determined to avoid or at least minimise. These were, in order of increasing repugnance, kitchen duties, latrine cleaning and solitary guard duty in the early hours. I'm pleased to report that I only experienced one of these, and the latrine in question belonged to the senior officers and was pretty immaculate anyway—so I used it.

The old army maxim, 'Never volunteer for anything', should not be too closely adhered to if one wishes to dodge latrine duty or standing around like a cold bottle of urine at 4.00 a.m. protecting the outskirts of an army base from the odd lost cow. Volunteering for certain team things can be an excellent way of becoming a protected elitist.

Although only a fair swimmer, I joined the company training team like a flash and spent many hours splashing around the fringe of Lake Nagambie. The Bren gun team was also a gem. Days could be spent lying on the ground, firing a quite comfortable-to-use Bren gun into a hillside. I must have fired 10 000 rounds of near time-expired ammunition into that hill. One day, it will probably be made into an open-cut mine.

There was a downside to these team activities. When my sister Tookie sailed for overseas on the *Oronsay* I was denied special leave to farewell her because of them.

Although the weather was extremely hot at times, military service was almost pleasant and certainly relaxing. A lot of bush was inspected and many friends made. We got very fit and, like all recruits, thought we had become reasonably good soldiers. At times, I quite fancied myself as a trained killer. Early one morning Alan Kerr, who was in a different platoon, appeared out of nowhere when my little band and I were all camouflaged and ready for a mock attack. 'Oh, Stru!' says Alan. 'If only

The author as an infant

First day at school

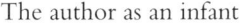

Camp Hill State School, brooding in the light of dawn in 1987
(Photo by author)

The author as a young man, viewing the world with confidence

Ready for a family picnic. My sister Diana holds the squirming infant Rosemary, who being a good sport, allowed use of this photo.

A good sized one-teacher class of 1946. The author is seen in the bottom right-hand corner. John Hial is third from his right. Jack Strellon is in the centre of the second row wearing a white shirt and Trevor Kennedy is fourth from the left, second back row.

your mother could see you now.' Corporal Kerr then wandered off into the mist like someone from *War and Peace*, leaving me uncertain as to whether it was a compliment or not.

A corporal seemed to have the best of both worlds. He remained with his mates, but got no rotten jobs and also got paid more. I resolved to reach Hitler's rank after basic training was finished and we went on to the Melbourne University Regiment.

As the three-month army camp neared its end, various optimistic rumours circulated that the medical faculty was re-assessing some of the failed candidates. Nothing came of this and I had little option but to accept my mother's offer of a loan from her tiny savings to pay the lecture fees. If I intended to eat, part-time work was essential. The other priority was to find a suitable shelter back in Melbourne.

St Richards' College for Young Gentlemen

Private Barry Vivian Thompson, one third-Bachelor of Arts and lover of cold baked beans, suggested I try his current bed and breakfast place in Parkville. 'Should be able to get in,' said Barry. 'Has quite a high turnover.'

To confuse some of the more snobbish students who lived in the university colleges my mates and I later christened our place of residence 'St Richards'. This enhanced the lowly status of our accommodation and occasionally mail would be addressed to this exclusive but unlisted college.

The sturdy, two storey, Victorian house which was St Richards still stands proudly at 34 Park Drive, Parkville. Should the reader pass this shrine of my well spent youth, several important features should be noted. Firstly, there is a separate entrance at the front for the upper floor where up to ten students could be accommodated. Secondly, the outer wall of the house abuts directly onto a cobbled alley. Finally, the high back fence is (or used to be) a challenge to anyone found locked out of Fortress St Richards.

Our hosts were Mr and Mrs Richards. Mrs Richards had considerable inner strength and gave the impression that everything occurring within 50 yards or so of her had been predetermined and that was the end of the matter. Early on we discovered this matriarch's initial was V. It took some time to discover that her first name was Valme and by that time she had been firmly labelled 'Venus'.

Venus may be described, I think generously, as a large, bony, fast moving, frightening, white haired lady with piercing blue eyes, a trap-like jaw and a voice with the range and the versatility of a Sarah Bernhardt combined with Ivan Rebroff. I, of course, was a skinny, pimply, shifty, impoverished, mediocre medical student endowed with a standard issue of native cunning. I think when our eyes first met, we instantly understood each other.

The Richards occupied the ground floor. Venus's partner in life was a timid man called Jack who had been a reporter on *The Argus* and was an acknowledged expert on shorthand. I got on well with Jack, perhaps because of a shared wariness towards a certain person. It may be relevant to note here that male spiders are generally relatively insignificant compared with their mates and tend to show little interest in life after the honeymoon. This couple had a daughter who was married to a motor mechanic called Harry and she in turn had produced a daughter. I remember seeing Venus single-handedly stripping down a Hillman engine one Saturday afternoon. I think she was showing Harry how to do it properly.

St Richards' breakfast cuisine

'The most important meal of the day is one's breakfast'—so I'd been taught along with 'early to bed early to rise makes a man healthy, wealthy and wise'.

However, after six weeks of the breakfast side of the 2 pounds, 5 shillings ($4.50) bed and breakfast deal, I voluntarily gave up breakfast at St Richards for the next three years.

Breakfast was eaten downstairs in a large dark room which was papered with an extraordinary wallpaper depicting a gloomy forest or at least the leafy side of a gloomy forest. In the centre of the table was a huge, water-filled bowl which radiated arum lilies. Next to it stood a jar of very dark jam made from some unidentifiable fruit.

The hungry student went down the external back staircase through to the small kitchen in which Venus officiated. Her narrowed eyes scrutinised each student as she handed them a bowl of Weeties, half an inch deep. Carrying one's bowl to the dining room, one passed her spouse meekly eating in a kitchen corner. A loud 'Morning, Mr Richards' was guaranteed to make him jump. Weeties are fairly insipid at the best of times, but with watered-down milk they definitely fell another gastronomic notch. Having finished his Weeties the student went back to the kitchen where the empty bowl was replaced with a plate containing a dollop of mince or tripe or brains or a boiled egg, depending upon the day of the week. Two triangles of thin toast, pre-spread with 'something' and a cup of tea rounded off this repast. No reading of papers was allowed because that 'caused blockages' according to Venus. These 'blockages' were apparently in the free flow of students rather than those students' bowels.

This system seemed to work until the day I asked for a tiny bit of butter to put in my boiled egg. This caused quite a flurry with Venus expending as much effort opening the fridge door as would be required to do the same to the vault door of a Collins Street bank.

Unfortunately, at about this time, I was required as part of the physiology course to keep a precise record of my food intake over a seven-day period. This was to determine the amount of potential energy I chomped through over a week. Amongst the lowest food values on my physiology table were tripe and brains. Naturally, with my usual desire to share newly acquired knowledge with interested parties, I provided Venus and my fellow breakfasters with instant estimations of the nutritional value of each item she had laboured so long and hard to prepare and presented with such tenderness. As the week wore on, there was a steady increase in the muttering and background clanging of saucepans in the kitchen.

One morning at breakfast, I was surprised to see a temporary girl resident from upstairs (and the pet of Venus) settling down to tackle a nice, plump lamb chop. I watched her demolish this with considerable interest and when she arose to collect her cup of tea and toast I asked if she could pass on to Mrs Richards that I 'could quite fancy a lamb chop'. She went off, smacking her own chops, and we all waited in silence. Soon we could hear our one girl resident going 'whisper whisper' in the kitchen. A loud cry came from Venus, 'I'll chop him! He'll have tripe and like it.' Clatter of dishes.

The reader might think I would then lack the courage to front up to my tripe or whatever. Not so! The hands of our fair Venus shook as she spooned out a third of a cup of minced bulls' ears. I had to face her anyway as the only escape route was through the kitchen. Things seemed to be going from bad to worse with everyone taking a special interest in our exchanges. Mr Richards seemed delighted to have the flak drawn away from his little perch.

A few days later, when waiting to bolt in to the bathroom when I next heard its latch being released, it occurred to me to give breakfast (and Venus) a miss and have a vacant bathroom almost guaranteed. Thereafter, breakfast consisted of a half-pint of milk in winter (placed on the windowsill last thing at night) or a cup of coffee in summer. This arrangement suited Venus and me perfectly and we settled into an uneasy relationship while life went on around us.

Sisters of the Holy Name
Venus wasn't keen on student drinking, although quite a nice little stack of empty sherry bottles seemed to accumulate each month in her own backyard. It was prudent, therefore, to keep our bottles of medium-dry sherry somewhere out of sight.

I'd been given an old gramophone by the Kerr family which I managed to connect up to an equally ancient radio. Both of these items could be described as 'bench models'. The front of the gramophone could be opened and two bottles and glasses stored comfortably inside it. Venus wanted to charge me extra for the gramophone when it arrived because, after consulting a friend's son, who was an apprentice electrician, she advised me that 'Radiograms are very expensive to run'. I disputed this because, firstly, I only had two records (both old 78 rpm) and, secondly, the machine was wound by hand. She reluctantly conceded that she had lost on a minor technicality.

Venus sprang into action a year later when I bought a second-hand 'reconditioned' radiogram which had been traded in at a city store. The day this arrived, like magic she appeared on my doorway and upped the weekly rate by 5 shillings. She then caused a panic by briskly moving to the old gramophone and, with a transient and extraordinarily pious expression, suggesting I donate it to the 'Sisters of the Holy Name'. I could sense the sherry bottle holding its breath as I prised her hands off that little cabinet. A month or so later, she carried it off in triumph, no doubt gaining a few credit points for the next life.

The following Saturday night as we made our way to Jack McCarthy's Fish Palace in North Melbourne we noticed a seemingly identical unit in the window of a second-hand shop nearby. The shop was closed and a closer inspection was not possible. I was tempted to purchase it and let it reappear miraculously at St Richards.

Things to do in a boarding house bathroom

A fresh human brain cost 5 shillings (50 c). The mortuary attendant pocketed the money before wrapping the brain up in newspaper and advised against squeezing the package too tightly. A proffered milk bottle was filled with formalin and two brains were nonchalantly transported to St Richards by a single pair of legs.

It was not my intention to serve up crumbed brains but rather to keep up with my colleagues. The chances of passing the Anatomy exam were enhanced by access to one's own private brain—or so it was believed. Information received led to the surreptitious deal near the post-mortem room.

All that had to be done was to cut the brain into a series of slices using a kitchen knife, and then stain them to highlight the organ's components. These slices could then be fixed in formalin. The bathroom was obviously the best place to carry out this process which was commenced in the bath.

The synthetic dyes developed in Germany last century had amazed all and sundry with their high degree of permanency. It is therefore not surprising that it took at least three hours of frantic scrubbing before any trace of whiteness appeared on the bottom of the communal bath on the first floor. Laid out on newspaper on the bathroom floor were a couple of dozen nicely stained brain sections, looking like pretty little blue lamb chops. They had taken up about 2 per cent of the dye; the bath and my hands soaked up the rest. I started to panic. Partial body snatching. Caught blue-handed. Venus will demand a new bath, etc.

The bathroom door was fitted with an engaged/vacant latch liked the ones used on the railways. Using sharp fingernails, it was possible to twist the mechanism either way from the outside. Then a dash was made to the local milk bar for gloves and the most abrasive cleaner then known. Fortunately, no queue had formed in the meantime outside the 'engaged' bathroom.

After some four hours, my arms felt as though I'd scrubbed my way to the centre of the earth and the bath was still pretty blue. The brain chops were packed away into two large jars and hidden in my room. The cleaning gear was placed in a nearby litter bin.

Next day, the state of the bath was a popular topic. Beside the bath was a neat little bottle from the local chemist labelled 'Gentian Violet: Apply daily for tinea. External Use Only'. It had not been opened. Before the culprit could commence the next stage of his or her treatment, a notice went up in the bathroom, 'Do not use foot paint in bathroom'. This was soon mutilated, just like its neighbouring sign, 'Please pull the chain allways [sic].'

For ten days, I had to wear a coat and gloves all the time and it was only after some six weeks of keeping my fists clenched, that I dared expose my now pale-blue fingernails. At the Anatomy oral exam, I was only asked a single cursory question about the external aspects of the brain. By that time, all at St Richards had gone down with tinea.

A month later the bathroom nearly attracted Venus's attention, but in a different way. Since all the students were male there was an understanding that if the shower was being used and an urgent need arose to use the toilet one could open the latch from the outside and stroll into the bathroom. One fellow did this and was having a quiet wee when he noticed a female partially covered with soap under the shower. His sphincters slammed shut and he retreated, shaken but unnoticed. When she had finished her uneventful shower our landlady's visitor made her way back down the stairs. This minor and innocent episode was eclipsed by what became known as the 'outside toilet incident' which I will now briefly summarise for the reader to savour.

It was midwinter and the upstairs toilet was out of action for a few days. Venus granted access to the downstairs external toilet, but not the downstairs internal one. Icy rain was falling when, for the third time, a large, gruff, engineering student wound his way down the stairs to find the toilet occupied again. After the fourth attempt, he was desperate with cramping lower abdominal pains. He seized a long-handled, stable broom, shoved the handle under the door and wiggled it briskly from side to side, shouting as he did, 'Come on out, you long sh.......g bastard! I'll do you!' In retrospect it is believed that this was the only time that Venus had been known to use the outside toilet!

The cleansing of St Richards'

All good times come to an end and the bath, or rather its contents, led to the 'great purge and cleansing' of St Richards. One of our number had graduated and after three years was to be sent far away by the education department. A pleasant little Saturday night party was organised with a few drinks upstairs, dinner in the city and then back for supper. The usual selection of bread, cheese, kabana and nuts were purchased and neatly set out on a table in the largest bedroom. Beer was placed in the bath and covered with ice and newspapers. About one-third was consumed before ten happy souls set off to the city for dinner. A good time was had and the guest of honour sat on the floor of the tram all the way back to Parkville.

The cheerful band ascended the stairs, anticipating stage three of the festivities. However, while we were each locating a glass or mug, the shocking discovery was made that the beer had gone. Not a bottle or so, but all of it. An empty bath. This was not an upstairs job because we had all been present. Gloom descended, especially as the pubs had closed hours ago at 6.00 p.m. Slowly the emergency grog of half flagons of port and sherry, etc., came out. We all cheered up and much later retired to our beds, slightly confused but generally filled with happy thoughts.

There was a special feeding routine on Sunday mornings. Instead of a downstairs breakfast, a cup of tea and two pieces of toast were shoved into each bedroom between 8.45 and 9.00 a.m. These were delivered sequentially by Mr Richards and it was a predictably quiet and uneventful operation. On this special Sunday morning, things went somewhat awry. Venus undertook the delivery and proceeded from room to room, waking up a series of potentially irritable young men. As she went from one room to another, someone said later, 'She seemed to be gathering up more and more steam.' My contribution was along the lines, 'Mrs Richards, we have had our property stolen.' She informed me, with a toss of her locks, that if I wasn't careful I might be asked to leave. From the safety of my bed, I replied that I might do just that. She slammed the door.

Things apparently went from bad to worse as she progressed. My long-term mate Alan Kerr was sitting up in bed reading *The Herald* and wondering what the door slamming was about when his door was flung open and, to his surprise, the cup-of-tea-less Venus shouted at him, 'You can leave, too!' and banged this final door. A puzzled Alan returned to his paper.

By mid-morning, we were having rather gloomy discussions and wondering why black coffee didn't have its normal mind-clearing effects. Someone was moodily looking through the Rooms and Flats Vacant section of the previous day's *Age*. 'Hello,' he said suddenly. 'Here's a cheap sounding place with a fridge and a kitchen, just around the corner!' A stir of interest. In fact, a flurry of interest. A tramp of feet down the stairs.

One hour later, a partial exodus began. This was despite the return of the beer by a hand-wringing Venus who pleaded, 'Perhaps you are all being a bit too hasty.' It took ten trips to move my gear by which time Venus had accepted the fact that she was losing her oldest resident. I'm pleased to report we parted amicably. By nightfall, the four of us were blissfully installed in a house and enjoying the benefits of a fridge, a kitchen and even a telephone. All this at very near the same price as St Richards.

The last time I saw Venus was some years later. I was a bustling prig of a novice intern and came across the whole family sitting in a row in Casualty. They had brought in a sick friend and I gave her my best Rolls Royce attention. Venus positively gushed with gratitude. She didn't mention the jar of pickled human brains I'd forgotten to remove from St Richards.

People Under the Microscope

Part of the Anatomy course involved a two-year study of human tissues as seen down the microscope. This gentle study, called Histology, was quite a pleasant and relaxing occupation. After surrendering a deposit against breakages, I was issued with a large box of slides each one containing a tiny sliver taken from some extinct human being.

No region was left out, as the slides represented tissues ranging from inside the nostril to the lining of the vagina. Hours were spent poring down the microscope learning to identify every tissue type and even to pick one region of gut from another. Tranquil hours could be spent drawing neat little representations of the different architectures to be seen.

Some of the pleasure was taken out of Histology by the knowledge that the same task must be performed with unknown samples under examination conditions. From a philosophical point of view, however, it was good to be able to study some of the basic building blocks of the body. Although Histology seemed fairly difficult at the time, Pathology several years later proved much harder because the student had to identify not only the tissue seen in the slide but also the particular disease present.

The brain is especially beautiful when looked at under the microscope. Special stains give the cells an inner glow and there is a gossamer of thousands of fibres which link one cell with another. When I looked at a little batch of someone's brain cells I wondered if that section remembered the name of the 1930 Melbourne Cup winner or whether it recorded the birthday of a family member. The structure of the back of the eye and parts of the ear are just as breathtaking. Obviously a great deal of thought and expertise had gone into the design and construction of the human.

Speaking of construction, a parallel subject to Histology was Embryology. This traced the development of the human from Day One right up until near birth. It was a bit like studying a cricketer's almanac because it involved learning dates when things happened and what else was going on at the same time.

Gestation is an amazing business and a lot of contortions are needed to bring all the bits and pieces together at the right time. As students, we just memorised the obvious sequence of events and received no insight into the extraordinary mechanism and what was driving it. Getting a man onto the moon is often claimed as the pinnacle of man's endeavours but what happens to an embryo is a million times more inspiring.

In the old Anatomy School in a tall specimen box there was a prosection of a pregnant woman who had been cut, perhaps with a circular saw, from the top of her head down to her bottom. She had been sawn all the way through her spine and the section showed the relationship of her near-term baby to all her insides, etc. It seemed weird to have two lives frozen together in time like that and I often wondered about their history.

There was a farmer who hired a hand who turned out to be grossly incompetent and stupid to boot. He only put up with him because of a severe shortage of labour. One day it got too much for the farmer who said, 'Are all your family as stupid as you?' The young man replied, 'No, in fact my brother is up at the University.' Curiosity got the better of the farmer who asked, 'What's your brother doing at the University?' The young man said, 'He's in the Anatomy School. He's got two heads and he's in a bottle.'

Oral Examinations

These were much worse than going to a dentist and being told to 'open wide'. They were a dreadful ordeal at which, on a predetermined date and time, several examiners were faced in the flesh. It was almost like a criminal meeting his executioner. Any student who actually looks forward to such an encounter must be an extraordinarily intelligent prig or a fool.

After six visits to the toilet in as many minutes the victim is ushered, with clean teeth, fingernails and shoes, into a large hall where four pairs of examiners sit behind small tables covered with selected remains. A batch of four students is directed to four different points of the compass. They halt at the appropriate table where they face the examiners with expressions which hopefully convey the right amount of integrity, humility and scholarly bent. A documentary on body language could be based on the subsequent events. Each little trio remains together for some ten minutes until the sound of a bell. Then, like a bee visiting another flowerbed, the student will move on to the next pair of examiners.

Death of a Sister

My next eldest sister Barbara who was known by the Maori diminutive, Tookie, died tragically in Cornwall at the age of twenty-three. On a perfect spring afternoon she and her girlfriend accepted a lift with a foreign diplomat. They were both killed when he rounded a bend on the wrong side of the road and struck a bus. Tookie was the sunniest member of the family, and my mother never fully recovered from her loss.

Her death occurred in May 1956 when I was still at university. A vacation job had fizzled out and I had decided to hitchhike home for a few days to get away from Melbourne. It was 11.30 and I had barely walked in when a telegram arrived from Diana who was in England with the news that Tookie had been seriously injured in a car accident. Another followed three hours later announcing her death. (In fact she had died almost immediately and Diana had cushioned the blow.)

Within minutes of the second telegram I found myself taking on the responsibility for caring and comforting the family, most especially my mother. There are four special memories I have of the next week or so:

1. Late that first night seeing by firelight that my father was silently shedding tears.

2. Gaining access to Beethoven's Emperor concerto. I had borrowed a portable 'radiogram' and records as a distraction for my parents and younger sister, Rosemary. Burl Ives cheered them up while Beethoven got into my brain and stayed there as my own source of divine inspiration. In 1996 I paid homage to Beethoven's grave in Vienna and found it a moving and satisfying experience. Brahms was next to him separated by poor little Schubert.

3. On the third night after my parents had gone off to heavy sedated slumbers I was left despondently studying the fire. In keeping with my new role as a responsible adult, I poured and drank a slug of the household whiskey. This first-time experience at the age of nineteen had the desired effect. I cried my eyes out, cleaned my teeth and slept like a top.

4. The last memory is of meeting Dr Sharland's daughter 'Poppet' in the street. She told me that her father had thought I had 'acted like an angel' in looking after my parents. This was praise from a man I greatly admired. He had removed a shocking appendix from Tookie three years earlier and I had run across Bendigo to summon his aid after the news of her death. I admired the whole Sharland family, especially Poppet, so seeing my approval rating elevated in her magnificent eyes did me the world of good. (Sadly a year later Poppet died in a car accident.)

Earning a Crust 1956 to 1958

For three years I lived a hand-to-mouth existence. Not many medical students did part-time work and my jobs were neatly divided into hospital and non-hospital employment.

The Myer Emporium

The Myer Emporium was a regular source of income. My first position was in the Men's Suit Department where I received five minutes training.

I learnt that the population was divided into three groups: midget, regular and portly. Bargain suits were advertised in the Saturday *Herald* and rows of these suits were lined up for the Monday morning rush. Our instructions were to get the customer into a suit coat as soon as possible, as its crisp newness would make his old one look like a crumpled rag. The top was easy to put on and the customer was rotated in front of a mirror while the slack was taken up from behind. Most also took a deep breath and stood upright for the necessary few seconds. Should alterations to the trouser length be required a couple of chalk marks were made and pins inserted. The whole operation, including a lay-by deposit, could be conducted in six minutes flat. If the customer dallied, a murmured warning that his wallet might be stolen from the changing cubicle worked wonders.

Supervising this circus was the rather rotund Mr Curnow who cruised around murmuring 'a nice fit' and 'a very good fabric'. When the special of the week was a charcoal number it seemed an ideal time to replace my ghastly dark blue suit. Mr Curnow had waxed enthusiastic about these suits, providing us with a variety of superlatives to recite by rote. Because payday was sometime off and the racks of charcoal suits were becoming markedly depleted I eased up to Mr Curnow with the request to put one aside.

My inspirational department leader saw no problem and asked who it was for. 'Me,' I said. 'Good grief, Sutherland, you don't want to buy rubbish like this. Material thin as blotting paper, wear out in no time! Haven't you learnt anything! Now if you want a good suit come and have a look at one of these ...'

Next I became a dab hand at selling gas stoves—in particular the English model 'New World' which was all baked enamel and stainless steel. My enthusiastic spiel about this appliance led to brisk sales, even hooking South Yarra matrons who had merely been 'passing through' the department. By this time I had learnt that the permanent staff gained benefits from meticulously recording their sales. Not so the casuals, so I did a secret deal with the floor manager. After recording my sale on his list, I could usually take a thirty-minute break in the coffee shop.

I had my best Myer job in 1956 at the start of the Olympic Games in Melbourne. I was placed beside the Enquiries Desk at the Lonsdale Street entrance. This was also the unofficial place for the directors and their families to collect or deliver items or messages, and to arrange transport. My first morning was rather unusual as my boss, Charlie Johnson, told me to roam around both stores for a few hours. This meant even more cups of coffee.

Most of the time a uniformed girl sat in the mini-booth labelled Enquiries whilst I leant against it dressed in my new half-paid-for charcoal suit. At times when the booth was empty members of the public would come and stand optimistically beside it. It was then up to me, depending on how I felt, to ask from my position of studied nonchalance whether I could help them.

Most questions were straightforward but sometimes I realised later that my directions had been wrong. Surprisingly few people came back to remonstrate with me.

One woman stumped me when she demanded the Complaints Department. She was particularly unpleasant and I was inspired to answer, 'Madam, we never get any complaints.' Somehow she located and stormed into the administration area way up in the clouds. After kicking up a scene she was escorted out of the store. I observed her departure as an innocent bystander partially screened by a column.

About this time Myer decided all departmental managers were to wear a white flower on their lapels. Charlie, who had wide and possibly secret responsibilities, was issued with one. He pinned it on me which gave me unexpected status, especially in the eyes of the junior girls. It also protected me further from the public because I fancy they thought I was a stranded member of a wedding party from St Francis's over the road. Speaking of which, on one Ash Wednesday I unwittingly pointed out to a customer that she had a dark mark on her forehead. This had been applied by the priest over the road minutes before.

Next day there was some excitement right in my territory. My daydreams were interrupted by a yell for help from the matronly looking store detective. She was hanging on to a large suitcase which a rough looking fellow in a grotty old overcoat was trying to shake from her grasp. They seesawed in front of me as the brute tried to reach the steps leading to Lonsdale Street.

With a courage which surprises me to this day, I sprang from my leaning position and firmly grasped the arm holding the suitcase which, of course, was towing the store detective. This halted the villain's progress. In the split second when I thought he would smash my face with the fist of his free hand, he gave me a look that Conan Doyle would have described as 'vehement malevolence'. Actually, he didn't hit me and, as we hung onto him and manouevred him around behind the Enquiries booth, his expression turned to one of extreme frustration rather than aggression. The detective tripped over the suitcase and accidentally the three of us fell to the floor. At this stage his coat flew off one shoulder to reveal a tiny deformed right hand which looked like a rubber glove as it was waved under my nose.

My bravery was not officially recognised by the Myer Emporium. However, the pretty girl who was selling wind-up ducks on the novelty table nearby gave me a duck which did not waddle properly. Fortunately, she had not seen the end of the fracas and at the time it was against my interests to confuse her with unnecessary details.

Every day a little violence occurred near the booth. Mostly it consisted of parents belting their children or people falling down the entrance stairs. One day someone fell up the stairs! A young mother was talking on the public telephone at one side of the Enquiries booth when she suddenly emitted a scream. Charlie and I, who were discussing our favourite subject (women), turned round and saw a man with his hand under the clothing of the woman's child. In a flash Charlie spun the man around and gave him such a crack on the jaw that he went backwards and upwards. This was impressive stuff and a few more shoppers than usual fell down the steps over the next few seconds. Charlie wanted to hit him again but I convinced him the man was dead. By the time the police arrived, my diagnosis had proved incorrect.

I felt quite self-important when dealing with the Myer directors. They didn't know me from Adam but I enjoyed dropping their names when doing their messages. This certainly smartened up the odd departmental head. If I was wearing my white flower it confused them even more as they had never seen me at a departmental meeting. On my final day at Myer I had to help a director, Tom Pettigrew, who was returning eleven bottles of a very expensive hair tonic. He was bald as a coot.

Hot Cross Buns—becoming a master pastry cook in eighteen hours

Each week the University appointments board issued a list of odd jobs, some of which offered unique work experience. The medical course was intense, breaks were infrequent and well paid fund-raising opportunities that could be fitted in had to be seized with avidity. One friend got a job which was discreetly advertised as manufacturing rubber items. Interest in his occupation fell when he returned home with some free pairs of rubber gloves. Whereas the most skilled job I had was assembling cuckoo clocks, the messiest was making hot cross buns.

Four assistant cooks were required for a straight run which lasted from midday Wednesday until 6.00 a.m. on Maundy Thursday. The pay was fantastic! Four of us reported to a funny little run-down bakery in Glenhuntly where they had set up extra tables for the mass production (by hand) of a few billion hot cross buns. No protective clothing was provided.

In a matter of minutes the four idiots had become dough-encrusted and were staggering around like a combination of the Michelin man and the Abominable Snowman. We started off in high spirits and enjoyed eating the first few hundred that emerged from the vintage 1810 oven. As time wore on the crafting of individual buns became mentally less stimulating. There was a tendency either to make them smaller and smaller or larger and larger. After six hours practice, some degree of uniformity emerged. This was individual uniformity of course, with each person's product clearly recognisable before baking. There were bucketfuls of a sugary brown glaze and 4 inch-wide paintbrushes to slop it over the cooling buns as the finishing touch.

As the afternoon sun penetrated into the back of the bakery the production area reached a nice comfortable temperature of about 50°C. Then a second oven arrived in a van and was plugged in near its bigger brother. It warmed up quickly.

In a hot operating theatre the well-clad surgeon usually requests that his or her perspiring brow be wiped. In films this is traditionally done by a junior nurse whose wide eyes above her mask show a mixture of adoration and respect. Apart from making the surgeon more comfortable, this manoeuvre reduces the amount of bacteria-filled sweat to drip into the insides of the poor unfortunate below.

Things got even hotter in this Dickension production unit but no-one was at hand to wipe our brows or anything else. The perspiration of four medical students flowed into and onto the buns. Shirts were abandoned as the furious tempo was maintained. No meal breaks were offered, the only interruption came when a van arrived, and, 'painting', loading and packing had to be done.

As dawn approached, fatigue made itself apparent in various ways. Both the galley slaves and their masters were getting snappy and becoming accident prone. Bags of dried fruit burst and more flour ended up on the floor. Buckets of brown glaze were kicked over or even stepped in. By now our shoes were grasped firmly by the floor which allowed one to lean at interesting angles.

Bees arrived with the dawn. Firstly in ones and twos and then in hundreds. Initially they headed for the varnish buckets but soon distributed themselves neatly like league footballers over the expanse of finished buns. To maintain uniformity the buns not covered with bees were to have a generous application of varnish/bee mixture. Production continued until the 6.00 a.m. deadline and by 7.00 a.m. the last van-load had left followed by the more enthusiastic bees. It was strange that bees were active so early and that so few stings occurred.

Four filthy, exhausted, but temporarily wealthy students then had to travel across Melbourne by train and tram. They and the occasional bee fell under the critical gaze of the early office workers. After a quick clean-up we made the 9.00 a.m. lecture after which we woke up feeling considerably refreshed.

For about three weeks my shoes smelt like a Christmas cake and I have rarely eaten a hot cross bun since. When offered one, I invariably cut it open and closely inspect its interior. Each blackened protruding lump has to be checked and identified as a sultana.

A glimpse of the inside workings of the Victorian Railways

One year I emerged from a post-exam hangover to find that jobs had suddenly become scarce. All the old faithfuls and stand-bys had disappeared. The Women's Hospital said, 'Sorry Johnny,' the Myers Emporium, 'Not in the foreseeable future Mr Sutherland,' and Venus said back at St Richards', 'Board payable in advance.'

Ever since this time job seekers have had my sympathy. Those who criticise the unemployed should spend a little of their time tramping from site to site as their funds dwindle. Job hunting makes one thirsty and hungry. A day of failure finds the searcher returning home in a prickly mood which is nicely rounded off by some forgotten bill in the day's post. After seven fruitless days the situation became desperate. As I lay in bed resting my throbbing feet *The Herald* informed me that the Victorian Railways was requiring staff immediately. By first light I'd homed in on the VR employment office.

Three days later eight of us stood naked in a line having deposited remarkably varied quantities of urine into the eight jars provided opposite us. (Someone with an ear for music could have produced a suitably rousing tune by striking these jars in an appropriate sequence.) When we were all blue with cold, a wizened and vaguely crippled old doctor was led along the line. He then inspected our back views. His eyes were thousands of miles away as he listened briefly to each heart. The Victorian Railways apparently considered the heart the third most important feature of potential employees after the genitals and bottom. The representative of the esteemed profession to which I aspired then shuffled off without having washed his hands at any stage.

Our group of eight were instructed to join other genitals and bottoms the following Monday for a full day's free training on the intricacies of the Victorian Railways.

It was a strange sensation at the age of twenty years, almost like being back at Camp Hill State School. The instructor explained that the engine was usually at the front end of the train and hopefully on tracks which led to its destination. Engines and guards vans carried colour coded discs which indicated different things. Fair enough. The colour coding of tickets revealed a great deal of information to the highly trained ticket collector, much of which was quite occult even to the most discerning railway enthusiast. By the end of the day we would be highly trained in 'state of the art' ticket inspecting. Again, fair enough.

We acolytes were irritated by a keen youth who was the nephew, godson or some other damn relative of the instructor. This oleaginous twit had been programmed to spew out answers. By mid-afternoon we were all jacked off with his performance and would have liked to kick this potential Chief Commissioner of the Railways to death.

By the time the long day drew to a close I wasn't madly keen to get cracking with a ticket clipper. Some things you quite look forward to having a go at after lengthy instruction, e.g. throwing a hand grenade or delivering a baby. The closing comments from the instructor caused considerable dissatisfaction amongst the rabble. He told us we were not to be paid for the day's training. We were then informed that we had to provide our own dark trousers which in my case was part of my only suit.

I was then issued with a cap and a badge no. 3635. My highly specialised duties were to commence as a ticket collector (trainee) in my good suit trousers at 5.45 a.m. next morning at Spencer Street Station. Clutching my cap and badge, I moodily tramped home in the cold night air to Parkville.

Joyous news awaited me at St Richards! A cleaners job at the Royal Women's Hospital was available from the next morning. Johnny would be back with his old mates in hospital issue overalls while his good suit pants rested in the wardrobe.

Over the next two years I received a number of impersonal letters demanding the return of the cap and badge. This headwear sometimes featured at parties. On moonlit nights the badge was a comforting sight as it shone in reflected light from the top of my bookcase.

One day, a singularly threatening piece of correspondence prompted the immediate return of this symbol of authority. The human skull it had sat on during its stay in Parkville looked quite bare thereafter.

The neutral ground of a hospital orderly

I first got my hands on patients at the age of eighteen. My days started at 5.00 a.m. when, exposed to a severe chill factor, I cycled downhill in midwinter to the Bendigo Benevolent Home. Even the army greatcoat seemed to shiver as the bike was parked. With chattering teeth I would scurry to my destination—the general ward, male, non-ambulatory. The occupants of this ward were basically as unaware of their impending departure from this world as they had been of their arrival into it (that's a quote from someone). The poor chaps were mostly incontinent and totally confused. The orderly's job was to clean 'em and feed 'em, clean 'em and feed 'em, etc.

This was a good introduction to the less attractive aspects of patient care and a stimulus to pass exams. On night duty the rows of slumbering and largely abandoned old men eerily vocalised their childhood fears in nightmares. At other times a demented patient would have a remarkably lucid period and, in the midst of an interesting conversation, go quite ga ga again. Interesting things brains.

In 1979 I was having a check-up from Peter Sutherland who is probably the most beloved physician in Melbourne. These tended to be social events, and generally his other customers would hear a lot of laughter coming from behind the closed door. On this occasion we were dealing with family and colleagues, including the recently deceased, and then hobbies. He described his hobby carpentry as 'noisy, messy and anti-social'. He showed me some photographs of beautiful furniture he had made.

Then he produced a neat little fruit bowl he had made. From the same shelf, he handed me a stainless steel bowl to wee into. I said I hoped he never mixed the bowls up, more laughter. I'll now come to the point of this story. As he finished prodding my belly, I said, 'I wonder what we shall die of Peter.' He replied, 'I don't really mind provided it's reasonably quick.' He then cocked his head to one side and said, 'Really anything but senile dementia. That way everyone finds out all about you.'

In these days when job descriptions are all the rage, the term 'health care worker' is wonderfully vague. Whereas Hospital Orderly at the Bendigo Benevolent Home was a dull, messy occupation, 'Orderly' at the Royal Melbourne Hospital was infinitely better. It became my prime source of income and allowed me to pick up a few clues at the same time. By the time I was a full-time student at the hospital, I was familiar with its geography, staff and pecking orders.

In a big hospital the orderly's position is often comfortably neutral if friction develops between the nursing and medical staff. In some ways like a medical eunuch, he may stand around and merge into the hospital furniture whilst waiting to perform some simple physical duty. Alternatively, he may spend time roaming the hospital on minor and often personal errands, or quietly help the junior nurse stack fresh linen in the utility room. An orderly who has been around for years generally has a serene relationship with the most fearsome of senior sisters. The latter may snap at interns and positively devour medical students but will accept the habits and quirks of the established orderly because he has cunningly integrated his work patterns with hers. For example, when her supremeness goes off to have morning tea with the other rulers he will be on hand to keep an eye on things. Anything of importance will be noticed in her absence.

Often the orderly is in the best position to enjoy hospital life to the full. The patients generally find him non-threatening. He's not going to puncture them with needles or shave off their pubic hair. Provided he doesn't drop them off the trolley or down the stairs, they accept him as fellow human.

After working as an orderly I reached the following conclusion. Many sick people once over their initial crisis would probably recover faster if they booked into a good hotel. It would be cheaper, quieter and offer better food and room service. The larger hotels usually have access to a capable doctor who will come at any hour. Also in a hotel room one is not woken every hour or so by someone dropping a stainless steel kidney disk on the nearby tiled floor.

Working as an orderly in the Plastic Surgery Ward was especially rewarding. At that time I was nineteen and had not yet started my clinical years.

Mr Benjamin Rank (later Sir Benjamin—it's amazing how many people I 'worked' with were later knighted) had a ward full of *proper* plastic cases. No face-lifting, tucking away of ageing skin rubbish, or making synthetic people look more synthetic. Many of the patients were long-term cases who required major repairs after severe burns or other accidents. The drama of the initial injury was long past but it took months and a series of operations to transfer full thickness skin, say, from their abdomens to perhaps their chin, where it would then be moved around and shaped.

Full thickness skin cannot just be lifted up and plopped on a bed of fat elsewhere because it needs to take a good blood supply with it as it moves. The plastic surgeon makes two parallel cuts on the abdomen (the patient is anaesthetised!) and then folds them together making a roll similar to a thick breakfast sausage. The ends of the sausage are still attached to the abdomen and from there obtain their blood. The gap is sewn up and the patient wakes up to admire the neat little handle that has appeared on his tummy.

After a few weeks when the handle has settled down and is obviously well supplied with blood, the patient goes back to the theatre. One end of the handle is cut free, moved closer to its final destination and re-attached. When the blood supply is obviously adequate from the new site it's back to the theatre and the other end is now swung a bit closer. This process continues until the rolled up living skin (called a pedicle) reaches the desired position.

Most of the patients tended to be healthy young men who naturally enough were bored silly by all the waiting around. Either they were getting a lot of attention from the nursing staff while having painful dressings changed or not enough attention once they were feeling better. The nurses' assessment of these young men's condition was reflected in how tightly they placed the screens around the bed.

This ward was on the seventh floor in the wing of the hospital directly opposite the Nurses Home. One afternoon I found all the lads that could manage it had wheeled themselves into a tight huddle in one corner of the balcony and were peering intently over the ledge. A young nurse was deliberately exposing her naked form to them as she lay on her bed in her room some 30 yards away at near eye level. As I was focusing on this interesting sight, the ward Sister 'sprang' the lot of us. She barked commands which would have done credit to an alarmed German Army Officer in World War II.

The lads managed to identify the nurse that had attracted their attention by bedroom window counting and other means. Sadly, by the time this intelligence was available they found she had abandoned her nursing career at the Royal Melbourne Hospital.

The Plastic Surgery Ward was unique in that late each Wednesday afternoon the patients were offered a beer issue. A little burst of alcohol is good for skin grafts, but Mr Rank was also a kindly soul who realised that, at least in the short term, there were few foreseeable pleasures for the patients to anticipate. I took to the store a special form signed by Sister and was then issued with a dozen or so cold bottles of beer to bring back intact to the ward. At the appointed time I went from bed to bed offering a beer here and a beer there. Most of the male patients would be sitting up in bed like Jackie smacking their lips in anticipation. For the next thirty minutes I was the most popular person in the hospital. I didn't drink alcohol in those days and enjoyed dispensing a medicine the patients relished.

One day my inexperience with alcohol caused quite a problem. It was Wednesday afternoon and I went around the ward pouring out the beer issue, but a change in routine due to a public holiday had severely affected my clientele. Most were either recovering from anaesthetics or had 'nil orally' labels over their beds. I pressed on with the few that were normal and it seemed logical regularly to top up their glasses and dispose of the beer issue. Their faces and the sausage-like grafts glowed very satisfactorily. Unfortunately, one fell out of bed and another sang happily until late at night. During visiting hours some unseemly behaviour was reported and the night staff recorded an excessive demand for urine bottles. Thereafter, the beer issue was in proportion to the number of approved recipients. The skin grafts never glowed so well again!

When Mr Rank did his ward round I would discreetly join his entourage immune from the student's fear of intimidating questions. From the outer circle senior students could be observed writhing with embarrassment as the great man canvassed the various possibilities of treatment. Then he would say, 'What else?, what else?, what else?' 'Do nothing,' was the final alternative he wanted. Many a patient enjoys a longer life because his doctor adopts this approach.

During one ward round a young nurse was instructed by Sister to pull back the dressings from a grafted area. She started to do so but suddenly shrieked, 'Christ, it's moving!' Craning our necks we saw the grafted area was covered with a cluster of maggots. We were all shocked except for the great man. He pointed out that the little creatures would only feed on dead tissue and so were in fact cleaning the region up. He advocated leaving the cradle over the patient's leg so that the poor chap could not see the activities at the far end of his body. Finally, he stressed that appropriate nursing practices would have ensured that such an event did not occur at the Royal Melbourne Hospital. 'Don't you agree Sister?' he asked rhetorically and then moved off to the next patient. As soon as he left the ward, Sister ignored his instructions and tackled the maggots.

When Mr Rank had suggested that the 'little chaps' be allowed to enjoy their meal I had observed his deputy's reaction. He turned around and looked out of the window with a face contorted with a mixture of revulsion and disbelief.

While I was in the Plastic Surgery Ward an Altona Refinery worker was brought in. He had burst into flames after being sprayed with molten phosphorus and then hosed down with water by a workmate. With full thickness burns covering some 80 per cent of his body, his outlook was hopeless. He took some ten days to die and during that period he exhibited all the various complications including gastric haemorrhage which would lead to the inevitable outcome.

As a child I had dropped a boiling kettle and scalded my foot. The pain was extraordinary but next day in hospital I had had spectacularly large blisters for all and sundry to admire. With this memory I naturally observed the poor fellow with considerable compassion. One aspect of plastic surgery is removing tumours which grew too well and I wondered about using modified tumour tissue in the treatment of severe burns. I sounded out my theory on an intern who in turn pointed out my ignorance. Nowadays, cells from the victim are grown up in sheets by tissue culture methods and then used as grafts.

Many orderlies have a high degree of expertise which is often self-taught. Theatre orderlies have had vast experience in the appropriate positioning of patients for various operations and the safe movement of unconscious patients. In emergency departments the older orderlies have a memory bank of literally thousands of regular and not so regular customers. Foolish indeed is the young intern who doesn't take notice of a comment about a particular patient whispered in his ear by an experienced orderly. A senior orderly can be a father figure to the staff in general. He can also be a great source of information and *dirt* on the senior hospital staff and what they got up to as either students or young interns.

I have a favourite story about a hospital clerk called Les Blizzard. Les recorded the names and addresses and presenting problems of the patients as they entered the foyer of the Casualty Department. In those days all the details were recorded in a large ledger. Each individual was allotted one line. Some might have to be listed as 'unknown woman aged ?18 DOA—City Morgue'. The good clerk must remain calm and in control of all situations. Les demonstrated this approach magnificently after there had been an accident in a large warehouse down near the docks.

Two men were shifting wool bales using large hand-held hooks. One swept his hook over a bale and embedded it in the forehead of his workmate. At great speed the injured man was taken to the hospital where strangely enough he walked from the ambulance up to the desk where Les was seated.

Les looked up and saw a patient standing opposite him with a large hook projecting out of his forehead with the handle nicely horizontal, some four inches above his head. After a brief glance Les gave a bored sniff, picked up his pen and looked at his watch. As he meticulously recorded the exact time in the ledger, he said 'Name?'

It's nice to report that this patient made a full recovery. Fortunately, the part of the brain that was damaged was a region he rarely used.

An Intern in Casualty 1961

Certifying the dead

She was a very large and very cold woman. She smelt of gin but she wasn't breathing out the fumes. Her pupils were dilated and fixed. The

Registrar shone a bright light into both eyes and got no response. With the help of the ambulance officer he raised one massive cold left breast, applied his stethoscope and listened for a mandatory 15 seconds. He nodded to the officer as he removed the stethoscope, signed the proffered form and nonchalantly strolled back into Casualty. The ambulance then sedately departed to the City Morgue.

Mortuary attendants tend to be nuggety little chaps of cheerful dispositions, with a habit of whistling as they go about their tasks. One I know has quite a brilliant repertoire from the most popular operas.

A cheerful, whistling mortuary attendant parked a trolley with an average weight stiff on it next to the trolley on which rested the body of the fat lady and it received a slight bump. As the attendant turned towards the door he heard a low, but definite, moan. He stopped whistling in mid-crotchet. With some curiosity he pulled the sheet down from the woman's head. In response, she emitted one long resonant snore.

The duty officer at the ambulance headquarters initially said, 'Good joke, ho-ho', when the call came for an ambulance to be dispatched with urgency to the City Morgue. Our heroine was wrapped in blankets and sent to a different hospital to the one which she had been parked outside the night before.

The Press had a wonderful time. She sold her story to *The Truth* which made up 98 per cent of the story, anyway. It turned out it was her sheer size which had allowed her to do her Lazarus act. The lone attendant had not even tried to fit her into the giant filing cabinet-like drawers that are used to near-freeze the newly arrived corpse.

Four months later this poor lady was dead again, this time permanently. On the second occasion a number of doctors were called in to confirm that the diagnosis was utterly and absolutely correct.

Naturally enough, this case concentrated the attention of hospital administrators, so that they were determined that it should never happen again. Of course, none of the said hospital administrators were going to be crawling into the back of an ambulance at 3.00 a.m. in the morning to determine whether or not an 18 stone woman was alive or dead. However, the new rules at the Royal Melbourne Hospital were not too bad and they certainly improved upon the routine of eye and heart examinations in cases requiring certification that death had occurred.

Within a week I witnessed a fellow intern flagrantly disregarding these rules. It was also witnessed by the acting Deputy Medical Superintendent whom we shall call 'Fred'. From his office this pain had an excellent view of the area surrounding the entrance to Casualty and this allowed him to wile away his time while watching the comings and goings to the temple of healing over which he was temporarily presiding. He was not keen on duty staff leaving the immediate building even in an emergency (discussed below), and in fact he would probably have preferred to have no patients admitted to the hospital whatsoever. Like many medical administrators of some years' standing, he would not have recognised a patient if he had fallen over one.

If one is going to have a motor vehicle accident or a baby, then 10.30 on a Monday morning is apparently the best time to do either or both of these things. A veneer of calm efficiency can be found in all corners of our great hospitals. They are fully staffed, mainly with people who have had a pleasant weekend and are suitably refreshed. The patient going to hospital will not be competing with peak hour traffic, pay-night or Saturday night drunks.

The scene therefore was 10.30 on a Monday morning in mid-summer. In brilliant sunshine a police utility van motored serenely down the drive and parked a discreet distance away from the main entrance to Casualty. Two constables alighted, one clutching a clipboard, and they strolled purposefully into Casualty. From his observation post, Fred noted this event. A few minutes later my colleague Jim accompanied the two policemen out to the utility. One policeman undid the zip of the cover a few inches and Jim looked in. He nodded his head, the zip was done up again and Jim signed his name on the document attached to the clipboard proffered by the other policeman. They chatted for a few more minutes before the police departed in their vehicle while Jim wandered back into Casualty after a brief glance at the glorious sunshine.

Within a thousandth of a second, Fred got himself from his first floor observation post into Casualty. Confronting Jim he demanded to know what the whole performance had been about. 'Certifying a stiff was a stiff,' said Jim. His worst fears confirmed, Fred now red-faced demanded that the police be immediately contacted so that the corpse could be brought back to be more thoroughly examined. 'Consider yourself suspended,' ranted Fred, and in no time the whole staff and assembled patients in Casualty were aware that Jim had not only disgraced the hospital but his medical future was near non-existent.

When Fred paused to see what effect this catastrophic verdict was having on him, Jim cocked his head to one side and said, 'Frankly, Fred, I'd probably do exactly the same again.' Even more blood perfused Fred's face but, before he could wind himself into a greater fury, Jim said, 'That bloke had been in a dam for four weeks, and the yabbies had eaten half his head away. I doubt if even you, Fred, could have resuscitated him.' Fred then withdrew promptly as if he had more pressing business to attend to. The rumour was that he subsequently kept a pair of binoculars in the drawer of his desk.

Poised for disaster
Sister May Spalding was in charge of Casualty for many years. She ruled it, she hovered over it, and when off duty gave the impression she did nothing but worry about it. In her absence, were the staff becoming too familiar? Yes. Were some of her tight standards and rules being varied?

Yes. Tall and pale, as though dusted with flour, she would glide silently from one part of Casualty to another. Beware the student chatting up a pretty nurse whilst stitching up some unconscious drunk. May would materialise from nowhere, sentence the poor inoffensive nurse to some ghastly task and give the suturer an exceedingly chilly glance

Several years ago May Spalding died. Efficient to the last, she rang for an ambulance explaining that she was having a heart attack. When the ambulance officers arrived they found the front door had been propped open and she was dead in the lounge. Had she expired immediately after their arrival her last words might have drawn attention to the state of their uniform or the cleanliness of their boots.

In a cupboard in Sister Spalding's office, were two large canvas containers labelled 'Emergency Packs'. The contents of these packs had been determined by, as you would expect, a Hospital Committee. It was not clear what particular type of emergency they were designed for.

A medical emergency tends to occur suddenly. Word came to Casualty that a lady was trapped in a city store lift and a doctor plus Emergency Pack were to be dispatched immediately. A taxi driver delivering a patient was told to wait while an Emergency Pack was obtained from Sister Spalding. This was done with some difficulty, because she insisted upon knowing more precise details of the emergency before her precious Emergency Pack was taken from her care. Finally off went my mate Jim, hemmed in the back seat of a taxi by a large canvas emergency bag.

Upon arrival at the store he and the bag were manhandled up to the second floor where a lady had her foot trapped in the lift. She was in considerable pain and by coincidence was one of my mother's card-playing friends. This fact had nothing to do with her particular predicament. Jim's priority was to give her pain relief so he tackled the Emergency Pack.

Unfortunately, none of the items in the bag were labelled. Soon he had unwrapped a variety of kidney dishes, stainless steel bowls and glass syringes. It looked as if he were running a medical Tupperware party. As he pursued this fruitless task my mother's friend made reassuring noises to him such as 'Not to worry dear, everything will be all right', etc.

Suddenly, as he squatted beside the near-closed lift, a gruff voice came from above. 'You want some morphine, do you Doc?' Towering over Jim was the then Police Surgeon, Dr John Birrell. 'I've got some in the glove box of my car,' said Birrell sailing off and leaving Jim to begin packing up the merchandise which by then radiated some 4 feet around him.

Sister Spalding was tight-lipped when she took possession of the bag again and muttered darkly of the need to re-sterilise all its contents. Apparently there were no drugs at all in the bag, and the only way it could be used to produce pain relief in a patient would be to knock him or her unconscious with one of the larger vessels or instruments enclosed therein.

This episode led to a proliferation of emergency bags for different types of emergency. It also led to a profusion of instructions as to their use, their accessibility, and a list of people who were required to give permission before they could be removed 3 feet from their position of storage.

My first professional house call

My turn came a few weeks later. I was perched on a stool behind the desk in Casualty chatting to a nurse. The phone rang and the Casualty clerk picked it up. 'No, it's Sutherland, I don't know where the other ones are. I'll tell him.' With bored indifference he turned to me and said, 'There's an emergency at Parliament House. A police car's coming to pick you up in a few seconds.' With a surge of excitement I said, 'I must get the right emergency bag.' The clerk spoke briefly into the phone and then turned and said, 'No time for that doctor, they'll have everything you need there.'

There was the sound of approaching sirens which abruptly stopped. I was hustled out to a police car, which set off with such force I almost went through the back seat and into the boot. In the four minutes it took to get to Parliament House I wondered why on earth they hadn't requested help from the much nearer St Vincent's Hospital. I then wondered what critical situation I might shortly have to handle in a masterly fashion.

If the Statesman I was about to resuscitate was a Minister then a gong in the New Year's honours seemed almost guaranteed. On the other hand, he might be a member of the Opposition. If he was a thorn in the side of the Government and I failed to resuscitate him, then this, also, might lead to some significant recognition.

I was led into Parliament House looking brisk, astute and ready to bark out orders. The first anti-climax came when there was absolutely no-one there to greet us. Then two breathless officials arrived from different directions, each wanting me to accompany them in the direction from which they had just come. The police sorted this out and the little band scurried down a corridor with one official alarming me by saying, 'It looks bad, doctor, it looks very bad indeed.'

I was hustled into the Women's toilets where a rather dazed cleaning lady was curled up on the floor. She held a white bloodstained towel against her forehead. Apparently she'd had some type of fit and cut herself in the process. The first aid kit was produced but, as an official said, 'There's not much actually left in it, is there?'

I patched her up as well as possible. Like the police, I was disappointed she was not a near-dead politician. Rather than leaving me to tramp back across the City in a white coat clutching a stethoscope, the police kindly gave me a lift back to the hospital. Curious colleagues inquired about the details of my expedition, and I remained suitably humble about the crisis which had been so ably handled at Parliament House.

Now I would like to leave hospital life in Melbourne, as indeed happened to me in 1962. Accompany me on a cruise or two after which we will find plenty of things to do back in Melbourne!

The Navy Lark—RAN 1962–1965 5

'Is my right honorable friend saying that Wrens' skirts must be held up until all sailors have been satisfied?'

Irene Ward, MP, in the Commons after being told that the supply of Wrens' uniforms would have to wait until RN uniforms had been dealt with.
(Wren: member of the Women's Royal Naval Service.)

Starvation or the Navy—A Bleak Prospect	108
The Interview	110
A Day at Sea	111
A Private Session with the Admiral	112
HMAS *Voyager*	114
In the Navy up to my neck	114
A few rules	114
My own little pozzy	115
A Naval contribution to the social life of Sydney	116
A painful start to Day One	118
Meeting my Godmother	118
Blossom	119
Meeting my Captain	120
The sick bay	121
The Navy gets down to business	122
On leadership	123
More on Blossom	124
Organ of generation	127
Socialising in Hobart	128
Pulling of rank and 'white-anting'	129
Chipping the decks	130
A trip up north	131
A hell of a fright	131
Hospital visits	132

The return of the Great White Traders 133
Marriage 134
HMAS *Cerberus*—The White Man's Grave 134
   Outpatients 135
   The Mammoth Ball 136
   A Senior Medical Officer and fun with stethoscopes 136
   Critical appraisal of a Surgeon Commander 138
   The Surgeon Commander gives me a fright 139
   An unfortunate death that led to hospital improvements 140
Out of the Frying Pan—Fun and Games in Darwin 141
   A secret report on the Defence establishments of Darwin in 1963 142
   Sport and death in the tropics 144
   A sick cat and Divisions 145
   An epidemic and the loss of a nice little business 146
   Things to do with 'natives' 147
   Goodbye, prickly heat 148
HMAS *Melbourne* 148
   Nonstop entertainment 149
   Saluting senior officers 150
   Keeping people asleep 150
   Going mad 152
   Going on a brothel raid 153
   Things sailors buy (caveat emptor) 155
   An unexpected medal for bravery 156

I served full-time in the RAN from 1962 to 1965 and travelled to the Far East on two occasions. The first time was on HMAS Voyager *and second, HMAS* Melbourne. *However, the views I express are those of a medical outsider who has not deliberately attempted to take the mickey out of the Senior Service.*

Starvation or the Navy—A Bleak Prospect

The financial outlook at the end of my third year of medicine was bleak. As the clinical years approached, vacations shortened or became non-existent and the opportunities for rewarding temporary jobs diminished. Not only were there the regular outgoings but there were also extra expenses, such as the cost of living-in as a student at the Women's Hospital. A few of my mates were in a similar position and we cast our eyes around for a quick solution to a fairly acute problem, i.e. a roof over our heads plus food and grog.

There were undergraduate schemes offered by the Armed Forces and the Department of Territories under which the student was supported for three years and, after completing hospital residency, he or she would spend four years in 'bondage' as it were to that particular organisation. We studied the various schemes and weighed the pros and cons of each proposal.

Four years straight in a remote area of Papua New Guinea seemed too dramatic a contrast to our current lifestyles and so the Department of the Territories was not given further consideration. We were left with the Army, Navy or Air Force. Choices were then made on the basis of either experience or non-experience. A fellow who had done National Service in the Air Force seemed unimpressed by their medical services and the one and only Army doctor I had come across during National Service was awful. He was a little bandy-legged bloke with rheumy eyes and may have been the twin of the doctor I had come across in the Victorian Railways. This left us with the Royal Australian Navy.

Having accepted the RAN by a process of elimination, the next stage was to convince the 'powers that be' that we were highly suitable and indeed potential assets to that organisation. The large form received required details of any previous military service. Corporal, Melbourne University Regiment, looked quite impressive but was offset by the blanks that had to be left under headings such as 'active military service', 'injuries', 'decorations', etc.

Glowing references were also required and this posed another small hurdle. Fortunately Drs Kerr and Sharland from Bendigo obliged with a pair of excellent references.

A near disaster occurred the night before the references were submitted. One had been passed around as a little light interlude during a poker game, different people reading different bits of it with different voices. All very hilarious. However, next day it was apparent that in the process the reference had blotted up a certain amount of port and/or claret. It also had a nice ring of red where a glass had rested for some time. It took a long time to remove these traces and the reference finally bore a strange circular watermark. As someone said later 'Stru got quite shirty about his reference didn't he'.

When everything was dry, off went the application and the waiting began. Finally, a letter arrived. A most memorable letter, since it was the only polite communication that I ever received from the RAN. It requested Mr Sutherland to attend at a particular time for an interview and sincerely hoped that this would be convenient. If it were not, the letter continued, no doubt some mutually convenient alternative could be determined. Most subsequent communications I had from the RAN were terse and to the point and ran along the following lines: 'The Naval Board hereby posts you to Her Majesty's Australian Ship such and such and directs you to repair to your duty—by direction of the Naval Board'.

The Interview

Surgeon Rear Admiral Lionel Lockwood was a pretty impressive fellow. I liked him from the start and he appeared to be kindly disposed towards me. With one exception he turned out to be the only permanent medical officer in the RAN who was to impress me during the period that the RAN and I were uneasy bedfellows. Some were quite dreadful. For example, the little Surgeon Captain who was on Lionel's right-hand side at the big interview was quite a pain. The interview had been going marvellously. By sheer good luck it turned out that Lionel had been at university with both Dr Kerr and Dr Sharland. Not only had they been in the same year and played football together but they had also been good mates. In no time at all Lionel and I were thoroughly enjoying ourselves and he painted a glowing picture of the role of the young doctor in the modern RAN. The third member of the panel was some senior public servant, horn-rimmed and taking copious notes.

My enjoyment of the interview began to wane as Lionel's medical colleague decided to have his penny's worth. I thought this was fair enough at first but his questions ranged from the ludicrous to the intangible. A couple of times I shot a glance at Lionel or the public servant indicating my surprise at what was being bowled across the table. I gave frank and honest answers and just as I was deciding to skip the RAN, Lionel turned to the Surgeon Captain and quietly said, 'Shut up pinhead.' He then turned to the public servant and asked him if he had any questions. 'No, I don't think so, no, no, not really, no, I think we have it all, no I don't think so.' Lionel then sprang to his feet, extended his hand and thanked me for coming. 'No doubt,' he said, 'this panel has been favourably impressed. You'll be communicated with in due course.' I nodded to the other two panel members and exited in what I hoped was a brisk naval fashion.

A week later a letter arrived informing me that, subject to a medical fitness report, I was a successful applicant. It was suggested that I present myself to HMAS *Lonsdale* two days later to be medically examined. HMAS *Lonsdale* was not some sleek grey ship of war resting uneasily at Port Melbourne, but a block of land whose nearest point to the sea was at least 65 feet from the high water mark. It consisted of a flagstaff with ensign, a large parade ground and a rabbit warren of buildings heavily painted as is the naval tradition. The sailors called me 'Sir' and a very retired Surgeon Captain conducted the medical examination in almost total silence. Three weeks later my landlady routinely shuffling through my mail was startled to find a large manilla envelope addressed to Sub-Lieutenant (U) O.P. S.K. Sutherland RAN. No doubt she held it up to the light and also shook it.

The letter explained that during the immediate future I was to be nurtured by HMAS *Lonsdale* which seemed ideal as that particular ship wasn't going anywhere fast. It also informed me that there were large bags of gold coins held at the said HMAS *Lonsdale*: I was entitled to back pay but also I was to buy uniforms with appropriate trimmings suited to a Sub-Lieutenant. These I could order from and pay for via the Commonwealth Clothing Factory in due course.

Discreet inquiries determined that the (U) in my new title stood for 'undergraduate' and 'O.P.' stood for 'on probation'. Further intelligence suggested that there was no urgency in regard to the Sub-Lieutenant's uniform and indeed most people seemed to purchase a dinner suit rather than their first naval uniform. At some time before graduation the Navy demanded that all students undergo a six week training course down at Flinders Naval Depot to teach them a wide range of naval matters. These included marching, P.T. (Physical Training) and surviving a walk through a smoke-filled chamber. This was definitely not a six week event to look forward to. By sheer luck I later became one of the few seagoing naval officers who had never done an introductory course. More of that later.

A Day at Sea

There was one occasion when not having a Sub-Lieutenant's uniform was an obvious advantage. From time to time the Surgeon Rear Admiral would arrange for an invitation to his undergraduates to inspect visiting ships, etc. On one such occasion I was invited to spend a day on HMAS *Melbourne* as it took families out and raced madly around Port Phillip Bay. The invitation said that 'Uniform was optional' and one was invited to bring a companion if one so desired. One of my mates, John Ipsen, was quite interested in seeing some of the workings of the Navy so I invited him to accompany me.

We arrived at the dock and found the 'officers only gangway' which led to the upper parts of this towering aircraft carrier. Ahead of us marched one of my fellow Sub-Lieutenants. Resplendent in his brand new uniform he threw a magnificent salute as he stepped on the deck. The officer of the watch returned the salute, there was a brief conversation and then we saw his portly figure shunted briskly to one side. John and I were not certain what to do so clutching our umbrellas and looking dignified we stepped on to the deck.

After a moment's delay the officer saluted and we acknowledged this response with a slight nod of our heads. He inquired whether we were the 'gentlemen of the press'. Before I could answer, John with a slightly pommy overtone, explained that we had been invited by the Surgeon Rear Admiral to observe the day's exercises. A young midshipman was produced from nowhere and instructed to take us to the main party.

After morning coffee in the wardroom while the *Melbourne* was steaming down the Bay we were taken to a splendid vantage point to see planes take off and land. Jammed in a crowd below us we could see the familiar face of one Sub-Lieutenant. At lunch the poor chap had sandwiches and cordial while we had pleasant pre-luncheon drinks and then dined quite comfortably on assorted seafood. Late in the afternoon when the *Melbourne* docked again it was raining quite steadily and some important person offered us a lift back to the City when his car arrived.

Quite a lot was learnt about the Navy that day. However, John showed no further interest in it and later developed a slightly embarrassing habit of sending me postcards when I was serving overseas. Late at a dinner party he would produce a blank postcard and pass it around for our mutual friends to scribble messages on. All very reasonable. However, the card would be addressed to Surgeon Lieutenant Stru Boy Sutherland, care of whichever ship it was, and friends' comments made after a few good bottles of wine were generally not suitable reading for the chain of people who handle Navy mail.

A Private Session with the Admiral

Early in November 1961 I was summoned by the medical Director General to his office at Albert Park. I approached this meeting with some trepidation. My residency year at the Royal Melbourne Hospital was drawing to a close and the thought of leaving mainstream medicine was not attractive. However, the choice had been mine and by the time I reached the navy office curiosity about my fate had replaced the earlier apprehension.

'I've got an interesting job for you, Doctor,' said the Surgeon Rear Admiral. He leant forward in the confidential fashion, later seen in the film *The Life of Brian* when John Cleese says, 'We've got an interesting job for you, Brian.' And Brian's job was the highly dangerous activity of painting 'Romans go home' all over the buildings of occupied Jerusalem.

'A very interesting job, Doctor,' continued the Admiral. 'One that's come up rather unexpectedly. How about a nice trip up north in a destroyer?'... 'Visit exotic places—Hong Kong, Japan, the Philippines. Wouldn't mind you getting me a silver candelabra from Hong Kong. A nice neat 6 month trip. You should have a grand time,' he concluded, as he leant back looking father-like and awaiting my response.

'What's the ship, Sir?' seemed an appropriate opener.

'*Voyager*, a nice destroyer. To accompany the *Melbourne*. Will leave from Sydney. Visit the southern states. Then link up with the Brits, etc. For SEATO exercises.'

'Who's the captain, Sir?' I queried.

'Wells,' replied the Admiral. 'David Wells. Damned good bloke. Very pukka though, very pukka.' (I was later to find on the open-ended 'pukka scale', David Wells scored about 5, whereas my admiral was 9-plus.)

'Why has this job come up so suddenly, Sir? Last time we spoke I thought I would start at Flinders Naval Depot.'

'Um,' said the Admiral, 'there was a little bit of a problem on *Voyager.*'

'Problem, Sir?' I queried.

'Er, yes,' said the Admiral, 'the doctor had to be, shall we say, removed.'

'Removed?'

'Court-martialled, actually,' said the Admiral. 'Court-martialled?' I echoed slowly.

'Yes, there was a little bit of a problem between the doctor and Captain Wells.'

'Problem?'

'Yes. Things were a bit crowded on board *Voyager*. The Captain wanted to bunk down a couple of junior officers in the sick bay. The doctor wouldn't have a bar of this, and with one thing and another, things got a bit heated.'

'I see,' said I, seeing my forthcoming naval career in a new light. I hadn't realised court-martials still occurred and had mentally banished such events into the mists of naval history, along with keelhauling, the lash and, to a lesser extent, sodomy.

'Now, Struan,' said the Admiral, putting his persuasiveness into top gear. 'I'm sure you'll be much more diplomatic, and I'll keep a bit of an eye on things, anyway.'

I was tempted to ask, 'Are you going to come along with me too?' but that didn't seem an appropriate response and could possibly lead to another court-martial.

'I'm sure you'll have a wonderful time. Finish your hospital year, have a couple of weeks leave, then off on a trip of a lifetime.'

'What about this six weeks' indoctrination course I'm meant to do?'

'We can't fit that in, Doctor,' said my friend Lionel, becoming more formal. 'Sides, you've been in the army, you know all about discipline and one thing and another. Wouldn't worry about that.'

Finally, with a cheery handshake, the Surgeon Rear Admiral dismissed me and returned to his task of shuffling a variety of Surgeon Lieutenants (square pegs) into the kinds of ships or bases (round holes) which might remotely reduce additional court-martials.

3 February 1962 saw me alighting from a taxi beside HMAS *Voyager* which was contentedly tied up at Garden Island Dockyard, Sydney. My hospital job in Melbourne had finished the day before as the proposed holiday had vaporised when a three-week job in Dr Ian Wood's Clinical Research Ward had appeared out of the blue. No regrets about that, but it made the change in my professional lifestyle even more marked. (The Admiral had a hand in my acceptance of the three week stint. Cunning bloke.)

HMAS *Voyager*

In the Navy up to my neck

Two suitcases were dumped at my feet and the taxi driver sped off. They were really heavy suitcases, one containing a collection of naval uniforms which had never been worn. Looking at the gangway and remembering my experiences of the HMAS *Melbourne* officers' gangway some years before, it appeared best to start off on a positive footing.

The sentry guarding the bottom of the gangway suddenly saluted and this I returned in an unobtrusive fashion. I then left the suitcases where they rested and proceeded up the gangway in an orderly fashion. There was no way I was going to commence my seagoing naval career by staggering uphill with two suitcases I could hardly lift. It was a wise and officer-like decision, which was obviously appreciated by the officer of the watch as he saluted me when I stepped on to my future home. He greeted me warmly, two healthy young chaps were despatched to collect the baggage, and I was swept off to the wardroom to meet my new mates.

On entering the wardroom cries of 'G'day, Doc' came from various figures lounging around in white and having afternoon tea. I was given a cup of tea in one of the most massive cups I've ever held in my life. A jovial and obese Lieutenant Commander said, 'We've got a big cocktail party on board tonight, Doc. Starts at 6 p.m. We'll need your help in showing the teeth.' A newspaper was lowered and a hawklike figure in a white boilersuit said, 'Hope to Christ you're not like your predecessor, Doc.' Loud laughter greeted this remark which I found out later came from the engineering commander, or, in naval parlance, Commander (E).

Sipping my cup of tea, I had a closer look at a few of the people who were to be my companions for six months or so. Some people say that first impressions can be incorrect. With this mob I found that those who looked happy, jovial fellows remained happy, jovial fellows, whilst those giving the impression of being uptight, withdrawn or bad-tempered proved, by and large, to be uptight, withdrawn or bad-tempered. I wasn't to know it at the time, but naval officers are generally far happier when stationed on a ship than attached to some land base, provided in most cases they are not too far away from their families. Wardrooms on a happy ship are in marked contrast to wardrooms in land bases. On a ship it's the social centre, whereas on a land base it's used less frequently since many of the officers will be married and live off the base. So a good wardroom on a ship should be like one's friendly local restaurant, always a welcoming place to enter.

A few rules

There were a few useful rules in regard to wardroom behaviour. These rules had developed over the centuries and some were very sensible and helped to maintain harmony on long trips. No gambling for money was

Uncle Angus, my dad's favourite brother

Uncle Ian looking cheerful

Jim Sharland, a fine Bendigo doctor and close friend of the author's father

Dr Keith Kerr resting in the bush near Bendigo

Uncle Charles and Aunt Marjory, photographed in London, circa 1953

The author as a horseman. Shortly after, the horse took off, scraping the rider against every tree as it went.

Three pals—the author, Dick Harvey and Alan Kerr

Outside 'St Richards', Parkville, 1956. The author, Bob Dalgleish and Barry Thompson

allowed and a foreign language could not be spoken. Appropriate phrases from another language could be employed, but this skill was not often demonstrated in the RAN. It's easy to see how heavy gambling losses might disturb the peace and harmony of the wardroom and, apparently, the suppression of foreign languages developed from a fear of mutiny.

There were two wardroom habits I thought rather cute: the Royal toast was always done with everyone seated. This was a leftover from the days when the deck heads were so low it was almost impossible to stand erect unless you were of small stature, like Captain Cook. The other interesting rule was that any officer entering the wardroom and not wearing the correct uniform for the season and time of day would have to make his way up to the most senior officer present who was correctly dressed. On reaching this officer, he would come to attention, gently, not in a barrack-ground fashion, and say, 'May my rig be excused, Sir?' It was grand when a most senior officer entered the wardroom in mufti and had to approach a junior lieutenant or even a midshipman to obtain such permission. It was particularly good if, say, the sub-lieutenant was entertaining a young lady; she was invariably impressed.

Another very sensible rule is that the Captain is not allowed to enter the wardroom unless invited by its head. This means that there is one place in a ship where the officers can collectively whinge about 'Father' without him tippy-toeing in and hearing the tail end of the conversation. On special occasions an invitation could be issued to the Captain

Back then to the wardroom. Eight of the total twenty-odd officers were present. The party that night was to thank the dockyard staff for refitting the ship and was being given by the First Lieutenant. The Captain would get back from Canberra either the following night or first thing on Monday morning, when we were to sail.

My own little pozzy

'I'll show you your cabin, now, Doc,' said Harry, the Captain's secretary.

'Lead the way, old chap,' I said, and Harry did just that.

Following Harry down a few ladders quickly established that most parts of a warship are made out of very hard steel which project outwards at the most unexpected angles, noisy motors are busily blasting hot air, etc., wherever it should be going, and some handrails were not handrails but steampipes having their insulating lagging replaced. Proper railing must be gripped very firmly, especially when temporarily concussed by a piece of projecting steel. This secure hold must be maintained until pain and confusion subside. Then one can continue ascending or descending without suffering broken legs, etc., due to the effects of gravity.

'Here we are, Doc,' said Harry as we came to a halt in the blind alley. Two camp stretchers were laid out on the deck outside the cabin door. 'They're for the sub-lieutenants,' said Harry. 'You've got a good cabin here, Doc. You share it with the Senior Engineer.' The cabin was about 6 by 8 feet, contained 2 bunks, 2 cupboards and 2 desks. 'I'd take the top bunk, Doc, because when Jack's pissed he's likely to stand on your face getting up into his bunk.' Harry looked around, 'Bloody good cabin this, Doc. Spacious,' he said. 'Stow your gear,' continued Harry. 'Guests arrive in 45 minutes time.' After asking him what should be worn I requested him, in confidence, to check my uniform prior to my first public appearance. Harry agreed to this and said he'd return in 30 minutes.

The cabin seemed a lot bigger with one person in it instead of two. It had an amazing number of cables running around its walls and the light switches and telephones would have to be classified as 'heavy duty'. It was situated just below water level and so naturally had no view of the local scenery. A year or so later when HMAS *Melbourne* chopped HMAS *Voyager* in half, the division of the latter went right through this cabin. Amongst those who died was one of the young sub-lieutenants who slept on the deck outside the cabin.

The Senior Engineer never trod on my face, in fact I hardly ever saw him. He spent his life in oil-stained white boiler suits and seemed to get by with about 4 hours sleep a night. If I woke in the middle of the night I could always tell if my cabinmate had arrived because of a distinct smell, like a service station that's just had a sloppy fuel delivery. I did occasionally have conversations with Jack, and these generally related to problems with the No. 2 boiler or leaks in the port gland space. (The gland space is somewhere near the shaft that drives the propellers.)

Halfway through the voyage Jack was moved to another inferior cabin and replaced with a Lieutenant Commander (L)—'L' being for electrician. This character decided that not only would he prefer my bunk but he also preferred my desk. Both desks were absolutely identical except one pointed in one direction and the other in the opposite. Since he was senior to me in rank, but not convincingly so in intelligence, I had little choice but to shift all my bits and pieces including myself from one region to another. This chap had a habit of staring at the wall for long periods, which was rather unnerving. Possibly he actually went to sleep with his eyes open.

A Naval contribution to the social life of Sydney
Thirty minutes later Harry returned as promised and found me resplendent in virgin white. With glistening gold epaulettes divided by the bright red band which indicated a medico, I might well, from a distance, have been mistaken for a gold-topped milk bottle. Indeed, I felt that a well polished trombone would be a more appropriate accoutrement to my new appearance than a stethoscope.

'Well done, Doc,' said Harry, 'the only trouble is your epaulettes are pointing the wrong way. Either that or your bum is arse-about.' Off came the white jacket, the strings were undone and the epaulettes reversed. Later, using a texta, I marked the undersurface of each epaulette with an arrow and the letter 'A' for 'Anterior pointing'.

'Follow me, Doc,' and off went Harry with me in hot pursuit, but wishing I could leave a blazed trail. As anticipated it took me something like 30 minutes to find the cabin later that night.

'Geez you're young for a doctor,' said the middle-aged lady as she stuck her sharp fingernails into my upper arm. 'Isn't he, George? Isn't he young for a doctor?' George, who was a dockyard foreman, was drinking deeply out of a strong scotch and ice and looked at us as if he couldn't care less. Halfway through the ninety minute party all was going swimmingly. The forecastle had been decked out in bunting and the barrels of the 4.5 inch guns had been perfectly positioned for supporting party lights. The fine display of naval glassware and bottles of duty free spirit glistened in the late afternoon sun. The First Lieutenant, who was a chirpy fellow known as 'Rocker', had given us a brief informal pep-talk, 'Show the teeth and do the best you can.'

On cue, at 1800 hours and without queuing, the thirsty mob swarmed up the gangway and headed straight for the forecastle. Every 2 minutes thereafter an empty spirit bottle would clunk into a box which contained its also empty colleagues under the makeshift bar. As a student I had learnt how to get stuck into free grog, but this mob were professionals.

I didn't know at the time that this was a PR exercise to try to improve the relations between the civilians and the naval personnel in the dockyard region. There had been industrial problems at the dockyard for years and the quality and reliability of the work carried out was a matter of great concern. Indeed, the inquiry after the sinking of *Voyager* found that some of the emergency hatches which should have been serviced at the just-completed refit were difficult to open.

Lieutenant Mike Hudson was an impressive host at the party. He made headlines a few years later when he had a clash with dockyard workers while he was Executive Officer on HMAS *Vendetta*. A dockyard worker, having completed some minor job on the forecastle, tossed his tea leaves over the side of Mike's spick and span ship. Most of them stayed on the forecastle so Mike instructed him to clean up. The worker refused, one thing led to another and the whole dockyard went on strike until Mike apologised to the worker concerned. Years later Mike became Chief of Naval Staff which meant that almost everyone had to salute him.

'Geez you're young for a doctor,' I was told for the hundredth time. Ten drinks later, the party was nearly over. I had met her 'closest friend' and been fed the most intimate details of their medical histories.

My upper arm was very sore from being clutched whilst I was manoeuvred around the forecastle. After shoving another naval savoury into her gob, she was halfway through 'Geez you are ...', when she suddenly sank to her knees. As she grabbed the wire railing a floodlit arc of vomit was seen to dive gracefully into Sydney Harbour. No-one took much notice. Soon the lights flickered meaningfully and, with a certain amount of shrieking and the odd tumble, the absorbers of naval hospitality headed off to the nearest RSL club to complete an enjoyable outing.

A painful start to Day One

When one has gone to sleep in an unfamiliar environment it is best not to sit up too suddenly when awakening. A slightly sore head from the night before faded into insignificance when I cracked my head on the cabin roof. In the semi-darkness I wondered where the hell I was. I needed to go to the toilet and lowered myself gingerly down from the top bunk. My foot landed on a warmish, small pillow which emitted a groan. It was the Senior Engineer's head hanging over the side of his bunk. One up to me, I thought. Returning to my bunk I managed to tread on his hand. This produced only a quiet moan.

Some time later a physical assault upon my person woke me and again my head struck the deck head. This was my first experience of being given 'a shake' by a steward bringing me a morning cup of tea. It may sound like living in the lap of luxury, but the purpose of the giant cup of scalding tea is to ensure that the officer wakes up and stays awake.

The morning wake-up appeared to be the only time a rating could lay hands on an officer without being charged. If the sea was rough the efforts to avoid collision between rating and officer or senior officer and junior officer became marked. When an admiral passed by below decks everyone pressed hard against the bulkheads, flattening themselves like huntsman spiders. Rough weather had another way of flattening people. Seasickness has no respect for rank and nothing's more ironic than a sick doctor at sea. Fortunately, seasickness never came my way and indeed I enjoyed rough weather, believing HMAS *Voyager* was invincible.

In stormy weather, water was poured over the wardroom tablecloths to prevent cutlery sliding off the table. If one already felt squeamish and came in to face a bowl of greasy soup, this ritual made the seasick victim review his immediate dietary plans. It was of great satisfaction, even a morale booster, when an unpopular senior officer experienced a solid episode of seasickness. Regular, unofficial bulletins of his lack of progress would sweep the ship. The medical officer would neither confirm nor deny these reports. If the officer was a particularly unpleasant fellow, the galley staff would make sure that cold, greasy pork chops were guaranteed to be on the menu and indeed, the only item on the menu, when at last, thin, pale and drawn, he re-entered the wardroom.

Meeting my Godmother

Godparents had not played a significant role in my life. My uncle Ian whom I met but once prior to him roaring off on his motorbike was one. The other was a Mrs Barclay. Throughout my childhood she had regularly reminded me that my birthday was at hand by forwarding welcome little postal notes of either 2/6d or 5 shillings. All I had to do to maintain this cashflow was, under mother's instructions, to write a note of thanks. These were practically identical from one year to another. Mrs Barclay and I hadn't clapped eyes on each other since I was about 3-years-old, and hence I could not pick her out from Eve.

Although the supply of postal notes had dried up some years earlier, curiosity took me to Neutral Bay to meet her on the morning after my first shipboard cocktail party. I got a warm greeting and a big hug. Mrs Barclay was a bright-eyed little character who, like my mother, had retained the interest and enthusiasm for life that is often found in 8-year-old girls. Whilst boys at this age are inclined to sulk and kick empty cans as they meander along, girls tend to skip and emit peels of laughter and chatter. If our ages had been closer, I would have fallen in love with Mrs Barclay almost instantly. It was a beautiful day and she led me down to the back of the garden where, surrounded by pots filled with a variety of bamboos, stood a marble-topped table and two chairs. On the table was a tray, two bottles of sherry and two glasses.

Pouring two healthy glassfuls, my godmother said, 'My daughter thinks I drink too much, and I probably do. The thing is, I can't stand drinking by myself. If there is no-one around, I'll wait outside the front gate and ask some passing person if they'd like to join me for a drink. This worries my daughter, as well,' she added brightly. 'She found me entertaining what she said was an old tramp, the other day. In fact he was a very nice gentleman. Cheers!'

With two bumps on my head from the cabin roof (or deckhead), a hangover from the night before and, in retrospect, residual damage from a series of farewells in Melbourne, a sherry was about the last thing on earth I wanted. One discreet sip made my salivary glands go into cramp-like spasms. An hour and a half later, both sherry bottles were empty and the bamboo plants surrounding me had surreptitiously received their aperitifs and were looking around for their entree. The impressive intake of sherry by my godmother appeared to have no effect on her whatsoever and, when I left her at the front gate, she was casting an eye in both directions in case a replacement for my company should hove into sight. I journeyed back to *Voyager* where I found a great increase in activity. Leave prior to sailing was finishing and her keen and not so keen crew were homing in from all directions.

Blossom

'Blossom's' surname eludes me but, for any RAN person of that era, the surname would be practically meaningless. The sick bay of an aircraft carrier has quite a large staff, that of a destroyer has a complement of one. The complement I had was Blossom. Blossom's full title was Sickberth Petty Officer (OP). The 'OP' implied he had been trained in operating theatre procedures 'to naval standards'. We had no operating theatre on *Voyager*.

Blossom was not of especially impressive physique. When first sighted he was shirtless, wearing a tattered pair of navy blue shorts and sandals which had seen better days. His physiognomy resembled young Warwick Fairfax when he was of similar age, that is, in his mid-twenties. Overall he did not meet my expectations concerning the appearance of a Petty Officer. A Petty Officer should, I felt, be a brisk, efficient machine ready to repel boarders, etc., at a moment's notice.

In the few hours before meeting Blossom, I'd been doing a bit of research with the Captain's secretary. No doctor had been carried by the *Voyager* for a six-month period, because she was either sitting in the dock or pottering around near the coast. During this period Blossom had been the medical supremo and he had resented the arrival of a medical officer a few months ago. He and my immediate predecessor had not hit it off but this conflict was rapidly overshadowed by the differences of opinion between the doctor and Captain Wells.

That doctor apparently had become increasingly annoyed about the proposal that officers be temporarily housed in his sick bay. The doctor's comments were passed back to the Captain by Blossom, possibly with some little extra flourishes. Blossom then relayed the Captain's responses back to the doctor, again possibly inflaming the situation further. When the doctor was removed from the ship to face disciplinary charges, the outcome was considered most satisfactory by Blossom. The inevitable appointment of another doctor prior to sailing overseas was a predicted eventuality, but when he found that the replacement was a 'new boy', he must have been somewhat reassured.

Meeting my Captain
Monday morning and we were to sail out of Sydney Harbour, with HMAS *Melbourne* and a few other hangers-on. There was great activity and everyone had a job to do but me. Unlike the army, there didn't seem to be collections of chaps sitting around waiting to be told what to do. I went up and checked how preparations were going on the bridge. The view was excellent and it was filled with many of the gadgets seen in the film *The Cruel Sea*. I knew this film backwards, in fact, I could take off some of the less demanding parts played by the late Jack Hawkins. First impressions that the bridge was an interesting place proved correct and I was to spend on average 2 hours a day at the back of the bridge for most of the voyage.

More and more urgent messages were coming over the amplifying system so I decided to leave the bridge and go down and check out the sick bay to see what Blossom was up to. By this time I fancied myself quite a dab at going down the near-vertical ladders at a rate not far removed from free-fall. I was doing this when something soft was struck. Further descent was attempted and strong resistance was met. Further pressure was about to be applied from above when, on looking down, I espied an officer's cap and a pair of epaulettes each with four gold bars. This was twice the number I had so I decided to yield and ascend from whence I'd come. The slightly stocky figure of Captain David Wells rose upwards and stood beside me. As he straightened his goldbraided cap which I had knocked askew, he inspected me with particularly intense blue eyes. His face was quite expressionless. I broke the ice by saluting. 'Ah, doctor,' he said. 'Welcome aboard. Talk to you later. Got to do some driving.' Well, I thought, he seems all right. He looked a bit like Jack Hawkins, too. Better Jack Hawkins than Noel Coward, I thought.

For decorative purposes, when the ship departs, any crew not required for immediate duties line the decks. This we did as at last we drew away from the wharf. The catering officer and I lined a tiny little deck which had an excellent but sheltered view. On the wharf relatives of the crew seemed either very upset by the departure of their loved ones, or somewhat relieved. It's interesting to note that one officer who was greatly distressed at farewelling his wife and young family had fully regained his composure by the time Hobart was reached a few days later. A very smooth young woman with an impressive sports car was waiting for him at the bottom of the gangway when he completed his duty, and he was not seen for a further 48 hours.

The sick bay
The fleet and Sydney Harbour were impressive sights that morning. Once out of the harbour we picked up what seemed phenomenal speed and raced hither and thither around the *Melbourne* like a madly excited whippet. I made my way down to the sick bay with caution and prepared to practise naval medicine.

The sick bay was situated immediately below the rear gun turret. Two long barrels extended from this turret. The first time I heard these guns fired was from inside the sick bay and they made a much louder bang than I had anticipated.

The sick bay itself was quite compact. Its main features were an examination couch, four metal-framed bunks, a boiling water steriliser, an ancient x-ray machine, a desk, a few chairs and, of course, Blossom. The desk was filled with a variety of naval forms. The steriliser contained a jumbled mixture of rusting instruments and old style re-useable syringes and needles. In the background, were cupboards which contained a variety of items I was quite unfamiliar with. There were jars of ointments and linctuses I had last seen when I was a chemist's lad. Most had never been mentioned in the medical course. A most attractive cedar box contained a collection of dental instruments. Sadly, I never had the opportunity to play with these on a suitable patient. The fridge in the corner was jam-packed with penicillin and vaccines. Strapped here and there were emergency lights, first aid litters, etc.

Blossom reluctantly rose from the best chair as I entered. 'Well, Petty Officer,' I said. 'Where are the patients?'

'I've dealt with them all, Sir,' says Blossom.

'What was wrong with them?' I said.

'Oh, this and that,' said Blossom.

'Let's have a look at the list.' 'Don't keep a list,' said Blossom.

'I think it might be an idea if we kept details of all the treatments.'

'If you wish to,' said Blossom, almost tossing his locks.

Blossom and I sat and looked at each other for a while. I then had a poke around the desk to see what was what.

Watched by Blossom, I had a closer stickybeak into the 'fridge, cupboards and steriliser. The 'fridge hadn't been defrosted for years and the vaccines in the freezer chamber like some long-extinct mammoth were encased in ice. Many items of an edible rather than a medical nature were neatly concealed behind the stacks of antibiotics. The sink and its contents were filthy and looked as though contact with them would result in an unpleasant social disease. A stainless steel bowl contained a half-eaten meal. Taking the better chair, I said, 'I think we need a bit of a clean-up here, don't you? How about you start with that steriliser.'

'Sir,' emitted Blossom.

Having apparently finished my day's medical duties, I then set out to learn further of the workings of a ship of war.

The Navy gets down to business

By my second day at sea, I realised I was basically a tourist carried as a form of insurance. Having got the hang of the situation the problem was how to keep occupied. Everyone, except Blossom and myself, was flat out performing sea-going duties.

Fresh sea air and the rolling environment meant that insomnia was no problem, indeed the return of the preschool after lunch nap became a distinct possibility. The Admiral was dead right, it did seem a chance of a lifetime. Far better paid than the other lieutenants, minimal duties, little chance of being involved in a good old naval battle, and provided with free accommodation. All the crew were medically fit and all appeared bent on self-preservation.

As we steamed south, *Melbourne* took on her aircraft which consisted of helicopters and really snappy little jets called Venoms. These jets were like sports cars with wings. They would shake and scream with rage on full throttle immediately before being flung off the aircraft carrier by the steam catapults. It is always fascinating to watch planes landing and taking off from an aircraft carrier.

On our second night at sea, the navy got really busy. *Voyager* was to be Resdes which meant 'rescue destroyer'. With the ships darkened, the *Melbourne* would tear along constantly changing position to face the wind correctly or to be inconsistent in the eyes of any lurking enemy submarine. The *Voyager*, which seemed particularly well darkened, had to position herself astern the *Melbourne* so that if a plane ditched either taking off or landing she was in a position to pick up any bits and pieces. Both ships burnt along in the same direction while planes were taking off or landing, but the moment the *Melbourne* changed position poor little *Voyager* had to go off in a sweeping arc, hence covering a greater distance, to resume the Resdes position. Whilst doing these ballet-like manoeuvres the *Voyager* would pound along at 28 knots or so and any item anywhere not firmly stowed would move very sharply to the left or right, depending on the particular manoeuvre. This often included the medical officer.

In the darkness the *Melbourne* as she ploughed on ahead could just be picked out from *Voyager's* bridge as a few dim lights swaying from side to side. Keeping the appropriate distance at night is quite difficult. A lot of the time the two ships have to be quite close—which makes radar barely satisfactory. In daylight a clever little gadget called a Stuart's distance meter can be used and it's pretty simple to keep the correct distance. The meter is set to the known height out of the water of the ship ahead of you, the desired distance is also entered, and when you're at the correct distance the visible portions of the ship ahead sit neatly in the view finder. Nelson probably used something like this when he was a lad.

Self-preservation is a strong instinct and, after several trips to the sick bay via the upper deck, I explored other routes. Had I not, I fancy that the Surgeon Rear Admiral's score of medical officers might have read as follows: Courtmartialled 1, Lost at Sea 1.

A warship is a bit like an ant's nest with the greatest amount of activity going on inside. During night exercises many of the crew will be off duty whilst others will be on their toes in case something happens. For the sailors there is practically no privacy and, to get from one end of the ship to the other in rough conditions, the path led right through sleeping and recreation areas, sealed off from one another by great hatches which had to be opened and closed as one progressed. Some areas would be brightly lit, with men writing, reading or playing cards. Others would be in gloom and care would have to be taken not to disturb a sleeper's arm or leg projecting from the side of the tiered bunks.

On leadership

A lot of the time Captain Wells appeared to have even less to do than me. I later realised that this is the impression a good captain gives, because the duties are appropriately delegated and his officers are not interfered with unless there is a clear indication that something is wrong. Captain Wells impressed me no end and the first officer's certificate I received from him also happened to be the best. Almost twenty years after I last saw Captain Wells I had reason to be grateful for his acquaintance under unexpected circumstances (see page 272).

If a leader is fair and respected, then the organisation he controls will generally be a happy one. His example will infiltrate right down the ranks and his consistency will tend to produce harmony. Those who are disgruntled by nature will find it hard to activate discontent amongst their colleagues and people are less likely to be victimised in one way or another. Good leadership should show encouragement and appreciation of all staff and there should be no hint of favouritism. A good leader should be welcomed and he or she should leave the staff feeling both refreshed and flattered. As the command of warships is changed every two years the reputation of a ship for being happy or unhappy can change quite regularly.

Whether the ship is happy or unhappy, a position that no-one would like to hold for too long is that of second in command. The First Lieutenant has to make sure everything is absolutely right all the time. His performance will usually decide whether he is promoted to the rank of commander and possibly later get his own ship. By the end of his time as a First Lieutenant, he will have a pretty clear idea of how far up the ladder he may rise. Whereas the captain is supposed to be the loneliest man on board ship, a newly arrived First Lieutenant is potentially the most disliked person. Some First Lieutenants slide comfortably into the role of being general smartener-up of everyone and everything, and later can't drop this mantle and turn out to be real bastards as captains.

More on Blossom
Considering my overall impressions of Captain Wells, the court-martialling of my predecessor seemed quite surprising. But then Blossom must be taken into account.

'Petty Officer,' I said, 'that steriliser's still filthy. There's bits and pieces floating around the water.'

'Been a bit busy, Sir,' said Blossom, as he emptied the contents of a glass syringe into a sailor's bottom, said 'Pull 'em up', strolled over and rinsed the syringe and needle under the tap, disassembled the syringe and dropped the lot back into the steriliser. He then dived his hand in and retrieved another syringe and barrel, placed it in a kidney dish covered with a towel and put it aside for use in the near future. To the departing rating, Blossom tossed the information that he would feel another little prick at 1800 hours.

'Petty Officer,' I said. 'I don't think anyone should get any more injections until the syringes and needles can be properly sterilised. I suggest you get cracking immediately, scrub the whole works out and replace those old well-used needles with a set of new ones. Do this before anything else.'

'Sir,' said Blossom.

An hour later I came back into the sick bay and found Blossom bent over my desk, obviously most preoccupied. 'What are you doing, Petty Officer.'

'Petit point, Sir,' came the response.
'Petit what?'
'Petit point.'
'What the hell's that?' I said.
He replied, 'A form of embroidery, Sir.'
'Embroidery.'
'Yes, embroidery, petit point.'
'I see. Steriliser fixed?' I inquired.
'Not yet, Sir, because I always do my petit point at this time of the day.'
'Petty Officer, I think we might have to bring a third party into the issue of the steriliser.'
'Sir?'
I wandered off, intending to seek advice from the Captain's secretary. Passing along the upper deck I espied Chief Petty Officer Coxswain Campbell, extracting his tall frame from his tiny office. This Chief was the oldest man on board, a World War II veteran, and obviously worth talking to off the cuff, as it were.

Five minutes later the Coxswain lowered his head to within one inch of my Petty Officer's glasses. 'Up to your old tricks, are you, Blossom?' said the Coxswain. He then concisely and in measured tones outlined the Navy's attitude towards the prompt obeying of lawful orders given by an officer to a rating. I found the Coxswain's burst very instructive.

After concluding with the exhortation, 'What the doctor wants, Blossom, do bloody quick!' he turned to me and said, 'Anything else for the moment, Doc?'

'Coxswain,' I said, 'I thought, say, a thousand word essay on why sterilisation is important on the sick bay of a warship could be a useful exercise.'

'You hear that, Blossom? Good idea. I'll have a copy too and I'll have it fast, Blossom. I'll have it by tomorrow morning, Blossom.'

Blossom's hands shook a little as he picked up his petit point and lowered it into the drawer which held the ship's supply of condoms.

For twenty-four hours Blossom's behaviour was exemplary. His essay was not particularly enlightening and may be summarised by the suggestion 'hot things have a tendency to kill things'. He was no doubt deeply wounded and affronted by his confrontation with the Coxswain and thereafter was no more helpful to me than was patently necessary. Some nice little tricks were tried like, if I was to have a day off in port, the sick list could be dramatically expanded by digging out earlier customers for review, or plonking a stack of routine annual medical examinations down for that morning.

As we steamed southwards, Blossom announced that he had read there was an outbreak of influenza in Japan and suggested that 'flu vaccine be bought to immunise the crew prior to heading north. I said I'd check this out in Hobart with the fleet medical officer. In those days 'flu vaccine was not considered especially effective and caused quite a few reactions. In Hobart I sought advice on this matter and was told it was not considered advisable or necessary. I informed Blossom of this fact. 'I see,' said Blossom. A couple of weeks later, as we were about to leave Western Australia, an expensive shipment of influenza vaccine was brought up the gangway. This had been ordered by you-guess-who with a verbal assurance that I had specifically requested it.

The captain's table next morning saw a divisional officer, namely me, having his whole division, namely Blossom, charged under an appropriate regulation. The best part about this was hearing Blossom getting a good dressing down from Captain Wells who, at the same time, gave the cold blue-eyed stare. He lost a couple of days' pay and after responding to the order 'Caps on', he marched off the deck, down to the sick bay in a seething rage. Frankly, I felt that not only should he have paid for the vaccine but he should have been injected with the whole 300 doses.

After the 'vaccine incident' as the Coxswain and I used to refer to it in Blossom's presence, he gave very little trouble. We even did a little operation at sea which required a short general anaesthetic. A sailor had a severe infection of an important tendon in his hand and urgent drainage was necessary to avoid permanent damage. The examination couch in the sick bay was used as an operating table and *Voyager* temporarily stopped going round in circles during this operation. In retrospect, we were both very lucky that things went so smoothly.

Late one afternoon I had difficulty opening the sliding door to the sick bay. Upon entering I found it in near complete darkness. Turning on a light I found Blossom caressing with a dampened towel the head of a handsome young sailor as he lay on the examination couch.

'What is going on here, Petty Officer?'

'He's got a migraine headache, Sir, I'm just giving his head a massage.'

'I wouldn't like to get the wrong impression, Petty Officer, but I'd suggest this need not be done in total darkness.'

Speaking of impressions, we were all surprised when Blossom bought a diamond engagement ring in Hong Kong. However, a year or so later all became clear when Blossom was encouraged to leave the Navy following the finding of a letter he had written to a young rating.

My final memory of Blossom is quite a beauty. The long voyage was ending and we were steaming south towards Townsville before arriving at Sydney where an extensive refit was to take place. Blossom astounded me by emptying all the cupboards in the sick bay and wiping the shelves and stacking vast numbers of Naval medical supplies in the centre of the sick bay floor. As alluded to earlier, many of these standard Naval stores were preparations with which I was not familiar. I left Blossom to this task after giving appropriate words of encouragement and returning some time later found an exhausted but well-pleased Blossom having a mug of tea. I opened a cupboard to admire his efforts, and found it completely empty. The same was evident with all the other cupboards.

'Where've you put everything, Blossom?'

'Ditched them,' said Blossom.

'Ditched them?' I queried.

'Ditched them.' 'Why on earth have you done that?' 'Well,' said Blossom, 'if we bring them back we have to account for them. Then they have to go back to the navy store and that involves a lot of paperwork. So if we ditch them then they've been ditched and we have nil return as far as Stores go.'

Archaeologists in years to come will be battling to explain the jars containing strange ointments that are distributed at regular intervals on the seabed just inside the Great Barrier Reef north of Townsville.

Organ of generation

When I was about 11 someone gave my dad a book called *Cameos of Crime* written by a former Deputy Commissioner of Police in Queensland whose name was O'Sullivan. In due course, any new book reached my hands and in this one was a description of the encounter of a convicted rapist with a venomous snake. Thirty years later I had a terrible job trying to trace this reference. Below is reproduced the section used in a chapter I wrote on snake bite.

> A dramatic incident, which must have occurred about eighty years ago, was related by O'Sullivan (1947). O'Sullivan was a former Deputy Commissioner in the Queensland police and he described the tale of a man whom he named 'Dale' for legal reasons. Dale had a past record of rape, assault and possibly infanticide. After several years in jail he appeared to be a reformed character and was soon engaged to the housekeeper at the homestead where he had found employment. The following is quoted from O'Sullivan:

> 'About ten days before the marriage was to take place, the woman went into Mitchell to do some shopping and provide the things suitable for the occasion. The very morning on which Dale was to leave his camp and join her was very hot and, practically naked, for the sake of coolness, he was lying on a blanket thrown on a sheet of bark. Just as he was getting up, a venomous snake bit him on the organ of generation. Then he was faced with an extraordinary position—there was only one way of saving life, and he must act instantly. He did so, and, seizing a sheath knife, he performed the act of amputation. It was the next afternoon before he reached the hospital at Roma, in a very low state. Septic conditions had set in, and the surgeon's knife was used to such an extent that Dale was completely emasculated. He was under treatment for about four months.'

My problem as an 11-year-old was trying to determine what on earth the villain's organ of generation was. My elder sisters were of no help and I had to arrive at the solution by a process of elimination.

The reason for relating this episode of childhood bewilderment will become clear soon. As the good ship *Voyager* steamed down the east coast of Tasmania an 18-year-old lad woke up in his bunk and, overcome by nausea, vomited into his cap, which rested nearby. This unusual start to the day was associated with an unpleasant ache in the lower right-hand side of his tummy. A little later in the sick bay this ache was smartly enhanced when I applied pressure on the region from above. Blossom and I exchanged meaningful glances. Further examination did two things. Firstly, the diagnosis of acute appendicitis seemed extremely likely (whacko the diddle-oh, an operation at sea!). Secondly, it revealed that the business end of the young man's organ of generation was tattooed with the most luridly coloured blowfly.

According to the chart we were about two inches away from the Royal Hobart Hospital and were scheduled to follow *Melbourne* into Hobart in some three hours time. Fortunately for the patient, who was now drugged and resting peacefully, an operation at sea could not be justified. Captain Wells told me that if we got permission to steam ahead of the fleet and this bloke was not at death's door when we tied up, there would be some very interesting questions, doctor. I checked the patient again and provided I didn't prod him in the abdomen he seemed quite content to lie very still. We duly arrived and the awaiting ambulance took our hero off to the Royal Hobart where he and his appendix were promptly separated. When the pain-killers had worn off and the patient was fully conscious, he became aware that there was a continuous stream of nursing staff sticking their heads through the curtains whenever his dressings were being changed.

When they saw him alert and conscious their disappointment was obvious and they promptly withdrew. He was blissfully unaware that during the hazy preoperative and postoperative stages practically every member of the nursing staff at the Royal Hobart Hospital had managed to have a peak at his tattoo. Details of this work of art had spread widely and staff who had been off duty for several days approached his bed with stealth at the first available opportunity. When he caught up with *Voyager* a few weeks later he admitted to having been embarrassed by the whole business, but on reflection said he wouldn't have minded showing the tattoo to certain of the nurses, one at a time.

Socialising in Hobart

Twice in Hobart I found myself close to very important persons. The officers of the fleet attended a late afternoon cocktail party in the grounds of Government House. Dressed in long summer whites we met the local community leaders while being blasted by a brisk wind from the Antarctic.

The chill factor was so severe that we were forced to drink neat spirits and huddle as close to the local inhabitants as possible. When we were all covered with permafrost, the band struck up God Save the Queen as the then Governor-General of Australia, Lord D'Lisle, appeared. At the conclusion of the anthem he mingled. As my commander-in-chief came within a foot or so of me I very discreetly came to attention, looking forward to having a bit of a chat. Without altering his pace as he passed, he swivelled his head round, said 'How d' y' do?', then looked again to the front and that was that. Five minutes later, having completed his colonial duties, His Excellency was back ensconced in the warmth of Government House. I had hoped to thank him for signing my commission.

Next day there was a chance to mix with another VIP. It was the day of the Hobart Regatta and *Voyager* had taken up the position of a glorified buoy in the middle of the Derwent. Senator John Gorton, then Minister for the Navy, later Prime Minister of Australia, came out by launch with his party to view activities from the ship.

For an hour the Senator and his party stayed in a discreet huddle in comfortable chairs, smoking and drinking while the yachts did whatever they had to do. He had a word with the Captain and the First Lieutenant. A few hundred of the rest of us had to be waiting in readiness in case our Minister wished to have a look around and say 'Well done chaps, etc.' This didn't happen and the last glimpse I had of Senator Gorton was of him boarding the launch with a large carton of cigarettes under his arm.

Later, I found that if one wanted to meet a VIP, it was better just to stroll straight up and start a conversation. The minders may be very cautious initially but usually if the Distinguished One appears to be interested in the conversation they relax a little. I learnt this trick from my young son when I saw him emerge from a crowd at the age of seven to shake hands with Prince Charles. They both appeared to enjoy the encounter.

Pulling of rank and 'white-anting'
Some officers made up for their inadequacies by overuse of rank. They were usually disgruntled passed-over lieutenant commanders who were suffering a premature mid-life crisis. Such officers took delight in telling a sub-lieutenant that 'I'll have that paper you're reading, Sub' or giving someone a particularly difficult time when they asked to have their 'rig excused'. This was not very common, but certain people became renowned for it. They were usually especially grovelling towards ranks much higher than theirs. A subtle form of rank pulling is described below. This was effectively countered by the 'white-anting' technique.

The Scene: wardroom of *Voyager* which is in port
I am duty medical officer and am curled up in a chair reading a book. Enter two sub-lieutenants with a pair of attractive local girls. After introductions I return to my book since they are brimming over with eagerness

to converse with one another. They are on their second round of drinks when in comes one large but jovial lieutenant commander. With considerable grace for his size, he manages to lower himself onto the sofa between the two girls. Within thirty minutes he has completely monopolised the girls' attention. He sprinkles the odd disparaging comment about sub-lieutenants into the conversation. The two sub-lieutenants hardly get a word in and take it in turns to collect drinks from the bar. He's a pretty good raconteur and the girls are almost convinced he's the captain of the ship. They accept the friendly pats he gives their kneecaps. To be even-handed, he throws in a few jokes about naval medical officers.

After sixty minutes or so, our lieutenant commander arises for a brief visit to the toilet. The girls watch his retreating frame, eyes shining with admiration. The sub-lieutenants glare.

'Isn't he a fascinating man?' breathes one girl.

'We are a bit worried about him,' says one sub-lieutenant confidentially, as he starts white-anting. 'We're delighted he's showing an interest in you girls, because there've been some problems in our last two port calls. He spent a lot of time hanging around the local primary schools and we're not quite sure what we can do about him. He hasn't actually been caught doing anything but everyone's been very concerned. We have the reputation of the navy to consider.'

'A very important reputation,' chimed in the other sub-lieutenant.

The wardroom door is flung open and back comes the jovial lieutenant commander rubbing his hands and smiling happily. He has no trouble finding room on the sofa because each girl has shrunk to her own side and each is earnestly talking to a sub-lieutenant. He makes a couple of courageous attempts to win back his audience and even offers to buy drinks. I find my book absolutely riveting. The girls invite the sub-lieutenants to a party onshore and leave the lieutenant commander attempting to strike up a conversation with me. He goes off to his bunk a bewildered man, leaving me to meditate on the usefulness of a little white-anting.

Chipping the decks
While plodding along at sea in calm days a lot of the crew spent hours chipping away at the decks with little hammers. This woodpecker-like behaviour removed layers of paint to expose the underlying metal. This was then treated with acid and various coats of paint re-applied. After several weeks of stepping around the lads so engaged I got the okay to join them for an occasional two-hour burst of exercise. By the time we reached Perth I was as brown as they were and could not be distinguished from them, being dressed in cloth cap, shorts and leather sandals.

One day I got an unexpected torrent of abuse from Petty Officer Sutcliffe for not adequately completing one region before moving onto the next. When I looked up, appropriately hurt, he slightly apologised.

Sutcliffe was a nuggety little bloke who absolutely adored being at sea. He maintained his marriage because he avoided land bases. 'Every leave's a honeymoon, Doc,' he said. 'and Mum's the captain of the ship. The mother-in-law's another matter,' he added darkly. I found Sutcliffe a great source of information and wisdom. (Most marriages would benefit from the partners having regular short breaks from each other, as this allows time for meditation and to appreciate the company of the other.)

A trip up north
For four months or so *Voyager* tagged along with HMAS *Melbourne* and visited exotic places. Generally speaking, everyone enjoyed themselves, did a lot of shopping and a fair bit of drinking, some got tattoos and others got clap. Exercises with other navies were carried out and this presented an opportunity to learn about international collaboration. After collaborating at sea, the sailors from the respective navies belted the hell out of one another as soon as they got back into port. Visiting some twelve ports over the four month period allowed the crew to make love to, fight, or bargain with a representative sample of all racial groups in the Far East except the Koreans. That year Australian diplomats in Korea had an easy time.

A few little highlights stand out:

In Guam, three of us had a very messy accident. Dressed in long whites we left the US navy officers 'happy hour' and in semi-darkness we took a short cut to where *Voyager* was at rest. *Voyager* suddenly disappeared out of sight as we disappeared into a deep and very smelly drain. The captain later made a wry remark about *Voyager* being boarded by three black and white minstrels. The worst thing was having to put on a smelly cap before walking up the gangway.

For a couple of days *Voyager* was allowed to escape from the company of HMAS *Melbourne* and steam down to Leyte Gulf to conduct a memorial service and visit the township of Tacloban. This goodwill visit was considered of great importance and a photographer from HMAS *Melbourne* was sent along with us. A brass band welcomed us at Tacloban and fourteen of us were put in seven open vehicles and cheered by a fair crowd as we were driven in procession down the main street. The hospitality was pretty impressive.

No official photographs were sighted because the photographer didn't come back to the ship. However, he did manage to get back to the *Melbourne* a few days later—minus all the camera gear and plus a good dose of Filipino clap.

A hell of a fright
In World War II HMAS *Perth* sailed happily through the Sunda Straits in what is now Indonesia smack into a massive collection of Japanese warships. *Perth* did not have a hope and was quickly sunk. As the peacetime Australian

fleet sailed towards the Sunda Straits heading north, President Sukarno belligerently announced the Sunda Straits were now closed to international shipping. Our beloved Mr Menzies said no it isn't and the fleet proceeded towards the Sunda Straits. Everyone was up very early and the gun crews were especially active. With their tin hats and anti-flash gear I hardly recognised them. No nice white uniforms were worn that morning and there was no hot breakfast either.

I was enjoying the novelty of this change in routine, when I noticed with alarm that live ammunition was being stacked around the anti-aircraft guns. Everywhere quite dangerous looking munitions with red tags and highly specific warnings were appearing from below decks. The large gun turrets were busily swinging from one position to another.

As we entered the Sunda Straits everyone took up their action stations. This meant that Blossom and I were safely ensconced in the sick bay immediately below the rear gun turret that bore two spectacularly noisy 4.5 inch guns. The idea of spending eternity entombed with Blossom on the seabed had little appeal.

After a few hours it became apparent that the Indonesian Navy had stayed clear of the region and we could relax. This meant that none of us would be getting medals.

The stacks of live ammunition disappeared from sight, leaving a very disappointed gunnery officer who was itching to blast something out of the water. The only injuries were sustained by a few ratings who dropped some of the larger rounds of ammunition on their feet.

Hospital visits

In practically every port we visited, I spent time at the local hospital and Captain Wells actively encouraged these expeditions. The variation in medical standards, equipment and common types of diseases was quite stunning. The doctor on a navy ship puts in a quarterly report; he is the only person on the ship other than the captain who can make such a direct report. The captain signs it and can make comments about it but is not entitled to alter any of the facts.

One of my reports included a summary of hospitals visited and this report was eventually reproduced in the *RAN Medical Newsletter* in its first and possibly only edition, dated December 1962. Reading it years later, the report has all the hallmarks of an innocent abroad. I was clearly impressed by a visit to the atomic bomb casualty clearing hospital in Hiroshima which was situated on a hill overlooking ground zero of the atomic bomb. Thousands of survivors regularly attend this hospital in a vast follow-up programme backed by the American government. The appearance of some of the patients could only be described as awful.

In the Peace Centre situated at ground zero there are photographs taken the day after the bombing. On the steps of one of the few solid buildings that survived, the outline of a body can be seen. This person vapourised by the blast could perhaps be considered luckier than many of the survivors. After leaving Hiroshima I read a book by Michihiko Hachiya, a retired Japanese doctor, entitled *Hiroshima Diary*. It is an extraordinarily moving account of what he saw and how little he could do in the days immediately following the bombing.

The return of the Great White Traders
As we jogged from port to port the tramp tramp tramp of returning crew clutching items ranging from transistor radios to carved camphorwood chests steadily increased the water displacement of HMAS *Voyager*. Having left the shelves of many electrical and other stores in Singapore and Hong Kong quite empty, the fleet found in Japan a variety of gizmos unheard of in Australia. When *Voyager* eventually turned southwards many of the crew had exceeded the duty free provisions of the Australian Customs Act. Not to worry, we all thought, we would swap, change and top up each other's lists so that we all averaged out near the maximum allowance. Soon it became pretty clear there were not many people who had minimal purchases.

A Customs Officer was to come on board in Townsville and remain with us to Brisbane, giving him time to check each person's Customs Declaration. By the time Townsville hove into sight there was a high degree of concern amongst many members of the crew. The sick bay had become a repository for quite a large number of bulky items and sheer panic developed amongst a few of us when it was announced that it would be necessary for the Customs Officer to sleep in the sick bay provided the ship's doctor agreed. The ship's doctor did not agree at all but could hardly give the real reason for not wanting that particular bloke in that particular spot.

When the Customs Officer stepped on board, the sick bay housed a surprisingly large number of cardboard cartons covered in thick brown paper and marked with a stamp 'Return to Navy Stores'. Some were suspended from the deck head, others crammed into the cupboards which Blossom had so recently emptied by tossing their contents along the edge of the Barrier Reef.

Rather than a Gestapo-like chap, the Customs Officer was a quite jovial fellow. Naturally he was treated with great respect by all hands. A casual conversation with him, as he tried to stow his small suitcase somewhere in the sick bay, disclosed that he was not a particularly good sailor. Regular mouthfuls of seasickness tablets with strong sedating side effects set the pace for the next couple of days. In the wardroom he always had a drink thrust into his hand the moment he seated himself. The weather roughened up a little bit and so the seasickness tablets were pushed to near toxic levels. By the time he climbed into his little bunk in the sick bay he could barely read the 'Return to Navy Stores' labels which covered a variety of cardboard boxes all around him.

Eventually the dear chap's unsteady feet touched dry land at Brisbane. The crew were left clutching their Customs Clearance forms with only a handful having to pay duty. One of these was the prize 'zonk' of the navy. He'd gone and bought a small black and white television set which had a very high rate of duty. Our friendly Customs man had noticed this on his declaration and said very pointedly and kindly, 'I think that's a small radiogram, don't you?'

'No, it's not,' said the fellow. 'It's a black and white television.'

A senior rating said, 'Son, what he's saying is that's not a television set, it's a small radiogram.'

'No, it ain't,' says the stupid one, and he promptly received a bill for at least two weeks' wages.

Marriage

A week after returning from *Voyager* I married my university sweetheart, Wendy Solomon. The wedding took place at St Peters, Brighton Beach, on 30 June 1962 and was one of the best weddings I had been associated with. Our first born, John, came into the world next year on 7 April and soon got into the swing of eating, growing and being healthy, etc. Our daughter Susie was born on 19 July 1965. Wendy was a good wife and an especially caring mother who was left to cope alone with the children when I was unexpectedly sent to sea again.

Although for some time after our marriage broke up in 1973 she was not my favourite person, I remember with gratitude her support and encouragement during the years since we had first met at the university in 1955. I had the privilege of bringing up son John and daughter Susie from 1973 until they flew the coop.

At the time of our marriage Wendy was an editor at Oxford University Press in Melbourne and later had a distinguished career as principal editor for Melbourne University Press.

HMAS *Cerberus*—The White Man's Grave

Upon our return from honeymoon we set up house in Frankston and I commenced a ten-month stint at HMAS *Cerberus* (Flinders Naval Depot) some 20 miles away. Like 99 per cent of those attached to this establishment my state of mind during this sojourn is best described as 'mildly disgruntled'.

Years ago some idiot decided to establish the main Australian training base as far as possible from the major sources of naval recruits. The isolated site is flat, windswept, sandfly infested and 40 miles from the flesh-pots of Melbourne. At least twice a year the weather is delightful.

Right from the start, I sensed an eerie, brooding feeling about this establishment. My impression was that for decades it had been inhabited by men and women who wished they had been posted elsewhere. Some people loved it, but many found the lonely weekends and being cut off from sea-going colleagues downright depressing.

New recruits from Queensland and Western Australia did not know what had hit them. Cold and miserable, exhausted, yelled at by everyone, they hardly got a chance to become homesick. The only refuge they had at hand was the Flinders Naval Hospital.

Flinders Naval Hospital consisted of a scattered but interconnected collection of one-storey buildings representing various architectural styles and ages. Entry into the hospital complex was via the 'nerve centre' or administration block. This housed the office of the Surgeon Captain, the ward master lieutenant, the small dispensary, the office of the duty petty officer and, most useful of all, the officers' toilet. The petty officers' office was manned some fourteen hours a day and was an informal smoke-filled social centre for many of the petty officers in the depot. It was always good for a free cup of coffee and after hours medical supplies.

There were three senior career medical officers at the hospital and four or five juniors, i.e. surgeon lieutenants. A variety of consultants visited the hospital regularly from Melbourne and were paid handsomely for their efforts. Serious illness was very uncommon and deaths practically unheard of.

The hospital was nothing like M.A.S.H., but staff and patients led a comfortable though boring life. Busiest and most important was the centrally placed kitchen or galley and the local pig farm was largely sustained by a regular supply of meals that had got nowhere near the patients.

Outpatients

Under some large pine trees situated discreetly away from the hospital was a ramshackle fibro building which was Flinders Naval Hospital Outpatients Department. This was the pits. Staffed by five ratings and one medical officer it was one of the busiest spots in the depot, and I copped it for ten months. They tossed in a bit of variety by rostering me for duty at the hospital every third night and every third weekend. Weekend duty from Friday afternoon until Monday morning gave one extra time to study the excitements of the depot. The outpatients staff were pretty good and one bloke in particular could pick someone that was really sick a mile away.

They got up to quite a few pranks, some not in the best of taste. Midmorning often saw the arrival of hot bread rolls which all and sundry fell upon at once. Even the patient on the couch would often be given a hot buttered roll. One morning I had removed a reasonably sized sebaceous cyst from a chap's neck. Pathology was not required and so it was discarded. After suturing up the small incision I went back to see the next patient. A few minutes later there was a hell of a commotion from the other end of our flimsy building. The sebaceous cyst had somehow or other found its way into a hot buttered roll.

The Mammoth Ball

Some twit dreamed up the Mammoth Ball as an ideal way of ensuring that all naval recruits would sleep soundly at night because of sheer exhaustion. The fact that they were already chronically exhausted did not appear to discourage the inventor.

The Mammoth Ball was quite heavy and about 6 feet in diameter. Its creator advocated the formation of two teams of thirty each with each team aiming to push the Mammoth Ball across the length of the other's territory. So far, so good. Once the game commenced, however, it became apparent by the trail of prone figures that, if the Mammoth Ball rolled over a recruit who was then trampled on by the boots of the propelling units, he was not going to notice he was exhausted. His condition became worse if, like a piece of passive dough, he was rolled backwards and forwards with the Mammoth Ball playing the role of a rolling pin. In wet conditions some recruits were rolled almost out of sight.

I was already busy enough in Outpatients without this new device causing almost as many casualties as a busy day on the Western Front in World War I. On my desk I had a small stamp which, if applied to the recruit's outpatient card, would bring tears of gratitude and relief to his eyes. It read, 'Excused PT and boots for XX days'. Soon many of the recruits were clutching little cards marked, 'Excused mammoth ball for 3 weeks, or 5 weeks'. Some of the more slightly built recruits had cards indicating they were excluded from Mammoth Ball for three years.

There were a few quite heated conferences between training staff and my leaders in the hospital. With great diplomacy I suggested to my seniors that if they wished to have Mammoth Ball activities continue they were welcome to come over and lend a hand in Outpatients. In time the Mammoth Ball disappeared.

A Senior Medical Officer and fun with stethoscopes

Surgeon lieutenants believe that, generally speaking, the higher the medical rank reached in the navy the greater the lack of competence to practise medicine. There are exceptions, but if you are not regularly taking direct responsibility for managing sick patients then not only do you get out of practice but your confidence subsides and a medical limbo is reached. In my four years in the navy I progressively forgot many of the finer details of medicine. It is different in the United States Navy where all aspects of family medicine are covered and many active specialists are found amongst the full-time medical staff.

One of my seniors was quite a dab at doing medical examinations which he conducted in a brisk, military fashion. The lower the rank of officer being examined, the more briskly was he examined. Quite senior officers got a far smoother performance.

This officer's stethoscope was unique. I found this out when entertaining my outpatient staff at home prior to Christmas. Among other gems of information I was told that some time back in an idle moment a sick berth attendant had jammed some cottonwool into the tubing of the medical officer's stethoscope. He tested it and not a sound passed up the tubing. Before he had a chance to take the plug out, the doctor bustled in, picked up the stethoscope and took it into a nearby cubicle.

He was intrigued to see the doctor listening intently to the patient's heart and then measuring the blood pressure and noting it down in the patient's history. For the rest of the morning the doctor, using a sabotaged and absolutely silent stethoscope, assessed hearts as being A1 and accurately determined blood pressure readings.

The cottonwool plugs were left in position for some weeks and then removed one evening. The day after, my senior colleague recoiled when suddenly the thumping sounds of a beating heart reached his eardrums. He was observed taking a patient's blood pressure three times with an expression of wonderment and satisfaction. A week later only one tube of the stethoscope was blocked and then the blockage was swapped to the other side. As a precaution the spare stethoscope was left in the fully blocked position.

No one was surprised when the doctor consulted an ENT specialist in the city and arranged to have an audiogram. The specialist, however, could not determine a cause for his fluctuating deafness. The sick berth attendant who perpetrated this was rather hoping the specialist would embark on some major ear surgery.

Stethoscopes are very useful gadgets. They used to be a clear indication that the bearer was a medical doctor. Clutching one was almost as good as walking round holding a brass plate saying 'Dr Fred Smith, MD'. Nowadays it's acceptable for even the most junior nurse in a hospital to wander round twiddling a stethoscope.

It has not occured to most patients that an important role of the stethoscope is to give their doctor a little peace and quiet. The voice of the garrulous, whining, whingeing patient can be reduced to a distant murmur by clamping on a stethoscope and thought processes can be helped while peacefully listening to the thumps and clunks of the patient's heart as the valves open and close. 'At least he's still alive; I can hear his heart.'

Patients may not realise that often when a doctor takes their blood pressure his thoughts drift away and the process has to be repeated again to note the blood pressure readings. However, some things are bound to get our attention. Once, in Darwin, I was listening to a fellow's heart as he lay on a couch on the second floor of the hospital at the RAAF base.

Although I was having a quiet think, I noticed the patient's heart rate getting faster and faster. That's a bit unusual, I thought. I looked at his top end and, to my surprise, he was wide-eyed and shaking. Screams and yells hit my eardrums the moment the stethoscope was removed. An earth tremor was shaking the building and from the window I could see people running in all directions. My reclining patient became an airborne blur, leaving his symptoms behind him.

Critical appraisal of a Surgeon Commander

I'm sorry about this, possums, as Barry Humphries would say, but I must get a few observations about this bloke off my chest. Although he had been a confirmed Surgeon Commander for at least three years when we met, Tim had never been to sea. He was a tall, thin overseas medical graduate, with beady dark eyes, a little bit deaf, a little bit querulous and even a little bit eccentric. He was titular head of the surgical wards, although the bulk of the surgery was done by visiting consultants at the hospital or up in the city. Tim got a bash at the occasional appendix or haemorrhoids. He had no postgraduate surgical qualifications and had a bee in his bonnet about certain types of therapy, such as the use of chlorophyll to promote wound healing. Chlorophyll ointments have been abandoned now for years.

Having just got back from sea and been appointed to Flinders Naval Depot, the Surgeon Rear Admiral had assured me that I would be able to attend clinics at the Royal Melbourne Hospital one afternoon a week. Arriving at FND I found Tim acting in charge of the hospital. 'Absolutely not,' was his response to my request to escape from outpatients and burn up to town to get a little medical instruction. This knock-back was not only very imperious but had an air of finality about it. 'Sir,' I said, thinking where on earth did you come from you old bastard. A week later the Surgeon Captain returned to the hospital and, after a little bit of stimulation from the Surgeon Rear Admiral in Melbourne, all surgeon lieutenants were allowed to spend half a day at the hospital of their choice in that city.

Young doctors expect their senior colleagues to be examples of industry and a source of great wisdom and advice. I should also add, cleanliness. Tim's fingers were permanently stained with tobacco tar (nicotine is colourless) and often the dirt in his fingernails could be seen through his surgical gloves. After a ward round in the morning, which might last ten minutes or so, inspecting the odd post ingrown toenail operation, he would retire to his office for a little light reading. Morning tea in the officers' mess, a five minute stroll away, would occupy a comfortable thirty or so minutes. Back in the wardroom at midday for a few leisurely pots prior to a standard three course naval lunch, with or without wine. Tim considered himself quite a good judge of wine (see below). Re-energised by the luncheon break and having caught up on the latest gossip around the place, Tim retired again to his office where, believe it or not, he settled down to assemble model cars from kits. Meantime I was in outpatients dealing with the latest massacre produced by the Mammoth Ball.

The surgeon lieutenants used to get together from time to time to have frank discussions about various matters and often disguised bottles of wine would be produced to make heavy drinking appear a little more civilised. One of these sessions involved various types of masked champagne and some cunning swine had put a bottle of dry apple cider amongst the other unknowns. On this particular occasion Tim was invited along but arrived quite late, by which time most of the champagnes had been drunk and the still-covered cider bottle picked as cider and left well alone. Tim strolled up and filled himself a glass of cider. After holding it to the light, sniffing it, rolling it slowly round his mouth and, thank heavens, not gargling, he announced it to be of French origin. The rapt audience encouraged him to be even more specific and he offered a choice of two most likely regions. When the bottle was unmasked he went into a prolonged sulk.

The Surgeon Commander gives me a fright

When Tim had exhausted himself with his latest post-lunch car modelling project he usually disappeared home quite early. On Friday afternoon he skipped the car modelling and, to everyone's satisfaction, the hospital was Tim-free. One Friday immediately after lunch I'd had a rather heated discussion with Tim about something or other and, to put it mildly, the matter was unresolved when we parted. About ten minutes later I found Tim approaching me in long whites clutching a sword. 'Hello,' I thought, 'this is a bit unexpected.' There seemed no way of escape and I determined to knee the old bugger in the crutch before he sliced one of my ears off. As he got into range and my lithe young body was ready to get in first, I noticed that the sword he clutched was still in its scabbard.

Before physically assaulting a surgeon commander I was wise enough to inquire, 'What on earth are you up to?'

'Divisions,' said Tim.

'What are you going to divide?' I said.

'I'm standing in for the Surgeon Captain at Divisions,' said Tim. (Divisions are a type of parade and inspection of all the different departments of the ship's company.)

'You shouldn't be marching around with a sword,' said I.

'Why not?' said Tim.

'Well, it seems wrong to see a doctor marching around with a sword, that's all.'

'Will if I want to,' said Tim, tossing his locks and purposefully heading to the parade ground. Hope you trip over it, I thought.

One way and another I didn't regard Tim as a godlike figure of medicine. A year or so later, I was appointed to HMAS *Melbourne*. Amongst other duties I was to administer the anaesthetics required by the Fleet Surgeon who was also based on the *Melbourne*. A month later Tim was appointed Fleet Surgeon! We spent an interesting six months or so on exercises up north with the fleet. Oddly enough our relationship improved and we worked quite harmoniously. This was just as well, as between the two of us we had the skills of one capable rural general practitioner.

An unfortunate death that led to hospital improvements

There was a flu-like epidemic and I'd spent Saturday evening packing the hospital with recruits suffering from sore throats and high temperatures. Once in a warm hospital bed they tended to cough themselves contentedly to sleep. Most of the night staff were sick berth attendants who had been trained at the hospital itself. Some were very capable, others were somewhat dim.

At 3.00 a.m. I climbed wearily back into bed having admitted the umpteenth near-identical case. A sick berth attendant tapped on the door and informed me that a patient admitted a few hours earlier had 'fallen out of bed and had some little bruises'. He didn't appear to have hurt himself. I asked him to check his pulse rate and temperature and bring that information back to me. They were both normal and so I said to keep an eye on him, and promptly went to sleep.

Before breakfast I set off on a quick round of the wards and to my surprise found this particular patient covered with a most spectacular rash. Although he was sitting up in bed looking reasonably well my alarm bells rang out loudly. The rash resembled a textbook picture of meningococcal septicaemia. This syndrome, which is a combination of meningitis and blood poisoning, would put the wind up any doctor.

Things then moved quickly. The patient was removed from the general ward into isolation and the diagnosis confirmed by lumbar puncture within the hour.

With this positive diagnosis my immediate reaction was to get him to Fairfield Hospital in Melbourne as soon as possible. Being a Sunday morning this could be done in a little over an hour. Permission first had to be obtained from the medical officer in charge of the hospital, a role temporarily being filled by the Surgeon Commander. He decided the patient should be treated at the Hospital and an undated protocol for the treatment of meningococcal septicaemia at Flinders Naval Hospital was produced from the front office. The civilian pharmacist was brought in to prepare a large intravenous infusion of sulpha drugs and cortisone. The Surgeon Commander in the meantime had returned home.

As soon as it was ready, I commenced the intravenous administration of the cocktail. The patient, who was a very pleasant young fellow, still looked surprisingly well and chatted on happily as I let the infusion drip, drip, drip in. Fifteen minutes into the infusion he suddenly died. I could not resuscitate him.

Every doctor has very good as well as extremely bad experiences. The dramatic loss of consciousness and death of this patient was one of the most ghastly moments of my life. As is often the case, the sooner an account of a particularly distressing episode is recorded, the better it is for all concerned. Within three hours of the lad's death I had prepared a full summary with recommendations after I had discussed the case with staff at the Fairfield Hospital.

Naturally to this day I regret not having got out of my bed to examine this recruit who had 'fallen out of bed and got some little bruises'. At that early stage the diagnosis may not have even been considered. But regrets remain.

The major benefit from this tragedy was the decision to staff the hospital with state registered nurses twenty-four hours a day. They brought new life into the place and the standards of nursing and patient observation dramatically improved. The problems I reported with lumbar puncture kits and resuscitation equipment were promptly attended to. The advice from Fairfield Hospital suggested that the sudden death was due to the fact that some cases of meningitis are highly sensitive to even a hint of fluid overload. The size of the infusion was sufficient to cause sudden compression of vital brain regions in the already swollen brain, which in turn resulted in death. This occurred even though only a portion of the infusion had been given.

Out of the Frying Pan—Fun and Games in Darwin

In 1964 the position of doctor in HMAS *Melville* at Darwin was up for grabs. There were some attractive aspects to *Melville*. It was on dry land and thousands of miles away from senior naval medical officers. The navy doctor looked after the army personnel as well and was thus responsible for some 300 enlisted people and 68 persons described as 'natives'. Communication with the incumbent doctor disclosed that medically the post was most stimulating, with private practice not discouraged, especially when it involved the care of navy and army dependents. The pregnancy rate was fairly high in Darwin so the opportunity for practising obstetrics and child care was higher with this posting than any other that the navy could offer.

On arrival in Darwin we found that the attractive scenario had changed literally overnight. The airforce, being desperately short of doctors, had decided to withdraw their man from Darwin and, as a temporary measure, it was ordained that I could comfortably take over the management of their 809 officers and men plus their 33 'natives'. It was suggested that the navy doctor might also look after the several thousand dependents of the RAAF. No doubt this sounded perfectly reasonable to a distant bureaucrat.

The first medical officer's journal I sent south was introduced by the following quote from the autobiography of the Roman emperor Tiberius Claudius, AD 18 (Carthage): 'I found the climate very trying, the natives savage, diseased, and overworked, the residents dull, quarrelsome, mercenary and behind the times, the swarms of unfamiliar creeping and flying insects most horrible.'

This quotation was very appropriate especially if 'the senior officers of the three Forces' replaced 'the residents'. Inter-service rivalry was rife and it was difficult to satisfy the demands of three masters. Neither the army nor the navy appeared to have forgotten the dismal performance of the RAAF during the bombing of Darwin some twenty-five years earlier.

One of the joys of Darwin was getting to know my Admiral's brother, the late Douglas Lockwood. For many years Doug had been the Northern Territory representative of the Melbourne *Herald* and was the author of a number of books including *I, the Aboriginal*. He had been in Darwin during the bombing and was a source of delicious gossip regarding that event and the behaviour of a number of well-known citizens. It was not only certain members of the RAAF who came off badly. The first book he published, *Fair Dinkum*, in 1960, includes descriptions of what he observed on the first day of the bombing.

A secret report on the Defence establishments of Darwin in 1963
The army was based on a beautiful little peninsula in the heart of Darwin called Larrakeyah. This was ruled by Lieutenant Colonel Miller, whose major interest was the breeding of Santa Gertrudis cattle. A crusty fellow who had come up through the ranks, he was named 'the Lord of Larrakeyah' by the proprietor of the local newspaper. The colonel seemed to give his officers a terrible time, but was all sweetness and light to me. Years later I found that he had been regularly complaining to Head Office that I was not giving his troops the attention I was giving to the navy and the airforce.

At the end of most days I did an evening surgery for Dr Ella Stack, who later became famous as Mayor of Darwin after Cyclone Tracey. I'd been doing this work for quite some time when I discovered that Dr Stack's evening receptionist was the Colonel's wife. This worried me, but I needed all the experience I could get. After a while, Dr Stack trusted me with some of her obstetrical patients.

Overall, the army was no threat to anyone, but did two useful things. It ran a joint officers' mess with the RAN and half the population of Darwin appeared to be honorary members of this most pleasant watering hole which looked out over the sea. Its other useful function was to run a canteen and bulk order service for servicemen and their dependents.

The navy shared many of the army facilities on Larrakeyah. This was an uneasy relationship because a lot of people didn't know who to salute first. The navy headquarters was in a fine old building overlooking the Darwin jetty. In this building was my main boss, the Naval Officer in Charge, Northern Territory. He was Acting Captain Keatinge, a tall, seemingly aloof fellow, who was very keen on encouraging sporting activities in the tropics.

Most serving officers are hypochondriacs, and so take a close interest in medical services. In effect, serving three commanding officers on a day-to-day basis was like being on a tightrope. There seemed to be a simmering personal animosity between the three that sometimes helped and sometimes didn't. Captain Keatinge certainly didn't muck around when he felt obliged to clarify the relationship between 'his doctor' and the other brother services.

The navy was great on communications and the Coonawarra Naval Radio Station some 10 miles inland housed and partially occupied the time of forty WRANS. The RAAF base that shared Darwin Airport, consisted of 800 plus officers and men, lots of dependents, and one aeroplane, which in twelve months I never saw airborne. I could never determine what the bulk of these people did, for when a major air exercise was on, the squadrons that arrived were preceded by a large supporting force. As well as perks such as trips hither and thither on routine RAAF flights, the RAAF troupe by and large had a better deal than those in the other forces. Their duration of appointment was less and their electricity supply was free. As a result, on a hot night the ground in their base vibrated because of the multitude of air conditioners all pounding away in unison.

The Group Captain was nicknamed 'Bonehead' as he was somewhat bald. He was also someone I wouldn't trust as far as I could kick. I never liked saluting him, and when it was absolutely unavoidable I gave him my 'slow, dignified' salute.

Right from the beginning I had a great time racing from one part of Darwin to another. Starting at the RAAF base, their sick list would be dealt with and hospital patients seen, then over to Larrakeyah for the Army/Navy group including inpatients, and later in the morning to the Darwin Hospital. Being able to treat service and non-service patients in the Darwin Hospital was great. Coonawarra would be visited after lunch and late afternoon locum work was always available. Obstetrics thrived, and a number of young mums wanted to name their babies after me, until they found out what my first name actually was.

Over the years there had been some weird doctors in Darwin, and a subtle testing routine was often carried out by the husband on behalf of the wife. The husband would present with an ache in the shoulder or kneecap and on examination absolutely no sign of any physical abnormality could be found. If the doctor's overall performance was considered satisfactory, then his wife might turn up as a new patient at the next family clinic. I learnt not to inquire after the husband's health as it only led to confusion!

Sport and death in the tropics

Darwin is pretty close to the Equator and for many months of the year the climate is revoltingly hot and steamy. Anyone, say from Melbourne, who feels a little exhausted after two days of a heat wave should not plan to spend too long in Darwin. Keeping people fit in the tropics can be quite a challenge and Captain Keatinge decreed that sport would be compulsory for all those under his command. Some people weren't too keen on this order.

After playing squash at midday my radiator practically boiled and it took an hour and a half or so before I was back to normal. A cold shower was out of the question because Darwin water comes via a long overland pipe. The sea water is tepid and for some months of the year it's inadvisable to swim in the sea because of the presence of Box jellyfish. Apart from these lethal creatures there are some very fine fish in the sea around Darwin and recreational fishing seemed a far more sensible occupation than playing soccer on the rock hard Darwin ground in blistering heat. The navy's workboat was available for fishing parties, and I spent many a happy hour fishing over the wreck of the USS *Peary*. Only recently did I realise that I had been guilty of fishing over and in an official USN war grave. Quite a few men were trapped on board the *Peary* when she sank. I am very sorry that I was so ignorant.

At a little party about four months after I first met him, Captain Keatinge suddenly dropped his authoritative and formal facade and proved quite enchanting company. He could mimic at least half a dozen of the late Peter Sellers best known characters. I warmed to him.

He was playing cricket after lunch two days later when he suddenly emitted a loud cry and collapsed. My chief petty officer was on the spot and commenced immediate resuscitation as he was transported to the Darwin Hospital. I left the RAAF base immediately and got to him in casualty before the duty intern appeared. Our failure to resuscitate him was made worse because his wife and family were seated on the other side of the curtains.

That week turned out to be a bad one for the navy. A couple of days later at Coonawarra an 18-month-old girl wandered around the parked navy bus. A little while later the driver got into the bus and drove off, oblivious to the fact he had run over the small child. When I got there she was lying still in the middle of the road, seemingly unharmed, but there was a telltale trickle of blood from one ear indicating a fatal fracture at the base of the skull. For legal reasons I had to identify this child later in the Darwin Hospital and I still remember the perfection of her waxy appearance.

Following the death of Captain Keatinge, Commander Sanderson assumed temporary command and *Melville* continued to be a 'happy ship'. The navy office soon thought fit to appoint their most junior captain, Captain Jeffrey 'Gladys' Gledhill, RAN, to the post. I think I would rather tell the tale of a sick cat and divisions than spend time further considering Gladys Gledhill.

A sick cat and Divisions

The wildlife in suburban Darwin is a study in itself. The green frogs are filled with a heavily pigmented green slime. If a frog is trodden upon at night the popping sound gives one a microsecond warning that one is about to slip sideways, or if on a flight of stairs one is about to go downwards. The greenness of the slime produces some quite artistic effects when a frog falls into a spin dryer full of sheets. Geckos are a useful tool for gamblers. Bets can be made as to the likely course taken by a gecko on a ceiling. Some Darwinians fancied crocodiles as handy suburban pets. One army sergeant had one in his chook run which had grown so fast it was C-shaped because of the confined space. He was a big, gruff fellow (the sergeant, that is) and no one seemed inclined to interfere on behalf of the poor creature.

Cats and dogs don't seem to do well in the tropics, and some very pathetic specimens roamed around as strays. One day an extremely emaciated cat took to sleeping under our house. It would hardly eat anything, it was incontinent and a swarm of flies followed it in a most determined fashion. After a few days it ceased deteriorating and just patiently hovered at death's door. After it had hovered for a further week it was suggested that I 'bring something home from the hospital to do something about it'.

At lunchtime next day I brought home a bottle of ether. Gently picking up the near moribund cat I lowered him carefully onto the bottom of a large galvanised garbage bin. He rested comfortably on an old towel and appeared to be dead lying there with his mouth open, exposing a set of chipped yellow teeth. 'Better make sure,' I thought, and so I discreetly put a cloth heavily soaked with ether beside the inert cat. A heavy lid was then placed on the bin with a brick on it.

I was just sinking my teeth into a corned beef sandwich when what sounded like a short drumroll floated up from underneath the house. I took another bite and the drumroll was repeated. I inspected the garbage tin and came to the conclusion that the cat had some spare lives to expend and was doing so by pretending to be a madly ricocheting bullet. It continued to do this in embarrassing spasms for some fifteen minutes. Occasionally there would be a single or even a treble drum beat, but finally all was silent. A cautious glance inside the bin disclosed the cat lying perfectly still with its mouth open.

My next problem was how to dispose of the body. Our garbage

appeared to be collected once every four months whether it needed collecting or not, and obviously an alternative had to be found. Digging a hole was impossible because the ground was so hard. The solution appeared to be the local tip.

I've always been interested in tips, and have often surprised the family by detouring off to inspect a municipal tip. It's a bit like my habit of following fire engines when possible. The Darwin tip was a ripper. Apart from its sheer size and languid burning stink, it had quite a variety of inhabitants domiciled on its periphery. These good folk lived either in substantial looking houses covered with galvanised iron or in various types of timber crates.

On this particular afternoon, as the blowflies buzzed dozily, a car bearing RAN number plates was seen to cruise slowly down to the 'Deposit Here' section of the tip. A naval officer alighted, opened the boot of the car and carefully removed a large paper parcel. It was firmly secured with perhaps an excessive amount of strong string. The parcel was carried over to a section of the tip which was smouldering in a brooding fashion. The officer put it on the tip and looked thoughtfully at the parcel for a few minutes. Then, after a look around which to the close observer might have appeared furtive, went briskly back to the car and drove off to Divisions at Coonawarra. His last view from the top of the hill disclosed four persons converging on the parcel from four different directions.

A possibility occupied my mind while following the Captain around at a respectful distance during Divisions. I wondered if the headlines of the local paper tomorrow would scream, 'KILLER CAT SAVAGES FOUR PENSIONERS. Police are investigating reports that a giant feral cat carried out an unprovoked attack yesterday at the Darwin Municipal Tip.'

An epidemic and the loss of a nice little business

The day after the 1963 annual RAAF ball held at the RAAF base, several hundred civilians and service people found they had something in common. All had nausea, vomiting and, later, diarrhoea. Some had very high temperatures and a number became extremely ill. Even those of exalted rank suffered from a good dose of cramping abdominal pain and for the next twenty-four hours the defences of Darwin were in disarray. The various 'chains of command' had a number of links temporarily missing because the links were sitting on toilets.

The same type of bacteria was grown from all the serious cases and attention was concerntrated on the RAAF kitchen where all the food for the ball had been prepared. Samples were taken from ninety-six cooks and messhands and four were positive for the particular bacteria in question. The toilet facilities at the mess were very poor and the messhands were responsible for cleaning their own toilets, which were in the immediate vicinity of the kitchen. Thanks to the help of Dr John Crotty of the Darwin Hospital laboratory, the nature and origin of this epidemic was

The author feeling the tropical heat

Steaming along behind HMAS *Melbourne*. Captain David Wells has just scrutinised his Medical Officer and his camera.

HMAS *Voyager* doing 'wheelies'

Three pretty boys in the Philippines. From left to right: the author, Mike Hudson and Richard Carpendale. Mike later became Chief of Naval Staff.

The Red-back spider (*Latrodectus hasselti*). Bites by this creature are the most frequent reason antivenom is given in Australia. At least 300 patients receive antivenom each year.

The Blue-ringed octopus (*Hapalochlaena maculosa*). This is the most dangerous octopus in the world a is common around Australia. There is no antidote its poison. (Photo Vern Draffin)

The Black House or Window spider (*Badumna insignis*). Now known to cause significant skin damage and urgent research on its venom is warranted. (Photo per kindness Dr Robert Raven)

Male Sydney Funnel-web spider (*Atrax robustus*)

The White-tailed spider (*Lampona cylindrata*). Bites by either sex can cause blisters which sometimes ulcera Problem needs more attention. (Photo Raymond Mascord)

quickly established. Although some senior RAAF people considered our investigation conspiratorial, the bulk of the personnel cooperated very willingly. The message finally got home to the RAAF administration that if such contaminated food had been served up say, at a family function, then it was quite possible that a small child would have died.

When the investigation was complete I gave an evening talk to all RAAF cooks and messhands to explain the findings and how such an occurrence could be avoided. There were a number of interruptions from a drunken catering officer, who turned up next day at the hospital to apologise.

The private caterers in Darwin often found their most reasonable tenders had been undercut and in time, some of these firms negotiated a mutually satisfactory arrangement with the mystery firm. The profitability of this firm was boosted largely because it had no wages to pay, the use of extensive kitchen facilities and large cold stores was free, and perhaps some of their contents were also available. A variety of vehicles could also be used for deliveries. For those involved in this nice little earner, it was a tragedy that the investigation of an epidemic of food poisoning put them out of business. The RAAF catering officers and men thereafter limited their activities to the RAAF base.

Things to do with 'natives'
My MO's journal disclosed the Armed Services on average had a total of ninety-nine 'natives'. They tended to wander round in twos or threes looking vaguely lost or poking away in a desultory fashion at the odd weed in the gardens surrounding the houses of very senior officers. Crisscrossing from one base to another, I could never determine what the other ninety or so were doing. All became clear when one was injured on mudflats a long way away from Darwin. I found out that their main task was maintaining a regular supply of large mud crabs, again for certain senior officers. It's wonderful to see one's taxes at work.

Although at this time the RAAF was having enough natural disasters of its own, it decided to have a practice plane crash disaster. The base was put on alert for a plane crash between 8 o'clock and midnight, i.e. exactly 10 p.m. The plane was to crash in the bush somewhere at the back of the base. Late in the afternoon a truckload of natives was deposited at regular intervals in the disaster zone. Each bore a large label indicating injuries such as 'broken leg', 'back injury', or 'dead'.

When the alarm went off several ambulances raced off to the disaster zone to identify and collect casualties. Although we had secretly obtained the plans for this exercise, it seemed important to treat the whole business as serious. Unfortunately, it was a pitch-black night and almost impossible to find the victims, unless they suddenly grinned or had been positioned with the label uppermost. Eventually most were collected, with some of the 'dead' ones hanging onto the back of the ambulances. It wasn't until later the next day it was realised that some of the labelled volunteers were still 'missing in action', lying patiently out in the bush waiting to be collected and brought back to civilisation. In the meantime the RAAF hospital had filled with RAAF personnel with real injuries. It's amazing what an overhanging branch can do to passengers in an open vehicle on a dark night! One fellow literally disappeared from the passenger's seat into the darkness after the jeep had hit an especially big bump.

Goodbye, prickly heat
If you have to work fairly hard, then for many people the climate of Darwin is downright revolting. However, there is something about living there which tends to promote better and warmer communication between people, and friendships forged in such circumstances tend to last many a year. Amongst my handful of closest friends are several I first met in Darwin. When the time came to leave, several useful things had been done. After a lot of resistance, safety belts had been fitted to all Service vehicles. Routine low toxicity spraying was carried out late every second afternoon around most Service homes in an attempt to control the sandfly population. My best memory was that of the mothers and their babies I'd been privileged to help deliver. Nature had been kind to all concerned, and no major complications had occurred in any delivery.

HMAS *Melbourne*

Whilst I was trying to be a proper doctor in Darwin, HMAS *Melbourne* had chopped *Voyager* in half. When I was appointed to the aircraft carrier my immediate thought was to make sure my cabin was not only well above water but preferably as far back as possible. This wish was surprisingly easy to fulfill because, when I boarded in Sydney, the *Melbourne* had been in dry dock for a month, most of the crew was on leave and an appropriate cabin was up for grabs. It was not far from the cabin of the Fleet Commander, a fact which proved useful at a later date when I wanted a source of 240 volts AC to run my tape recorder. Such a supply was only available at a few points in the ship and an obliging electrician tapped the one from the Admiral's cabin. Admittedly it quivered a little when his air-conditioner went on or off.

Life was not dull on the *Melbourne*. There were lots of things to see and do and, with over one thousand crew, there were plenty of people to talk and listen to. Even if I could have gone back to the *Voyager*, it would have seemed boring after the much more 'upmarket' *Melbourne*. New and interesting people were always appearing in the wardroom. There were observers from other navies, flying crew, fleet chaplains and dentists, etc.

A number of these intriguing characters also had special skills. One was the late Laurie Matheson who was fluent in Russian. Laurie had come up through the ranks and claimed that when he was a rating on Manus Island he had had a captured Japanese Rear Admiral to clean his boots! The night Prime Minister Harold Holt disappeared in the surf at Portsea, Lieutenant Matheson gave a very impressive performance on national television as co-ordinator of the search.

As well as the variety of companionship, the food was much better on *Melbourne*, and she gave an invincible feeling as she surged along. For the film buffs she carried a stock of 100 000 old cowboy films, and also Hitchcock's *Psycho*. I couldn't help noticing the day after *Psycho* was shown most people did not pull their shower curtains across!

Nonstop entertainment

Watching planes taking off and landing on an aircraft carrier is one of the most fascinating of live entertainments. It's a toss-up which is the better—the steam catapult launching the plane off the deck as the plane's engines scream furiously—or the variations of landing technique shown by the different pilots. If the pilot attempts a landing and fails to hook any of the arrester wires he is obliged to abort the landing and come back for another try. Some pilots were chronic 'last-wirers'. Others were either 'wobblers', or 'thumpers'. The former wobbled as they came in to land and the latter landed rather heavily. Sometimes thumpers also became 'bouncers', with the plane's hook missing the arrester wires on the bounce. Some came in low and mean and looked as though they were going to fly straight into the quarterdeck.

There seemed to be an infinite variety of things that could go wrong in the simple process of a plane being blasted off a flight deck and some time later coming to rest again on the same flight deck. Accidents ranged from the catapult crew suffering steam burns to flight deck crew being injured when 'something came loose when a plane landed'. It's fascinating to watch a device that has been designed to explode on impact bouncing across the deck and disappearing over the side!

At night it was difficult to see the faces of the aircrew, which are often quite fascinating. In some planes the observer faced the tail of the plane and all he could do was count the missed arrester wires as they flashed past. It was always interesting seeing someone get out of the observer's section if it had been their first flight. One fellow was shaking so much he seemed to make the whole flight deck vibrate.

Some of the pilots or 'birdies' were pretty impressed with themselves and they were a skillful and pretty cheerful bunch. Few people would care to step into their shoes, particularly considering the hazards of night flying. Landing on a faintly illuminated aircraft carrier at high speed at night would have to be classified as a dangerous occupation. Fortunately, during this trip only one fatal crash occurred. The plane went over the side and the injured observer was rescued but the pilot still sits in his plane at the bottom of the sea.

When word got around that the *Melbourne's* 2IC, Commander Bailey, was to do his annual compulsory take off and landing at sea, he observed as he got into his plane that every vantage point was packed. It was noticed with some satisfaction that the seas were quite rough and the flight deck's movements reflected this fact. After giving the assembled crowd a thoughtful glare, the Commander moved his plane to the catapult and shortly was airborne.

By the time he made his landing approach some time later the weather had become rougher. He lowered his wheels and many pairs of eyeballs watched closely as he adjusted his trim and descended onto the lurching flight deck. His hook neatly picked up the first arrester wire and he landed as gently as a butterfly. The fellow next to me turned and said thoughtfully, 'He's not just a bastard, he's a skillful bastard.'

Saluting senior officers

If one came round a corner and nearly collided with a commander or a captain then one did not hesitate to salute. With senior medical officers, however, it seemed reasonable not to overdo the salutations.

One chap who didn't get saluted very often was Acting Surgeon Commander Treloar. When it was announced that I was to 'pop off up to Sydney' and go wherever HMAS *Melbourne* went for six months or so, there was a sharp intake of breath by some of my colleagues. It was generally agreed that Stru and Treloar would not hit it off but oddly enough we got on quite harmoniously. Treloar at that stage was finishing his appointment as Fleet Surgeon and we never actually sailed together, because he was replaced shortly after *Melbourne* left the dry dock. I rationed the salutes so one a day sufficed. If we passed in the dockyard he never knew whether a salute was coming or not. Now and again I would throw him a real bobby dazzler of a salute, but generally an absentminded nod was all that was offered. When Tim arrived from Flinders Naval Depot to take over as Fleet Surgeon it was even more important not to overdo the saluting business. When I threw Tim my special salute with the steady unwavering eyeball, he would usually return the salute with a smile of great happiness on his dial.

Of course, the really serious non-saluter concentrates on the ploy of being without a cap whenever possible. In uniform but without a cap one may not salute or return a salute.

Keeping people asleep

Compared with the sick bay on HMAS *Voyager*, that on the *Melbourne* was like the Mayo Clinic. It had an operating theatre, several examination rooms, plenty of bunks and even a small laboratory. It had a carefully selected and well trained staff of fourteen ratings, the Surgeon Commander and myself.

The way the Chief Petty Officer organised a routine mass immunisation of the crew was a delight to see. There is little time for niceties or polite chitchat when a thousand or so men require a shot of cholera vaccine in one arm and typhoid vaccine in the other. The recipients paused for only a split second between the two syringe-covered tables. It was sheer poetry of co-ordinated motion to pick up a syringe, inject, put down, pick up a syringe, inject, put down.

There was the occasional hitch to this system. Sometimes the pressure from behind would slacken off and some twit would stand perfectly still for too long. After he'd received two or even three doses in each arm, the lack of movement would be noticed. One fellow somehow or other got back in the queue and was only saved from hyperimmunisation because the earlier needle puncture sites were bleeding happily. At regular intervals the whole process would come to a halt when someone fainted at the sight of the assembled syringes. Almost invariably, this would be a huge young stoker who took quite a bit of shifting. There was also the problem of getting the volunteers back into the queue after he had been removed.

As a general rule no-one was told what they were receiving, and those that inquired received the information, post injection, as they disappeared around a corner. Occasionally someone would pipe up 'Excuse me, I'm allergic', but by this time the injections would have been given, and all that could be done was to lead the recipient aside for interrogation. It was always busy for a day or so after these mass immunisations as many of our customers came back with significant local or general reactions.

The naval architects no doubt had a sound reason for situating the operating theatre immediately below the steam catapult. Being in the theatre when the catapult was in action was akin to being in a metal biscuit tin which is regularly struck by a metal bar. My job was to make the patient oblivious to these sounds as soon as possible. As I put him to sleep his last thoughts were usually of a certainty of impending death. Occasionally, a particularly powerful crash by the catapult caused a collection of cockroaches, alive and dead, to tumble from the deck head onto the operating table. Between the firing of the catapult the good ship of course was constantly changing its direction. In the theatre the only component firmly fixed was the operating table. At regular intervals everyone except the patient grabbed hold of some part of this table to stay in the action, as it were.

Although the anaesthetic trolley had self-locking wheels this did not stop it vibrating itself away from the patient. Since there were tubes from the anaesthetic machine connected up to the patient there was a limit as to how far the machine could go walkabout with safety. It was all quite good fun. Whilst Tim was probing around the patient's appendix I'd hang on to the operating table with one hand and restrain the anaesthetic trolley with the other.

We and the patients would not have been so lucky if major surgery or extensive resuscitation had been required. At the time cockroaches were landing on us, a Buccaneer jet was taking off from the aircraft carrier *Ark Royal* a few miles away. The pilot aborted the take-off because he could not obtain full power from the starboard engine. The reason for this malfunction was that one of the flight deck crew had been sucked into the jet as the pilot gave it full revs. Some of our plasma supplies were sent by helicopter to try and help this poor fellow.

Going mad
Even in peacetime there is a lot of stress and anxiety evident amongst the crew of a naval ship. Some of this is due to homesickness or a feeling of helplessness in regard to some major domestic problem. Fatigue can be an important problem in the tropics, especially towards the end of a long exercise. Some people just bottle up their cares and woes and then suddenly one day do the most surprising things.

'Medical Officer to the Bridge', seemed a harmless enough invitation, as *Melbourne* steamed along placidly on a beautiful tropical evening. It had been a beer-issue night for those off duty and I passed plenty of amiable faces as I made my way up to the bridge. I couldn't help but hear that quite a few other persons and specific groups were being invited to my destination with some degree of urgency. 'At once' followed their invitations.

The centre of attention from the bridge was a spotlighted figure which had jumped onto the rotating main radar antenna. It was a huge wild-eyed stoker who had managed to clear the high tension carrying devices in the area and was shouting abuse to all and sundry as he went round in circles. As he came round again he announced that he'd pissed in the officers' water supply. That made me stop and think. By now the power supply was being cut off in stages and the screen ground to a halt.

He was then invited to come down and this he did with great determination, promptly pitching into the special sea duty men awaiting below. It was on for young and old and the scene was not unlike a group of ants attacking a large wasp. Back and forth they went but after a while steady progress was made towards the prison cells on a lower deck. The Medical Officer followed this procession at a respectful distance. Finally the stoker was pushed into a cell, the pushers having great difficulty breaking the half nelsons and headlocks which had been applied to him. The last one finally squeezed out minus boots and the door was locked.

Ten minutes later the banging and thumping inside the cell became less frequent and after a conference with the stoker's divisional officer it was agreed we should initially take a gentle approach. The door was opened slightly and the divisional officer popped his head in. This produced an almost inhuman growl and one boot flew straight over his left ear. 'The Medical Officer just wants to make sure you're not hurt.' This was news to me, but it produced a slightly less terrifying growl. Next thing I knew I'd been popped into the cell and the door was closed.

Sitting on the bunk my companion was methodically and thoughtfully pounded the wall with someone's boot and produced a surprisingly King Kong-like rumbling sound. My little heart went pitapat. He suddenly turned round and said, 'Gday, Doc.'

With a very dry mouth I said, 'You've had a bit of fun, haven't you?'

'Doc,' he said, 'I'm fed up with this—ship,—navy, and—divisional officer.'

'Well,' I said, rather benevolently, 'I think we can say that you're pretty likely to get back to Australia well before any of the rest of us.'

'D'y reckon?'

'Guarantee it, mate. I'll go off and get some pills for you and I don't want you to say anything to anyone or do anything till you've had 'em.'

'Right, Doc.' Signalling to the eyeball looking through the hole in the door that I was ready to leave, I stepped gingerly over a pair of boots and two caps.

A week later, doped up and in a childlike trance, the promise was kept. The stoker flew south to deal with the problem which had triggered off his behaviour and which he had learnt of in a letter from home. It was little consolation to find that the officers' drinking water was not separated from that of the rest of the crew.

It's amazing how wrong one can be! A couple of weeks later there was a young lad spending a few days in the brig for some moderate misdemeanour. Any bloke in a service lockup has to be seen by the Medical Officer once a day and such visits are as much social, or even pastoral, as they are medical. This sailor seemed to be taking his three day stint almost contentedly but late in the evening of the second day he was being taken out for a bit of a walk and an airing when he suddenly dived straight over the railings into the pitch black sea. The search was futile and that was the end of him.

Going on a brothel raid

In most ports a naval ship either hires, borrows, or offloads a vehicle or two to transport their naval patrol. An important function of such patrols, although naval regulations may not put their duties quite so succinctly, is to scoop up sailors who are getting into trouble before the local police grab them. Some places are out of bounds, a fact that attracts sailors like ants to honey, and these out of bounds places have to be checked to make sure they have in fact, remained out of bounds.

Sailors seem to like letting off steam even if later it costs them a week's pay and stoppage of leave. A few even fancied a good fight to round off an evening's revelry. They would fight anyone although preference was usually given to the crews of another ship or, better still, members of other navies. The unwise might even tackle their own naval patrol. They invariably came off second best with this latter group because the naval patrols were sober, wore helmets and carried large, heavy sticks.

I took up an offer to accompany the naval patrol on one of their forays in the Phillipines with enthusiasm. It seemed a safe way to visit some of the battle grounds that had led to me sewing up split heads in the early hours and an ethically acceptable way of satisfying my curiosity about certain highly unsavoury dives.

The party roared off late in the evening, everyone except me armed with a baton resembling a small baseball bat. An empty paddy wagon followed us in what was a well planned tour of the nightspots. The first stop was the notorious out of bounds Yellow Bar which was packed with people of all sexes, sizes and shapes. The noise and smoke was terrific and in single file the ten of us went from one end to the other and out again. On the little dance floor a tall black American sailor was demonstrating how he was going to make love to Mrs Kennedy when he got home. He was using a chair to represent Jackie Kennedy. On the way out two officers from our own ship were spotted in mufti entertaining two of the local girls. The younger of these officers got a friendly bop on the head with a baton by the officer in charge of the patrol. The Yellow Bar was only out of bounds to ratings, and of course no officer would consider going near such a place!

After strolling arrogantly through half-a-dozen bars and discreetly checking the reception areas of some brothels, we filed into the rear of a small and very grimy theatrette. No charge for the patrol. Under a red light in the centre of the stage, two figures are wrestling quietly. When my eyes became accustomed to the dim light it was apparent that they were in fact two females going through various manoeuvres. A senior officer is recognised in the front row. At one stage he takes his glasses off, polishes them, puts them on and leans forward for a better look. A series of spasms and moans indicate that this couple's performance is over and there's a rather embarrassing silence as the two ladies pick up their clothing and exit.

The two girls are replaced by quite a pretty girl and a somewhat washed-out, gaunt teenage youth. The audience gives them a warm welcome but a little while later when it has become quite obvious that the youth is, shall we say, a 'slow starter', some derisive comments start coming from the audience. I felt a bit sorry for the poor chap because even the girl starts chivvying him. In the long run he did a reasonable job. It was strongly rumoured that one of the officers on the *Melbourne* had participated in such a public performance at a Mediterranean port. Apparently, the inadequate efforts of the paid performer proved too much and he climbed onto the stage and took over. This was witnessed by a number of officers of the Mediterranean fleet. A year later he married the daughter of a Rear Admiral. I often wondered if these two events were connected. He was a big racy fellow and I lacked the courage to ask him!

When another pair of studiously bored females began to undress one another we moved out into the night to continue our surveillance work elsewhere. The drunk in the paddy wagon had sobered up somewhat and was being teased by a couple of hundred Filipino children. He didn't know where he was or how he got there but he sobered up even faster when the paddy wagon took off. The driver considered a naval patrol should always proceed at a very business-like speed, especially around corners.

The power of the human mind to cut itself off from a difficult situation was demonstrated during our final brothel raid. These of course were not raids in the true sense but rather stickybeaking. Instead of being greeted on this occasion by the usual sullen diplomacy, the proprietress seized hold of our leader's arm beseeching, 'You come, you come, trouble.' Not wishing to be left behind we followed them closely up the stairs. From behind a half open door a female was giving someone the rounds of the kitchen or in this case the rounds of the bedroom. A gruff male voice was occasionally responding to the outburst. Madam pushed open the door and we all trooped in. This made the room very crowded.

Sitting naked on a messed-up bed was a fellow I knew quite well. His expression was vaguely serious but not threatened, like a monkey watching humans file past its cage. Standing beside him was a small woman of uncertain age who was naked except for a flimsy wrap of revolting pink. Because of the shouting match the normal rules of evidence were not followed. However, it appeared that our shipmate, whom we will call Little Jim, owed the lady some money. When he had the opportunity to speak Little Jim indicated that he had not received everything that he was expected to pay for. This brought an angry response from the lady in the pink wrap who pointed with vigour at a small damp patch in the middle of the bed.

Little Jim gave the impression that the whole matter was best left to be debated between the two women and the officer in charge so he removed himself from the bed and sought some fresh air by the window. A few minutes later, when attention was again turned on him, he was nonchalantly having a wee through the slats in the venetian blind onto the courtyard below. Shortly afterwards, he was taken down the stairs a heck of a lot faster than he had come up them. Next day he claimed his brain had no record of the whole episode. This attitude was supported by his disbelief a few days later when he came down with a solid dose of the clap. Whenever someone snaps venetian blinds closed I recall the naval patrol and especially that crowded bedroom scene.

Things sailors buy (caveat emptor)

The sailor had paid a lot of money for a huge diving watch. It was so complicated it could probably even detect sunken wrecks and his left hand almost dragged along the ground because it was so heavy. Back in the dockyard he decided to try it in the swimming pool. When he sank to the floor of the pool he was pleased to see all the little dials seemed to be doing the right thing. He swam to the surface and was admiring the watch in the sun when he noticed there was a fluid level in his waterproof watch.

He and his mates hunted down the watch seller and, after threats, received a replacement, some money and a few other freebies. A further test in the dockside pool proved quite satisfactory. Three days later the watch stopped at sea. When the back was taken off a dead cockroach was found harnessed to tiny little pedals which drove the mechanism.

Sailors will buy anything. In the Philippines large live eagles can be purchased. Getting them back on board posed a problem. Alcohol is not allowed to be brought back on a ship unless it's inside the sailor. It follows that when a rating comes staggering up the gangway clutching a large sack it attracts the immediate attention of the duty watch. As I passed, the following conversation was heard, 'What've you got in that bag son?' said the Chief Petty Officer.

'An eagle, Chief.'

'Don't be smart with me, son.'

'It's a bloody eagle, Chief.' Chief plunges hand into bag. Howl of pain and Chief takes hand out of bag. Fortunately the injury is minor.

Half an hour later, an inebriated sailor managed to get up the gangway and found a large eagle peering closely into his face. After frightening the hell out of returning crew all night, the eagle was untied and was last seen strutting in a dignified fashion into the centre of the US navy base.

An unexpected medal for bravery

The war games had been going for weeks and it was decided that all hands deserved a little bit of a picnic barbecue on dry land. A suitable beach on a deserted spot of the Malay Peninsula was chosen for this event. A few British and Australian ships anchored close to the shore and cutters ferried a few thousand cartons of beer and a few thousand naval personnel onto the beach. There, we gave the poms some decent steak to eat and proper beer. The poms were also shown how to play proper football and towards the end of the festivities, some were shown how to have a proper fight. At sunset everyone was safely back on board, tired but happy. Only a few minor deliberate or accidental injuries had occurred. The whole exercise was a brilliant example of organisation and efficiency.

One Saturday morning a year after I had left the navy my front doorbell rang. Standing there was a Commonwealth driver who handed me a small package about the size of a cigarette packet which he asked me to sign for. The packet contained a medal for Campaign Service on the Malay Peninsula. Attached to it was a very attractive green and purple ribbon. Some mistake here, I thought, and wondered what the charges were for fraudulently obtaining or displaying such a campaign medal.

Investigations suggested the following explanation for the arrival of this medal. Sometime after the barbecue the Captain's secretary on the *Melbourne* realised that all who had participated had in fact stepped onto what was officially a zone of hostilities. At that time British troops were engaged in direct action against the Communists on the peninsula. It followed that all at that barbecue were entitled to receive (indirectly from the Queen) the Malay Peninsula campaign medal and ribbon. The remorseless process of medal distribution was then activated and in time my front doorbell was rung.

A proud relative kindly suggested a miniature of this medal be obtained to wear on formal occasions. I ignored his advice, which is just as well, as in the thirty years since no invitations have been received to a medal-wearing occasion. The opportunity has never arisen to tell fellow guests how a simple barbecue could turn into the greatest beach assault of the whole Malaysian campaign!

When people were not playing war games, late afternoon at sea was always most pleasant and reflective. On *Melbourne* little groups of friends could be seen strolling around the flight deck. Almost as a routine the Roman Catholic chaplain, the instructor Lieutenant Commander and I would stroll back and forth the full length of the deck. We would have pleasantly philosophical discussions with a question being put to one of the three, which was considered in silence down one length of the flight deck. It would then be answered as we returned to the starting point.

I emerged from four years in the navy with an overall respect and indeed a fondness for naval personnel. The experience was broadening, particularly so when on HMAS *Melbourne*.

Part 2

A Research Career, 1966–1998

Chronology of Main Events 1966–1998

1966	April Commence at CSL.
1967	Appointed Foundation Head of Immunology Research.
	Commence work on the toxin of the Blue-ringed octopus.
	Reactivate the Sydney Funnel-web spider antivenom project.
	Supervise projects relating to CSL vaccines, etc., for the next fourteen years.
1970	Major snake venom project starts.
	Isolation of main toxin from Tiger snake venom.
	Radioisotope laboratory established.
1972	Successful detection of minute quantities of snake venom in human tissues and fluids by radioimmunoassay.
1973	Isolation of further snake neurotoxins including textilotoxin.
1974	Major survey of effects of Australian animal toxins and care of the envenomed patient.
	Establish a department of pharmacological research.
1976	Survey of the properties of antivenoms and possible causes of adverse reactions.
	Major articles on management of snake and spider bites in *Australian Family Physician*.
1977	Study of movement of venom in animals commences. First aid measures to be investigated.
1978	Pressure-immobilisation technique developed.
	Survey of 2144 cases of Red-back spider bite published.
	Small-scaled snake venom found to be the most toxic snake venom ever described.
1979	Article on Pressure-immobilisation type of first aid published in *The Lancet*.
	Survey of the major snake venoms in the world completed.
	Study of the effects of all major Australian snake venoms in monkeys completed.
	Survey of the use of antivenom in Australia over a twelve-month period.
	First issue of snake venom detection kits to Australian hospitals.
1980	January Experiments suggest pressure-immobilisation first aid suitable for Funnel-web spider bites.
	7 February Suspended after dispute with Director over staff cuts.
	21 February Reinstated.
	13 March Collaboration with Drs Duncan and Tibballs to explore possible treatments for Funnel-web spider victims.
	April Renewed attempt to produce a Funnel-web spider antivenom.
	24 June First sign of a possible Funnel-web spider antivenom.
	July Experimental Funnel-web spider antivenom shown to be effective in monkeys.
	December Funnel-web spider antivenom issued for clinical trial in NSW.

1981	1 February First clinical use of Funnel-web spider antivenom. 17 June Forty-fifth birthday. 18 June Paper clip incident. 7 July Suspended from CSL over the paper clip incident. 16 July Round one of hearing. 29 July Round two. 5 August Round three. 26 August Final round and back to work.
1990	January Ant venom project expands. 30 June Exit of Neville McCarthy from CSL. 1 July Dr Brian McNamee new Manager Director. October Venom research laboratory re-established. 12 November Dr Ian Gust appointed R and D director
1991	Venom research in full swing. Projects include White-tailed spider, platypus venom, box jellyfish venom, new tests for allergy to venoms and the development of alternatives to the use of laboratory animals. June Author declines to sign work plan devised by Ian Gust.
1992	June Under duress author signs work plan.
1993	June Gust and author cannot agree on work plan. 5 July Gust ups the ante, author unsuccessfully appeals to MD to intervene. 12 July Author releases an open letter to MD: produces some relief. 30 July The seeds of AVRU are sown. 1 September CSL offers author voluntary redundancy; negotiations commence. 25 November CSL offers AVRU $60k.
1994	1 February CSL offers AVRU $150k and loan of equipment, removal costs, etc. April Approaches Victorian Government and Mr Stockdale, Victorian Treasurer, for support for AVRU. 1 April University appointment 5 April Kim Beazley announces float of CSL will occur in May. 9 April 'McNamee's serve.' May CSL public float. 28 June Give final research seminar. 13 July Leave CSL for AVRU.
1995	January $100K announced by Marie Tehan. She asks Dr Carmen Lawrence to match it. October Dr Ria Leonard hired/AVRU fundraising brochure printed. December Mrs Tehan announces further $100K. In press release condemns lack of support from Dr Lawrence.
1996	January Caroline Wiltshire starts lab work. March Dr Michael Wooldridge becomes Minister for Health and Family Services. 1 July Dr Ken Winkel starts as Deputy Director (part-time).

	9 September Dr Gabrielle Hawdon starts as Research Officer (part-time).
	December Dr James Tibballs appointed an Honorary Senior Associate.
1997	January Drs Winkel and Hawdon take over the doctor's advisory service, SKS continues as backstop.
	Appointment of Professor David Warrell as an Honorary Senior Associate.
	October AVRU moves into new labs in Medical School.
1998	February Dr Anna Young joins the Unit.
	June $178K equipment grant from Victorian Government plus two-year salary funding.

Snakes, Spiders and Test Tubes 6

'How many goodly creatures are there here!
How beauteous mankind is! O brave new world,
That has such people in't.'

William Shakespeare, *The Tempest*

Prologue	163
Seeking an interesting career	164
Starting at CSL: Some first impressions	166
Learning the ropes	169
My own little lab	172
Animal toxins; an introduction via the Blue-ringed octopus	175

Prologue

Most of my time at the Commonwealth Serum Laboratories was exciting and only rarely did I not have a feeling of pleasurable anticipation as I made my seven thousand or so journeys there and back. On the negative side, I felt at times subjected to less than sympathetic treatment by some of my seniors and indifference by federal authorities. Hopefully, others so-treated may find some solace in this tale and the odd self-serving bureaucrat may even gain some enlightenment!

Although as a student I sometimes dreamed of a career in medical research, it never seemed feasible. I could count on one hand the medical researchers in Australia I knew about and they all seemed too brilliant for words. However, some years later a series of fortuitous moves gave me the confidence to ease myself discreetly into this field. A modern medical researcher usually has to face formidable entry hurdles, undergo precarious funding and, in my opinion, work in a highly stressful environment. At least, initially, I had it pretty easy and it was a great way to earn a crust.

Seeking an interesting career

1965 was my last year in the navy. Bouncing along on the ocean blue I spotted an advertisement under 'Medical Research' in *The Medical Journal of Australia*. 'Medical Officer required by the Commonwealth Serum Laboratories, Parkville, Melbourne', 'Ample opportunity to pursue independent research' and 'Position may lead to top Executive status'.

This was of immediate interest, for six months earlier my post-RAN plans to enter general practice had been stymied. My potential partner, Dr Philip Stretton had unexpectedly had no choice but to sell his Frankston practice. Philip was a wise and skillful doctor, we got on well and he had made the option of general practice very attractive. This was in contrast with some other practices for which I had done locums in the Frankston area.

General practice being out for the time being, I was left to admire a string of beautiful tropical sunsets whilst contemplating alternative career paths. For a while I had flirted with dermatology and became a bit of a dab at diagnosing the less common rashes. Dermatologists are sometimes denigrated, but their patients rarely die of a particular disease and they are infrequently called into hospital at night. However, some patients suffering acute skin diseases may become very ill and great skill is needed in their diagnosis and management. After all, one's skin is the largest organ of the body and the main barrier between our insides and the outside world.

Whenever in Melbourne I had popped into various skin clinics and found the best one for my purposes was at the Royal Children's Hospital. The young patients tended to suffer from acute but potentially curable conditions. The downside was that at times the patients' noise was almost unbearable. To commit oneself to forty years of high-pitched screaming from little patients was, I'm afraid, not a gloriously attractive proposition. The arrival of Susie on 19 July of this year enabled her to fill in the gaps left in some of John's vocal performances. When their dear little reddened screwed-up faces emitted a pain-producing sound like a thousand cicadas even geriatric medicine appeared attractive.

Another serious possibility was endocrinology. The treatment of thyroid disease, diabetes and other hormonal disturbances was becoming more scientific by the day. After some correspondence, I got myself attached to the Endocrine Clinic at my old hospital and read with great enthusiasm the latest publications in endocrinology. However, after further study, the subject seemed too complicated and potentially boring. For me there was little excitement in the routine management of diabetes, for example.

I seriously considered obstetrics as I'd found delivering babies in Darwin highly satisfactory. On the negative side I would be starting from scratch some four years behind my contemporaries.

There was an increasing number of articles on virology and, to a greater extent, immunology, in the run-of-the-mill medical journals. In the midst of articles on the latest way of treating ingrown toenails or injecting varicose veins, intriguing articles were popping up on tissue transplantation and how the body made antibodies to various vaccines.

Immunology in Australia and, more specifically, in Melbourne had been given a great boost when Sir Macfarlane Burnet shared a Nobel Prize in 1960 for work on immunological tolerance. I'd been a keen observer of 'Sir Mac' as he was called for many years and I usually received an absentminded nod when we passed in corridors at the Royal Melbourne Hospital Walter and Eliza Hall complex. For a few weeks in early 1962 the Admiral had let me do a casual locum in the Clinical Research Unit and this had brought me closer to the workings of the Hall Institute. There was a small but beneficial spin-off some time later. A number of the patients suffered from liver failure and large quantities of fluid had to be removed from their abdominal cavities at regular intervals. This was done by pushing a large trocar and cannula about the size of a biro through their abdominal wall. (The trocar is an unfriendly-looking sharp-pointed perforator which slides into the cannula enabling it to penetrate tissues. The cannula is a particularly wide-bored needle, so much so that its lumen would punch out a cylinder of tissue were it not neatly filled by the trocar during insertion.). It may be all right to do this once or twice, but the patient understandably finds it's most unpleasant when it's being done more frequently and at different sites.

I remembered using a neat little gizmo called a Middlesex needle when studiously trying not to kill patients with anaesthetics. This needle could be left in place and, through its rubber bung, fluids could be injected or taken out without causing patient discomfort. For example, after part of the needle had been inserted into the vein on the back of the patient's hand it would be strapped in place. Drugs could then be given intravenously by injecting through the rubber bung. With permission, I used these needles to remove the abdominal fluid and they worked reasonably well; more importantly the patients thought they were a great improvement. The Unit Head, Dr Ian Wood (later Sir Ian), suggested I write it up for *The Medical Journal of Australia*. This seemed straightforward but writing one's first article is as difficult, for many, as it is for others to have a first baby.

Two prompting letters from Dr Wood saw me scribbling my way through ten drafts which resulted in what I thought was a splendid definitive paper. A month later at the end of a naval exercise on *Voyager*, the re-written, practically unrecognisable paper was returned by Dr Wood. The only original line was the concluding sentence, which said in part, '... with the abovementioned technique the procedure ceases to be an ordeal to either doctor or patient'. Dr Wood declined to be a co-author and arranged for the paper to be sent to *The Medical Journal of Australia*. Later that year my first publication appeared under the Original Articles section impressively headed 'Simplified abdominal paracentesis with the Middlesex needle', Struan K. Sutherland, from the Clinical Research Unit at the Royal Melbourne Hospital and the Walter and Eliza Hall Institute of Medical Research, Melbourne.

Bedazzling as the title may have been, the article itself consisted of a mere five paragraphs! However, it was a start and in job applications looked quite good. It may even have helped when I applied for the advertised job at CSL.

Starting at CSL: Some first impressions

Perhaps unfortunately for some of the later hierarchy at the Commonwealth Serum Laboratories (hereafter called CSL), I was hired as a Medical Officer. The date of commencement was 18 April 1966. Between storing my navy uniforms carefully away in case of World War III and starting at CSL I took stock of the history of my relationships with difficult people. As a result in February 1966 I enrolled in a Dale Carnegie course in the city which included 'Fundamental Techniques in Handling People'. The course was also helpful in teaching how to communicate. It was wonderful to see at the end of the public speaking section that some people who earlier were too nervous even to announce their names, had become quite eloquent.

Wendy and I put a deposit on a fine old house at Brighton Beach. It was, as they say in the trade, 'a renovator's challenge'. It was a fluke we had the opportunity to take up the challenge. Wendy's father was having lunch with his old mate Justin Hancock who informed him that his sister Marjory's house was up for sale because she wanted to move into a unit. Marjory didn't want it auctioned but wanted to screen likely purchasers. Miss Hancock was won over by young John's bright red hair. She said he reminded her of her nephew, Arnold Hancock, who was on the board of CSL. Wheels within wheels! I felt an immediate affinity with the house which has strengthened over the years. It allowed each family member to have his or her own territory which, combined with solid brick walls, was a great aid to domestic harmony.

Thus, refreshed from a few sea voyages, freshly trained to get on well with my fellow men and women, and absolutely brimming with enthusiasm, I arrived at CSL.

In 1916 the Commonwealth Serum Laboratories was established in Melbourne because World War I threatened the shipment of supplies of vaccines and antiserum from the United Kingdom. It moved from its first site in the original Walter and Eliza Hall Institute out to Parkville in 1919. As it grew from strength to strength, so did its impact on the health of the Australian population.

An admirable history of these Laboratories entitled *Committed to Saving*

Lives has been written by A. H. Brogan (Hyland House, 1990). This official history meticulously details the ups and downs of the establishment. It describes the important overseas discoveries which triggered off flurries of activity at CSL and the local production of new biologicals such as insulin in 1922, the first snake antivenom in 1931 and the production of penicillin in 1944. The organisation can be proud of a number of other achievements, especially the production of poliomyelitis vaccine in 1956.

Against this background of human endeavour were occasional upheavals. The most famous one was the clash between the dynamic Director Dr Val Bazeley and the then Prime Minister Mr Robert Menzies in 1961. This led to the appointment of a Commission to oversee CSL's functions until the full privatisation of CSL in 1994.

For much of CSL's existence its hierarchy had been left largely to its own devices. Indeed Dr Jim Forbes, who later became Chairman of CSL, said that when he was Federal Minister for Health he seldom gave CSL a thought. As a result: any unexpected ministerial directive would result in furious 'all stops out' activity; any adverse media report would really stir the ant's nest; and while CSL issued indignant press releases the Minister would make calming and reassuring noises. This cosy arrangement could, in my opinion, cover up ineptitudes on both sides.

When CSL was making a profit, which it almost invariably did, there was little political interference. The Government funded activities like blood fractionation on which CSL had, and still has, a monopoly and from this secure position CSL could venture forth to increase its share of the marketplace. Not only did the Federal Government look after CSL's interests but it regularly poured large sums of money into the organisation for capital works programmes and research.

Management thus had the best of both worlds. If things got tough it could cry for help under its National Interest obligations and public service links. Since it was competing with the outside world, management also seized all the lush trappings of corporate life.

On this planet, CSL is unique. It sprawls over 11 hectares on the side of a hill. Abutting the site are hockey fields, two Mental Health establishments, the Melbourne Zoological Gardens, Turana 'Youth Training Centre' and a very large geriatric hospital. The site is covered with a variety of buildings which architecturally range from the quaint to the grotesque. Apart from the new Parliament House in Canberra, possibly no other piece of land in this country has had so much Commonwealth money sunk into it.

It is a fascinating place which never fails to intrigue the visitor. Steam pipes hiss as they meander from building to building and Vaccine Production Units try to contain the vilest smells (they say there's money in muck). Electric trucks scuttle around as they transfer stainless steel vaccine containers that look like little space modules. Day and night, machines hum away busily while experimental animals chew their pellets thoughtfully. The odd strange pong invariably escapes and drifts through the wards of the elderly demented and then over the hill into the homes of the good Burghers of Brunswick. Strange mechanical sounds abound and false alarms bring the Fire Brigade en masse at least once a week.

When asked how many people worked at CSL, a jovial employment officer routinely replied, 'About half the buggers.' This was tongue-in-cheek because he knew that most of the 1200 people employed took their jobs seriously and many were and are extremely dedicated.

People usually worry about their first day in a new job. Generally, after an initial introduction and run-around, they are dumped in a corner, given something to do and ignored. They have to look intensely busy, remain on edge all day and be ready to show intelligent keenness should any of the big bosses come along. Day one is often only marginally productive but extraordinarily exhausting.

Installed in an office in the administration building I observed people scuttling in and out of the adjacent offices. Phones rang incessantly and the row of typists outside the offices clacked away at their typewriters. If my phone rang it was either for the previous occupant or a wrong number. Some passers-by scrutinised me with expressionless stares which I returned. In retrospect two of these were the first sightings of men who later made my life difficult.

The prestigious position of Director had always been held by a medical graduate. I was later to learn that certain ambitious technical staff members felt a latent hostility towards 'medical doctors' . The former were better paid and usually had more input into policy matters. At the time I was blissfully unaware of how this chronic antagonism would later affect me.

CSL had encouraged me to retain my honorary appointment to surgical outpatients at the Royal Melbourne Hospital. As a result, for many years I had direct and often ongoing contact with patients. It is always essential to try to keep up-to-date and also to be reminded what a hard slog medicine can be. In these clinics I kept my eyes open for patients who were obviously hard done by and deserving 'Rolls-Royce' treatment. This might involve personally taking them to the X-ray department, talking to the radiologist, and generally pushing things along. I got great satisfaction out of being a proper doctor at these clinics and remain grateful for the backing of CSL.

Tea breaks and lunch offered interesting learning opportunities. 'The senior officer's dining room' was the meeting ground for a most diverse group of people. I could plonk myself down beside a senior engineer, a veterinary officer, or someone who had been involved in vaccine preparation for thirty odd years. Little by little, I accumulated information about CSL and my fellow workers. I realised for the first time that others could get as much satisfaction out of their professions as I did from medicine. They also tended to be more humble.

Veterinarians particularly impressed me and it was the start of a fruitful relationship with this profession which continues to this day. Vets need a highly practical approach towards problems and they face economic and communication restraints regarding patient care which I find refreshingly different. One Monday morning, at morning tea, I was listening to a vet describing to two of his colleagues an episode witnessed the night before. He had just entered a milk bar when a young man collapsed and had a type of fit. The young vet was speculating on what was the likely cause of the fit. After listening to this conversation for some time the most senior vet, Dr Don Oxer, took his pipe out of his mouth and remarked, 'Why didn't you knock him on the head and do a post mortem?'

From a medical point of view this was a startling attitude but obviously it became acceptable, and sometimes necessary if some strange disease in a herd had to be identified.

Unlike the medico, the vet can't retreat into anonymity behind a large hospital facade and an animal's owner seems more likely to query the fees than a doctor's patient. The vet's patients are quite at liberty to bite, scratch, or stamp on his or her foot. Another occupational hazard is being kicked. For years, I have thought it would be sobering and highly educational for all medical students to spend some time working with their local vet.

Seating was quite random in the dining room, so it was an efficient way to learn the ins and outs of CSL. I recall affectionately a number of distinguished but unassuming CSL veterans who went out of their way to help and guide newcomers. In particular, Dr John Graydon and Merv Hinton emphasised the importance of CSL's role in public health rather than as a profit-making organisation. I still believe in a policy of reputation first, profits second.

Sadly, the senior officers' dining room was declared elitist in the mid-1970s. Thereafter all staff ate in the main cafeteria. That dining room was my classroom and after visiting it I would make little notes of things to find out more about. It was a great place to take visitors so they could meet relevant colleagues. However, at the canteen one queued up to the typical cacophony of a canteen with little hope of finding the desired people.

In school holidays sometimes one of my children would 'come to work with Daddy', and be put in a white coat and whisked off to play with baby rabbits, etc. They loved the canteen because the staff spoiled them silly.

Learning the ropes

'What do you do at CSL?' curious friends and acquaintances would ask. During the first four months or so this was a difficult question to answer.

Reading and listening seemed a fair summary. I was next to the offices of Dr John Trinca, CSL's Deputy Director, and Dr Peter Schiff who showed great patience as I popped in and out all day seeking information. They could not have been more helpful, inviting me to impromptu meetings on a host of subjects. A year before, Peter Schiff had completed a PhD in Biochemistry at the Australian National University and had spotted the same ad for a medical officer at CSL. At the time he was working as an Obstetrical Registrar and was hankering to get back to research. He was hired after an interview in Sydney with Dr Bill Lane at which Peter said Bill did all the talking. Fortunately, when I applied for the same position they had decided another medical officer was needed anyway.

My in-tray was regularly topped up with items to absorb and/or comment on. These ranged from reports of strange reactions to vaccines to new types of penicillin or scientific journals I had heard of but never read. Initially anything I wrote usually needed major surgery. This was gently attended to by John Trinca.

On my arrival, Peter's laboratory was nearly complete but because of the usefulness of his administrative abilities to CSL he remained desk-bound. This fact is mentioned because by the end of 1966 I had swiped for 'safe-keeping' all the new equipment from his laboratory. Within twelve months of coming Peter had been appointed Chief of Research, a position he held for the next twenty-five years. Highly intelligent and eloquent, Peter taught, advised and supported me admirably during the next few years.

Peter was particularly helpful in bringing me up-to-date in biochemistry. The interests and expertise of John Trinca, who was training us both, were far wider. He was an expert in immunisation, especially against tetanus. Indeed, it was this subject that had brought about our first meeting three years earlier. When a Naval Rating at Flinders had suffered a horrific local reaction to the normally benign tetanus toxoid, I visited CSL to discuss the matter. At the time, working at CSL was not on my agenda, but John later told me that their impression of me back then had helped their decision when I responded to their advertisement. Another factor, of which I was unaware at the time, was that John, who was CSL's expert on allergy, had been strongly supported by my Uncle Charles at an important time in his career. More wheels within wheels!

In 1966, CSL had hundreds of products. The combinations and variations of the queries which arose about many of these biologicals were phenomenal. I became a reasonable dab at replying to letters complaining about particular vaccines. The reply often went along the lines, 'We have examined the protocol of that particular batch in question, and found it to be quite satisfactory. Over 100 000 doses of this particular batch have been issued and no significant untoward reports have been received.'

Later, adverse reports were looked at far more thoroughly, but in those days it was a comfortable and common practice to hit the ball straight back at the bowler. Soon I had a collection of paragraphs from which I could select those appropriate for a particular written query.

The phone was a different matter. Often there was an acute problem which required an immediate answer. In my first week I had a call from a doctor in a blind panic because instead of giving the Sabin oral vaccine by mouth he'd pulled it up in a syringe and injected it. The other doctors' offices were all empty and so in a calm voice, disguising my state of panic, I told him that it was unlikely any untoward effects would develop, to keep the patient under observation; and I would seek further information. 'Yes,' said Dr Trinca, after I had breathlessly tracked him down in the library, 'They quite often do that. Apparently not much usually happens.'

I soon became aware that, although every vaccine sent out had a most comprehensive leaflet accompanying it, the doctor or the sister ringing up often had a problem which was not covered by that leaflet. Twenty-eight years later I was still being floored by the diversity of the queries. Vets accidentally inject themselves with cattle vaccines when suddenly kicked from behind. Teenagers are lined up and inadvertently given whooping cough vaccine which is only meant for infants.

After about four months I felt that I had got the hang of the office work and was even shaping up well as a public servant. The working week was comfortable but not especially exciting. I was beginning to enjoy my bi-weekly hospital clinics more than life at CSL. Then a couple of interlocking events occurred which led to my having what for years I considered the best job in Australia.

In the early 1960s, not only was research into immunology about to take off, but hospitals were establishing clinical immunology units. Of particular interest to me was the study and treatment of children who responded poorly to vaccination and were prone to far more infections than 'normal children'. When the sera of some of these children were examined, they were found to be markedly deficient in the group of proteins which have antibody activity. A practical way to help these children was to give them monthly injections of other peoples' antibodies. For years CSL had produced immune gamma globulin, which was used, amongst other things, for protection against what is now known as hepatitis A. The dosage required by some of these little kids was quite high and a monthly intramuscular injection was extremely distressing.

The best way around this problem was to inject the diluted material intravenously. Since the standard preparation given into the blood-stream caused quite severe reactions in some patients, a project was commenced at CSL to prepare gamma globulin especially for intravenous use. I was drawn into this exercise basically so I could get the experience. I soon became familiar with many aspects of the research, ranging from the details of manufacture to the laboratory tests used to investigate these unfortunate children.

One of the tests I saw at the Royal Children's Hospital fascinated me. I remember being passed a glass microscope slide on which a drop of a child's serum had been separated into its various components. From the central application point exquisite pink-stained feathery arcs radiated out in both directions. Each of these differently shaped arcs represented a separate plasma protein. In the immunodeficient child, some of these arcs were diminished or totally absent.

This technique, known as immunoelectrophoresis, could be used to study any type of antibody as well as the component it was directed against, called the antigen. I found this as striking as the the reaction between immune rabbit serum and its antigen described by my lecturer in zoology years earlier (page 76). So impressed was I with these slides that I was loathe to hand them back. Later it was found that an intravenous gamma globulin preparation worked effectively in these children and the more important missing arcs would reappear in the patient's serum.

My own little lab

Something of great importance occurred two weeks after I had seen those impressive slides. I was sitting behind my tidy desk surreptitiously admiring a new typist across the way, when John Trinca strolled in with a proposition. He explained that the College of Allergists and CSL were proposing a joint project and he needed a list of all the research equipment and specialised skills available at CSL. 'Wander round,' John said, 'and check things like the analytical ultracentrifuge and the amino acid analysers.'

This sounded like being back on the Enquiry Desk at the Myer Emporium but over the next three weeks I progressed from being a total ignoramus to having a reasonable working knowledge of the purpose and limitations of the various techniques available at CSL. Soon I had a fair idea of the time taken to perform electron microscopy or amino acid analysis. At the same time, I established contact with a network of biochemists to whom I could ask basic and often naive questions without them taking the mickey out of me. In subsequent years I asked many times for advice or to have my calculations for complex solutions of chemicals called buffers checked by these good people. Merle Gilbo and Joan Pearson were splendid in this regard and the late Mr Jan Birner turned an ignorant medico into a reasonable thin-layer chromatographer.

Some biochemists are doomed, at least in the short term, to endless repetitions of identical tests on a succession of product batches. These results should be predictable and practically identical. One was a paper electrophoresis technique which was carried out on various serum fractionation products. After adding this to my list I asked Mrs Emma

Polacsek, the biochemist-in-charge, what the gadget was in the corner. She said it was an immunoelectrophoresis apparatus which hadn't been used for some time. 'Gee,' I said. 'I'd love to have a go at getting that working.' 'Feel free,' she said. 'There's some bench space in the corner you can use.' From that day on, life at CSL became vastly more interesting.

A white coat, a little bit of bench space and helpful people soon had me as happy as a sandboy. Even though I've never discovered why a sandboy should be particularly happy. Within a couple of weeks I was obtaining reasonable results from the immunoelectrophoresis apparatus but it took months to produce technically perfect results. When problems could not be solved locally they were soon sorted out by Xenia Dennett of the Children's Hospital. What started off as a minor challenge soon became absorbing. As word got around I was asked to analyse various samples from the production departments and research divisions, and when some of these results proved particularly useful I found myself spending more and more time in the Laboratory. (Bow ties became part of my uniform after a couple of accidents in which my ordinary tie dipped into liquid-filled beakers.)

In no time I was whipping through the paperwork at a rate of knots just so I could get out of the Administration Building and back to the bench. I found many letters could be dealt with by a quick phone call. This was not only more effective but was also cheaper than having a letter typed out, one which was often only of academic interest when it eventually arrived. Later when fax machines were introduced they became invaluable for sending out backup material after the telephone call.

I also took steps to wriggle out of some of CSL's many routine meetings which often seemed self-perpetuating. Initially I found them instructive and restful, and they were even a pleasant way to while away the time between lunch and afternoon tea. Some were boring, repetitive and always attended by certain windbags, so when possible I arranged with the chairperson to appear only when required. In return, I promised to give the promptest attention to any submissions required which would be typed up for distribution. My attitude was that a meeting should only be held when necessary and the person calling the meeting should have a clear idea what recommendations he or she wished the meeting to endorse.

In my first year at CSL my duties also included participation in clinical trials. In one of these I accompanied the late Dr Alan Duxbury to St Joseph's babies home in Broadmeadows where Alan injected a number of babies with the new 'split' influenza vaccine. My job was to immunise other infants with routine childhood vaccines. Alan, who was a delightful and greatly loved man, exuded an air of disorganisation. A cheerful phone call at 10.00 p.m. was sometimes the first notification that we were 'on' at St Josephs at 8.00 a.m. the next day. After a number of days had been stuffed up in this fashion I managed to participate less frequently.

As the months went by, my work in the lab continued to be fun and became more productive. There was much to learn and no excuse for not learning it! I had to catch up and then try to keep abreast with the latest relevant scientific literature. Every day brought little triumphs. New skills like estimation of protein nitrogen and the making up of buffer solutions were learnt. I was taught how to photograph the pretty lines developed on glass slides. These slides, or more precisely the way they were kept clean, nearly led to a disaster.

Before they were used the slides had to be extremely clean and free of any grease. In retrospect, alcohol would have done the job as well and have posed less of a threat. However, a large glass jar filled with glass slides and highly inflammable acetone had been left by the previous user of the equipment. Prior to use, the required number of slides had to be removed from the acetone, laid out on a holder and polished briefly with a tissue.

The gels used to coat the slides were then heated up in a beaker of water over a Bunsen burner which was at the other end of the bench to the large jar containing the inflammable acetone. One day a storeman made a delivery to the lab when I was absent for a few minutes. He moved the large jar to the other end of the bench, stacked various items from his trolley on the bench and left.

Fifteen minutes later a sharp crack attracted my attention to the Bunsen burner. My eyes stuck out further than usual as they sighted the jar of acetone as close as it could possibly be to the lighted burner. As I looked further, heat induced cracks appeared in the sides of the jar. I turned the gas off and stealthily retreated into the main laboratory, which was empty. Should I sound the alarm?

It would either blow up during the next 30 seconds or there would only be danger from acetone leaking onto the floor. 'I wouldn't go in there, mate,' I said to an approaching biochemist, and the two of us waited 10 minutes out in the corridor. The warmish jar was then gingerly lowered into a bucket where it gracefully and safely fell apart. After this lucky escape I made sure that no potentially dangerous substances were used unnecessarily and, if they had to be, never without stringent precautions. Indeed I have been known to 'do my block' when a laboratory worker has endangered the life of others through sheer carelessness.

There were two spectacular accidents while I was at CSL. One was inexplicable, the other patently wasn't. The first involved steam.

I've always been wary of steam and never feel comfortable near steam sterilisers or even pressure cookers. I loved steam engines as a kid but although I spent much time admiring them up close I was never terribly happy about the giant sterilisers (autoclaves) which were regularly opened behind my back in operating theatres. I know they have safety valves and are checked rigorously but sometimes things still go wrong. When I first arrived at CSL they had two of the biggest steam autoclaves I had ever seen. Either would comfortably enclose a small family car and still have room to spare. One of these brutes was housed in a single-storey extension to the main penicillin production block. In a conducted tour during my first week, I observed this fellow in operation with a number of heavy duty gauges with their arrows dancing around indicating pressures. Charts also recorded pressures and temperatures. The whole caboodle was undoubtedly fail-safe.

A week later a little after 8 o'clock in the morning it blew up . The building was demolished, but fortunately no-one was injured. This was quite incredible as two attendants had been checking the dials some three minutes earlier. The twin of the disintegrated autoclave was later moved to the ground floor of the two-storey building in which I worked for six years. I wasn't sorry to see it replaced by a nattier, less threatening machine.

I had a box seat for the next explosion. One morning, attracted by a booming sound, I looked out of the window as a mushroom cloud rose seventy or so metres above the nearby insulin production building. Five minutes earlier a welder had been given the okay to do some repairs in the ceiling of the building. Below him was a huge stainless steel tank covered by thick plastic sheeting and inside the tank was a large volume of absolute alcohol. Welding sparks lazily spiralled down, burnt their way instantly through the plastic sheeting and ignited the alcohol. Those who saw this impressive blaze will never forget it. Fortunately no one was injured.

An introduction to animal toxins via the Blue-ringed octopus

An interest in animal toxins just crept up on me! I know exactly when marine toxins first attracted my attention. The front page of the Melbourne Age on 22 June 1967 carried a short report of a twenty-one-year-old soldier who apparently died quick smart after being bitten by a small Blue-ringed octopus in Sydney. He had only joined the Army the day before—which must represent one of the shortest military careers ever recorded. 'Poor bloke,' I thought, and promptly forgot about him and turned to see what was on television that night.

Late the following day, after most of the 1200 staff of CSL had left, I was doing some final tinkering with the odd test tube. A security guard wandered in and said, 'Doc, I've got an esky from Sydney and no-one knows what to do with it. It's not addressed to anyone. Last week I got in the poo for not putting a sample in the freezer—and then on Monday I got hammered for putting an unlabelled package in the freezer. I'm fed up.'

Together we opened the esky and from the pelleted dry ice extracted some containers filled with blood and other samples. Wrapped in plastic was a rigidly frozen envelope. When it became warm and pliable it was opened and revealed that the samples enclosed were from the young soldier who had died earlier that week in Sydney. The writer wanted CSL to detect octopus toxin in the samples.

I put the samples away in my freezer and the following Monday tried to find out who at CSL should investigate the problem. Within a couple of days I had discovered that no-one was either interested in or trained in the business of venom or animal toxin detection. This was communicated back to the forensic people in Sydney and it later appeared no-one in Australia was involved in this type of work.

I was stuck with the samples and reminded of the problem they posed each time I opened the freezer. It seemed extraordinary that a tiny octopus could kill a fully grown man so quickly. A visit to the library disclosed that a similar octopus had killed another serviceman in Darwin some years before. The octopus seemed extraordinarily selective as the only documented deaths in Australia were both members of the armed forces! Although this was coincidental there was no doubt that it was the only octopus in the world which had the potential to fatally poison people. Furthermore, it was one of the commonest octopuses found around the Australian coast. The problem was too tempting not to be investigated. Within a matter of weeks, a collection of these exquisitely coloured little creatures was darting about in an aquarium at CSL, oblivious to the fact that they were shortly to be sacrificed in the cause of science.

Although it was only part-time work the octopus project was a real eye-opener and established in me a lifelong interest in marine biology. Contacts I made with marine biologists, many of whom have an extraordinarily interesting life, continued thereafter. Pottering along an ocean beach is always delightful but if you're looking for certain specimens and being paid for it, then you are a very lucky person indeed. I fondly recall a beautiful day collecting forty or so surprised Blue-ringed octopuses on the mudflats at Port Albert in Victoria. It is something I could happily do once a week forever!

Early on we explored how the tiny beak of the octopus could penetrate skin and force in the toxic saliva. A simple if spectacular experiment was proposed by the late Dr Frank Warburton after I had described our aims. 'Why not get a rabbit,' said Frank. 'Shave its back, put it in a restraining box and then plonk the octopus on it. If the soldier died after carrying the octopus on his arm for a few minutes you should be able to reproduce the situation with a rabbit.'

Fine-toothed electrical clippers were used to shave a patch of fur about the size of a small saucer off the rabbit's back. It looked slightly surprised when a cold octopus was placed on the shaved area. The octopus attempted to crawl off twice but long forceps were used to push it back to a central position. Alarmed at being out of the water its brilliant blue colouration intensified and it appeared to squat hard onto the rabbit's back. Two minutes later the rabbit gave a shiver, went completely limp and ceased breathing. The octopus marched off to one side and was gingerly netted and placed back into the tank.

This fearsome demonstration by an intact octopus left the witnesses as ashen-faced as the rabbit. For a venomous bite to produce paralysis at such speed, not only must a lot of toxin be injected but it must be extremely potent. The Blue-ringed octopus certainly had both these capabilities.

A similar toxin to that produced by the Blue-ringed octopus was found in creatures such as the Toad or Puffer fish. In Japan, the Puffer fish toxin accidentally kills dozens of fish-fanciers each year. As yet no commercial antidote to this toxin is available and all these years later there is not a satisfactory method for detecting the Blue-ringed octopus toxin in the serum of patients.

In humans, the paralysis produced by the Blue-ringed octopus will wear off in time provided adequate artificial ventilation is given and this may have to be done for some hours. There is a well known incident in which a paralysed schoolteacher was aware of everything going on around him except when the first aiders slackened their efforts; then he lapsed into unconsciousness due to lack of oxygen. The patient recalled that during a conscious phase he heard a first aider observe 'Looks like this poor chap has had it.' The first aid which might have saved the lives of the two servicemen is described on page 374.

I made one fascinating observation when studying octopuses in a tank—I saw it only once and am unaware of any similar reports. When I dropped a live crab into the water containing a hungry Blue-ringed octopus the normal response was for the octopus to attack the crab and devour it. Sometimes the crab would injure the octopus. On this occasion the octopus glided over the fairly large crab spraying saliva as it went. The octopus then waited at one end of the tank while the crab 'inhaled' the toxin, becoming paralysed. The octopus then swam back and proceded to pull the crab apart. This observation appeared to explain why the octopus needed so much venom and how it used it to avoid being injured. It is not known why the octopus is not affected by the toxin it releases into the water.

We established that the killing potential of the octopus was tremendous. In mouse studies it was found that the average octopus appeared to have enough toxin in its salivary glands to paralyse as many as ten men. First I obtained some good advice from Dr Everton Trethewie and Dr Shirley Freeman who had separately explored the toxin several years earlier. They had found that with experimental animals the toxin caused paralysis and did not directly affect the heart.

My decision to do a couple of days work on the octopus toxin turned out to be a momentous one. I realised that much interesting work was waiting to be done on the detection of marine toxins and this later led to the challenge of toxin detection in general. My work on this toxin introduced me to a whole lot of techniques highly relevant to the study of venoms and the preparation of antivenoms. It also led to my first publication on toxins which was co-authored by Dr Bill Lane. This seems to be an appropriate opportunity to introduce Dr Bill Lane who greatly influenced my career at CSL and to look at the project which most occupied my time at that establishment.

lding Day with my first wife, Wendy

The author's second wife, Megan

and Susie, 1989. John has just been admitted to the Bar.

Calling it quits. Mrs Marjory Davey at her farewell, June 1994. 'Fourteen years, and we never had a row', said Marjory. 'I was frightened of you!', was my response.

A bright young bunch. Photograph taken by the author after AVRU meeting, 15 June 1998. Left to right, Dr James Tibballs, Dr Anna Young, Dr Ken Winkel, Dr Steven Pincus and Dr Gabrielle Hawdon.

Mates for fifty-five years. The author with Alan Kerr, then Administrator of Norfolk Island, 1996

7 The Sydney Funnel-web Spider

'From ghoulies and ghosties and long-leggety beasties
And things that go bump in the night,
Good Lord, deliver us!'

Anonymous (Cornish)

Prologue	182
The Department of Immunlogy Research	182
Getting the Okay	183
The Challenge	183
Seeing the Enemy	185
Milking Spiders	185
Doing Tricks with Venom	186
Growing a Beard	186
Back to the Drawing Board	187
The Hunt for the Real McCoy	188
Finding a Suitable Laboratory Animal	188
A Funnel-web Death	189
Becoming Obsessed	190
Some Important Personal Events	190
How Does the Funnel-web Spider Kill People?	192
Getting Close to Monkeys	193
Complacency and No Antivenom in Sight	195
The Death of Mrs Christine Sturges	195
The Final Assault	196
The Funnel-web Strikes Again	198
Success at Last	198
A Bite to Remember	214
How to Confuse the People of Sydney	215
Out of the Woods	216
Funnel-webs Still have Secrets	217

Main events in the search for an effective therapy for Sydney Funnel-web spider bites

1927	First recorded death. A two-year-old boy dies ninety minutes after being bitten by a male Sydney Funnel-web spider at Thornleigh, Sydney.
1934	Dr Charles Kellaway at the Walter and Eliza Hall Institute of Medical Research in Melbourne conducts the first research on the venom. He fails to demonstrate toxicity in common laboratory animals.
1957-1961	Dr Saul Wiener, working at CSL, fails to find either an antivenom or a drug which would neutralise the venom but expands knowledge of the venom.
1964	Miss Merle Gilbo and Dr Normal Coles publish details of their isolation, at CSL, of what was considered to be the main toxin in Funnel-web venom.
1967	Permission given to try new techniques to make antibodies to the main toxin in rabbits. Succeeds, but found 'main toxin' not responsible for most important effects in monkeys and, presumably, man.
1970	A seventeen-year-old woman, six months pregnant, dies after a Sydney Funnel-web spider bite.
1972	Proven the main toxin, designated Atraxotoxin, binds to glassware and, without special precautions, is lost.
1972-79	Atraxotoxin purified and its actions explored. Antibodies made but ineffectual. No antidotes found. Extensive ongoing collaboration, every new avenue explored.
1979	January Article on pressure-immobilisation type of first aid for snake bite published in *The Lancet*. Death of Mrs Sturges, aged thirty-one years, following a Sydney Funnel-web spider bite.
1980	January Death of James Culley, aged two years, due to a Sydney Funnel-web spider bite. Experiments in monkeys suggest pressure-immobilisation first aid suitable for Sydney Funnel-web spider bites. February Author suspended during 7-21 February after dispute with Director over staff cuts. March Long-planned collaboration begins with Drs Alan Duncan and James Tibballs at the Royal Children's Hospital. Using monkeys, the aim is to explore drugs which might dampen the venom's affects. April Monkey work exceeds expectations. Greater understanding of the syndrome obtained. Pharmacological antidotes were shown to be feasible, but were complicated, risky and required prolonged invasive monitoring. Results were uncertain and so drugs were not an ideal solution. Renewed attempts to make an antivenom as even a feeble one might help drug therapy. Rabbits commence intense schedule using best male venom. (Rabbits are not affected by this venom.)

June Very weak antibodies were detected in a rabbit serum. A new process from overseas allowed harvesting and concentration without loss of activity of this first, albeit weak, antivenom. The immunisation of all rabbits was intensified.
July The antivenom is shown to be dramatically effective in monkeys.
October *The Medical Journal of Australia* carries three articles which deal with the antivenom, the syndrome in monkeys and first aid for Sydney Funnel-web spider bites.
First production batch of antivenom manufactured.
December Funnel-web antivenom issued for clinical trial in NSW.

1981 February First clinical use of Funnel-web antivenom. The patient is Gordon Wheatley, the clinician, Malcolm Fisher.

Prologue

A fairly accurate assessment could be made of the length of time spent in the presence of Dr Bill Lane by counting the number of cigarette butts appearing in the ashtray. This habit unfortunately led to his premature death in 1974 and the arrival of a new director, Dr Neville McCarthy. I found Bill Lane supportive and inspiring, and Neville McCarthy just the opposite.

Bill Lane was a fairly big fellow with beady brown eyes behind thick spectacles. When I first arrived at CSL I was perturbed to find that he more or less ignored me. Later I realised he had just taken over as Director and was preoccupied with more important matters.

Bill had quite a presence and was an excellent off-the-cuff public speaker. He had a sense of humour, took an interest in people and is the only person I have ever known with a photographic memory. It was possible to have a healthy argument with him and there was no hint of acrimony whether one won or lost. He was an entertaining dinner companion and I liked and respected him very much.

Bill would often go for a stroll around CSL and listen to people's whinges and expectations, leaving an appropriate number of butts behind. He developed a habit of coming into my small area and perching his tall frame on a laboratory stool while he tried to determine what I was up to.

One thing led to another and in August 1967 he decided I was to head a new department of Immunology Research. I was somewhat surprised at the speed of these initial developments. Bill Lane muttered something about 'always room for a talented amateur', an observation I never knew quite how to take.

The Department of Immunology Research

I was at least twenty years younger than the other research heads but they were very supportive if approached with humility. I had much to learn! I started with two offsiders but, by the time Bill Lane became ill in 1974, I and about fourteen staff had occupied a larger laboratory for five years. Although it was the smallest research department at CSL, it was often responsible for almost one-half of the scientific papers published each year.

Over the fourteen years I headed the department most of our collective energies were directed at 'non-venomous' research. Projects included the preparation and testing of antilymphocyte globulin intended for immunosuppressing human transplant patients and we dealt with diverse problems ranging from the surveillance of rabies-suspect samples from Darwin to novel experimental veterinary vaccines.

One very useful exercise was the establishment of an assay to measure the activity of Rh(D) human immunoglobulin. This immunoglobulin is given to an Rh-negative mother within seventy-two hours of delivery of a Rh-positive baby and will prevent the development of Rh antibodies.

All these projects were interesting, but in my mind lacked the special excitement I found in hunting down unique Australian problems—like the Funnel-web spider.

Getting the Okay

'You want to what?' said Dr Bill Lane. It was late 1967 and he had been passing the office where Dr Peter Schiff and I were discussing the then defunct Funnel-web spider project. 'Wouldn't mind trying a few immunological tricks,' I replied. 'Listen, young Sutherland,' said Bill Lane. 'I spent a great deal of effort closing that project down and it'll stay closed.' He ground out his cigarette, settled into a chair and glared at me fixedly. He also threw a glare at Peter for good measure.

Bill lit another cigarette and continued, 'When I was research director the hardest thing was not starting projects, but stopping them. It would be easier to stop the tide coming in than to stop some of these projects that have gone on and on and on. The Funnel-web's dead and that's that. Finito!'

'Struan just wants to do a little one-off experiment,' said Peter Schiff. 'He thinks the carrier protein technique might produce an antivenom.'

'He does, does he?' said Dr Lane, unwillingly showing a little interest.

'It's only an idea, and if it doesn't work it'll be dropped within a month or so.' Grudging permission was given with the proviso that it was truly a one-off experiment.

This one-off experiment managed to extend itself to some fourteen interesting and sometimes quite frustrating years. I wouldn't have missed it for quids!

The Challenge

Six weeks before this meeting I had been looking through an old file of case histories. It contained details of a two-year-old girl who had died shortly after being bitten by a Funnel-web spider which had been lurking in her slipper. The horror of this tragedy came alive in the statements prepared for the Coroner. The rapidity with which the child became unconscious and died, despite prompt medical care, was most extraordinary. At the time my daughter was the same age as the victim and I wondered how I would cope with such a situation. A bit of probing around revealed reports of quite a few other cases, all of which involved either women or children.

Hunting around and talking to people unearthed a number of interesting facts. The male Funnel-web was the spider responsible for deaths. No deaths or serious illnesses had occurred after bites by the female spider. The spider was not known to be dangerous until the first recorded death occurred in 1927. Ten known deaths had been attributed to this spider up to 1961. This did not seem a particularly high death rate, but the spiders certainly put the wind up the two million people who lived in the area it inhabited. There were other types of Funnel-webs elsewhere in New South Wales and Queensland which were also thought to be highly dangerous, but the fact that the Sydney Funnel-web spider killed children in less than two hours clearly made it the most dangerous spider in the world. Yet cats and dogs appeared immune.

In 1934 Dr Charles Kellaway, working at the Walter and Eliza Hall Institute of Medical Research in Melbourne, conducted the first research on the venom. He failed to demonstrate toxicity in common laboratory animals.

Dr Saul Wiener, who worked at CSL from 1952 until 1961, succeeded in making a Red-back spider antivenom and a Stonefish antivenom but unfortunately had no success with the Funnel-web. As with every one he studied, he did excellent work on the Funnel-web venom. Saul established that most laboratory animals except monkeys were extremely resistant to the effects of the venom. He tried a variety of ways of immunising animals to raise antivenom and had no success. As a final attempt, he immunised a horse with as much female venom as was available, but the resultant product failed even to neutralise the female venom. Saul also investigated a wide range of drugs as possible antidotes to its effects, but had no positive results. Unfortunately, at least for venom research in Australia, he became a Collins Street allergist.

After Saul Wiener left, two biochemists, Merle Gilbo and Dr Norman Coles, isolated and characterised two components from the spider venom. It was assumed that one of these was the most important component of the venom. It was quite toxic in mice and it was felt that its relatively small size explained why it failed to stimulate an antivenom response in immunised animals. When this biochemical work had been completed, Bill Lane firmly brought the shutters down on the Funnel-web project, which had run on and off for some five years.

By 1967 there was a method available which made various small molecules more likely to promote an immune response by attaching the smaller molecule to a larger carrier protein. This technique had enabled the making of antibodies to important small molecules such as insulin. This, in turn, led to methods which measured the minute amounts of insulin circulating in the body for the first time. It had also resulted in a few Nobel Prizes.

With the details of the little girl's death fresh in my mind and the knowledge of the work that had gone on beforehand, it seemed reasonable to try this technique on the Funnel-web spider toxin. I proposed attaching the toxin isolated by Gilbo and Coles to a carrier protein and using it to immunise a rabbit. It was pretty straightforward; it would either work or it wouldn't.

Seeing the Enemy

The first time I saw a live Funnel-web spider in 1967 I recoiled in horror. Serious doubts arose as to the wisdom of having anything to do with such an evil black brute. They are big spiders, with legs some 3 cm long and the female has a particularly fat abdomen. They are fearless, have a filthy temper and their gait varies from a swagger to a brisk march. I've been told that when stamped on their tough skin makes a cracking sound like a chicken bone breaking. When annoyed they rear up with their fangs raised as high as possible. Venom can be seen dripping from the tips of these hollow fangs as the spider displays footwork that would do any boxer proud. Once the target is in range the spider drives the fangs down in a narrow arc with such violence that sometimes they become firmly embedded. Having a big black spider actually stuck on to you would cause even the calmest person to panic—especially if they were in the Sydney area.

Milking Spiders

First we had to establish a colony of spiders at CSL. Special metal boxes were made to transfer Funnel-web spiders to Melbourne from the Sydney area and the Australian Reptile Park at Gosford. These boxes went backwards and forwards by air for years, transferring spiders on their one-way trip and reducing Sydney's arachnid population by thousands over the years. There were a couple of problems which had to be solved.

The travel-weary and irritable spiders had to be moved from their small containers to the larger holding jars which contained a layer of sand. Experience confirmed that an irate and much-travelled Funnel-web spider could move very quickly and rarely played dead as some other spiders do. Sometimes two spiders were put in the one jar and, because of the cotton wool they were packed in, this would only become apparent when one spider had been safely transferred and a movement was detected out of the corner of one's eye. This would precipitate a little frenzied activity.

Secondly, Funnel-web venom was collected by sucking the drops of venom off the tips of the fangs, using a Pasteur pipette attached to a suction line. When the lid was taken off the jar the spider reared up and the tip of the pipette could be maneuvered from side to side over a minute or so until no further venom appeared.

The problem with this method was that sometimes the spider had made a fine web around the side of the jar which it could run up quickly and escape. Occasionally a spider would blatantly march straight up the pipette. In Dr Wiener's time, two laboratory assistants were bitten by spiders taking the latter route. Fortunately, they were female spiders and no significant illness developed.

Playing around on Anzac Day 1968 I found that the simplest and safest way to milk the spiders was to have a single, clearly-marked lid with a central hole in it wide enough to allow entry of the pipette. When a spider was to be milked this lid replaced the normal lid on the jar. For ease of capture in case of escape, the transferral of the spiders and all milking operations were carried out in a large, shallow box placed on the bench.

The rules were quite simple: no spider work was to be carried out other than in the box on the bench; only one jar was to be opened at a time; and no-one was allowed to distract the attention of the technician when he or she was transferring spiders or removing or replacing a lid. Over the years a number of technicians did a wonderful job maintaining and milking as many as 300 spiders. Following these rules, no-one ever got bitten despite the efforts of the occasional arachnid Houdini.

A discreet notice on the laboratory wall told of the procedures to be followed when bitten by a Funnel-web. From memory this gave the following advice: Note the sex of the spider and return it to the jar. Apply an arterial tourniquet if possible (at the time this was the accepted type of first aid). If Dr S. is not available, arrange to be transported to the Royal Melbourne Hospital taking the nearby plastic bag which contains information on the effects of Funnel-web spider venom.

I often wondered what these 300-odd spiders brooded about while they rested in their jars in a locked cupboard. Initially I had quite a few nightmares involving them, as did most of the technicians when they first commenced work.

Doing Tricks with Venom

It took months to collect sufficient venom for the first experiment, because each milking when dried yielded less than a fifth of a milligram. After about six months we had a reasonable quantity of a purified component, similar to the one which Gilbo and Coles had found to be highly toxic to mice. A skilled biochemist, Allan Telford, helped link some of this component to bovine serum albumin as a carrier protein. To try to raise antibodies, several rabbits were then regularly injected with the combined material over a period of several months. At the same time, other aspects of venom research were expanding and Dr Lane made the comment that the Immunology Research Department was like a demonstration of Parkinson's law before his very eyes.

Growing a Beard

In the Navy permission to grow a beard had to be obtained from the Captain. If permission was granted, the initial time for being unshaved was a minimum of six weeks. At that stage I was not certain whether I could grow a reasonable beard and, as a general rule, the less one seeks permission to do anything the better. So in the Navy I remained beardless.

At the end of 1967, amongst the 1200 staff at CSL there were only two people with beards. One was a large Russian who worked in the Serum Fractionation region and the other one was a woman technician in the Quality Control Section. At various outside research meetings I had noticed beards beginning to appear here and there and so over the Christmas break of 1967 I sprouted whiskers which resulted in Bill Lane calling me 'Frobisher'. The reason I discuss beards in the context of this research is that over the same Christmas period Allan Telford also grew a beard. The first day back at work we roamed around looking for one another, failing to recognise each other when we passed in a corridor.

Back to the Drawing Board

Although our beard-growing experiments may have been successful, things were not so good on the Funnel-web scene. The serum taken from the immunised rabbits gave the mice only marginal protection against the toxic component. This was a depressing finding and my despondency deepened as other related experiments failed one by one. Clearly the problem was more difficult than the 'new boy' had thought.

There was something queer about the toxin that had been isolated by Gilbo and Coles. It killed mice and rabbits very quickly. On the other hand, Saul Wiener's mice appeared resistant to the whole venom and only died when given massive doses and then some hours after the injection. The toxin obviously needed closer examination, although it would be a painstaking procedure to purify.

Fortunately I went to Dr Gordon Ada at the Walter and Eliza Hall Institute for advice. He made some useful suggestions about alternative ways of presenting this annoying toxin to rabbits. As I left him, almost as an afterthought he said, 'Perhaps you might use some of the ultrafiltration chambers which are just coming on the market. They should allow you to separate your low molecular weight venom components faster and with a far better yield.'

Thanks to Gordon's advice, six months later milligram quantities of the purified Gilbo and Coles toxin were available. The choice of a species on which to test this toxin seemed extremely important. Saul Wiener had found most animals other than monkeys highly resistant and there were no records of any domestic animals dying from Sydney Funnel-web spider bites. I felt there was no option but to test the purified toxin in a monkey.

A nice little monkey watched me inject a dose of the toxin through a very fine needle into one of his veins. Absolutely nothing happened. Next day the same monkey received 3 mg per kg of the toxin which was a massive dose. Again nothing happened. Another monkey, which had received a minute dose of crude whole venom, developed the full syndrome of muscle twitching and gross production of saliva, and became comatose. It was clear that the last eighteen months of effort had been directed at a minor toxin and in fact we knew nothing about the major toxin or toxins.

The Hunt for the Real McCoy

Once we realised that the toxin we had isolated from crude Funnel-web venom was the wrong one, the hunt was on to find the real McCoy. We named it 'atraxotoxin' because the Funnel-web is of the genus *Atrax*. By mid-1973 we had established that most of the standard purification methods resulted in a mysterious loss of the activity we were looking for. Atraxotoxin was absorbed out of solutions of venom by many materials such as paper, cellulose and, in particular, glassware.

The discovery that much of the atraxotoxin painstakingly collected over the years had been left stuck on glassware was devastating. Our washing up lady, who kept the glassware meticulously clean, had washed the atraxotoxin down the sink! Meanwhile we had battled on exploring a venom which became lower and lower in atraxotoxin activity each time it came into contact with more glassware.

We solved the problem by dissolving the venom in a weak acid solution and treating all glassware with a thin layer of silicone. These measures had to be in place from the start. Spiders were milked using internally coated pipettes and the collected venom was rinsed out in the dilute acid solution into bottles that had also been treated.

These methods improved both the yield and the toxicity of the venom collected. Soon atraxotoxin was isolated and, although it only represented 10 per cent of the venom by weight, it had all the activity we were interested in. While the harvesting of this toxin improved, other important investigations were underway.

Finding a Suitable Laboratory Animal

Even today, work on venom and antivenom unfortunately has to use experimental animals. As stated earlier, common laboratory animals were relatively resistant to Funnel-web venom and only limited use could be made of monkeys. Apart from monkeys, man was the only other animal known to be highly sensitive. Believe it or not, some people did volunteer their bodies for venom experiments.

Our numerous venom and antivenom studies required large numbers of experimental animals and we had to find a creature which was cheap, convenient to use and needed only minute amounts of venom to demonstrate a reproducible effect. An amazing variety of little beasts was tested for susceptibility to venom. They included earthworms, mealworms, slater beetles, crickets and even tadpoles. Any small critter which caught my attention when gardening on Sunday afternoon had a one-way trip to CSL next morning!

In the end, we found that newborn mice were the ideal test animal. Each week at CSL, hundreds of these newborn mice were culled anyway to maintain optimal litter size.

Newborn mice became the backbone of the project. With practice, it was possible to inject them with as little as a millionth of a litre of venom. They were exquisitely sensitive to the effects of the venom for the first twenty-four hours of their lives. After this their resistance rapidly increased and they were no longer suitable.

A Funnel-web Death

On 28 December 1970 near Nowra, a seventeen-year-old woman who was six months pregnant was bitten by a Funnel-web spider. The spider was brushed from the branch of a tree and bit her on the breast. Sixteen hours later she was dead.

The day after she died happened to be a public holiday. I was at CSL processing some venom and at the same time applying fresh wallpaper to one wall of my office. The phone went and it was a journalist who said he had just discussed the Funnel-web problem with my Director, Dr Bill Lane. Bill had told him that the chances of success in the preparation of antivenom by me, or anyone else, was extremely remote. Although I was inclined to agree with him, my responses to the journalist's questions tended to be a little more optimistic.

Following this death, some higher authority decreed that the progress I was making should be closely examined. Accordingly a few weeks later I was told that Professor Roy Douglas Wright (A.k.a. 'Pansy' Wright) would like to come out and discuss the project with me in detail. At the time he was Emeritus Professor of Physiology at the University of Melbourne.

I was pleased rather than worried about the scrutiny. In fact I was delighted, because like any fervent researcher I welcomed the chance to bore anyone silly with my major work interest. On top of this, Pansy was immensely wise and knowledgeable and I felt privileged that my project was to be assessed by such a man. He was almost the only person I knew from whom I would accept a decision to pack it in and let someone else have a go.

The great man came out to CSL and for two and a half hours was presented with charts, fractionation patterns and reports of studies in animals. I also inundated him with overseas correspondence on the project, since many venom samples were being sent to other centres for collaborative studies.

Throughout all this he remained silent, except for an occasional grunt of acknowledgment. Finally, I said, 'Well that's it Pansy.' (I didn't actually say Pansy, it was more like 'Well that's it Sir, that's the situation.') I then sat back and waited for his reaction. After some thirty seconds or more of meditative silence, he raised his eyes and in his very deep voice drawled, 'I would just keep doing what you're doing.'

Becoming Obsessed

Pansy cheered me up to some extent, but did not suggest any new or blindingly simple avenues to explore. The Funnel-web project occupied much of my thoughts day and night. My principal hobby was the renovation of our home but, once the thinking aspects of the day's handywork had been planned, my thoughts returned to the project. For a while I found a good form of escapism: while spending a weekend at Mildura at a motel situated beside a golf club I had observed obsessive golfers playing on in pouring rain. I enjoyed the challenge of golf for a while and three of my five strokes were absolutely brilliant. However, I was slowed down by the need to inspect venomous creatures during my many visits to the rough.

Twice I was wounded by stray balls on the battlefield that was Brighton Public Golf Course. Concerned relatives maneuvered me into membership of the Victorian Golf Club. This proved counterproductive because I could ill-afford membership of such a beautiful club and I found fanatical golfers off-putting.

Eventually I decided golf was an expensive way to ruin a nice walk and went back to regular bike-riding or brisk walks along Brighton beach both for recreation and problem solving. I soon became permanently addicted to solitary tramps beside the sea for therapeutic and creative purposes. From time to time projects like the Funnel-web spider were nudged forward by an idea conceived on the beach which is but five minutes away from my home.

Some Important Personal Events

This seems a good spot to relate some events that changed my personal life.

At 3.00 p.m. on Tuesday, 9 October 1973, Margaret Malloy walked into my Laboratory and, as they say, into my life. Slightly built, quietly spoken and with features along the lines of Meryl Streep, she was delivering some library journals for me. Being in the middle of a lively conversation with my colleagues, I had the brilliant idea of asking her to wait in my office on the pretence I had something to return to the library. Five minutes conversation with her and I was completely hooked. She was twenty-two, I was thirty-seven. It was years before I looked at another woman.

At the time I was going through a fairly rough patch as I was in my third month as a single parent. A stressful time it may have been, but I had the consolation of maintaining close contact with John aged eleven and Susie nine. I unwittingly went through the normal cycles of disbelief, anger and a spot of despondency from time to time. It was sometimes midnight before I got the washing out on the line and then went to bed with a strangely pleasing feeling of martyrdom.

In the hope of eventually gaining custody of the kids I kept a comprehensive diary including meals served, etc. Early on some of my catering lacked imagination but was grand on efficiency. For example, the lunchtime sandwiches for the week would be prepared on Sunday and then frozen. John later said he always knew it was the middle of Tuesday when he was eating a recently thawed sardine sandwich! Fortunately, as I was attacking the problem of atraxotoxin vigorously at this time CSL's management was highly supportive. On the home front I had the services of seventy-year-old Miss Ismay Pickett who babysat the kids after school. Miss Pickett became a bit of a fixture and as she aged I was able to repay her for some of her many kindnesses.

A week before I met Margaret Malloy, Dad had gone into the Bendigo Benevolent Home. At eighty-nine he became progressively vague and a short spell in an institution that he had been associated with for many years seemed a good idea. A few years earlier funds for the construction of an impressive steam laundry to service the Home and the hospital had been collected largely owing to his efforts. I think he was secretly proud of his resultant nickname—'Steamboat Sutherland'. It reflected his persistence. Thus I assumed he would be well looked after at a time when I was having my own, hopefully, temporary difficulties.

On Friday, 12 October, I had a wonderful lunch with Margaret Malloy in Carlton. Since I had never cared for Margaret as a name, I audaciously asked her if I could call her Megan. She agreed and some months later formalised the change. I remember at lunch asking if I could touch her and she said 'of course'. I reached over and touched her cheek.

We made our separate ways back to CSL and my offsiders remarked on my cheerfulness. That night family catering was easy because the tennis club across the street was having a pie night. I went to sleep with happy thoughts about Megan, wondering if she would get on with the kids. She was the eldest of seven which seemed a good background. At

11.30 the phone rang and I was told Dad had just died. I had no idea he had suddenly deteriorated in the previous twenty-four hours. I lay awake thinking about my father's body cooling down and the fact I had been falling profoundly in love while he was dying. Next day I left the kids with my sister Diana and drove up to Bendigo to arrange the funeral. All the way up I kept thinking how wonderful it was that it appeared that, as I was losing one loved one, I was finding another.

At Bendigo a particularly dreary looking undertaker asked me if I wanted to view Dad since he was 'only next door' and it would 'only take a jiff'. I declined because I felt, and still do, that he would have preferred not to be seen when he wasn't looking his best. There was also another reason; my friend John Ipsen had recently overdosed and I had been called on to identify him at the morgue. After this experience, I preferred to avoid such stark visual evidence that a loved one was indeed very dead.

I was certain I would cry at Dad's funeral, but because there was such a large turnout I felt only pride. However, it was some years before I revisited his grave because I could vividly imagine what was going on in his coffin. Later on, after the first of many visits to pour a dram of whisky on his grave, I was moved to compose a poem entitled 'First Visit to a Father's Grave'. *The Age* declined to publish this very fine poem!

Two days after Dad's funeral found me lunching with Megan and her mother. A week later it was Megan's turn to be scrutinised at dinner by John and Susie. Fortunately they all got on splendidly. Megan and I were married on 9 April 1976, my divorce being one of the first processed by the new Family Court.

Megan was extremely supportive and inspirational. Unfortunately my single-mindedness and work habits took their toll. My time with Megan was the most fruitful and emotionally serene time of my life.

How Does the Funnel-web Spider Kill People?

It may be helpful at this stage to describe the effects of this venom on humans. Most venoms cause death by paralysis; this prevents the creature's prey from escaping or putting up too much of a fight. A paralysed animal can't breathe and, without outside help, soon dies.

After a Funnel-web spider's bite the illness doesn't resemble any other type of poisoning and there is no suggestion of paralysis. Fortunately, nine out of ten people bitten by a Funnel-web spider develop no illness because an inadequate amount of venom has been injected. However, any bite by this large black spider is in itself a frightening experience. This was especially so before antivenom became available and if a death had recently occurred.

Most bites are painful because the fangs are large and the venom is acidic. Sweating develops around the bite site, hairs in the region stand up and neighbouring muscles develop irregular, uncontrolled twitchings.

Within minutes strange taste sensations may be experienced and twitchings of the facial muscles be observed. Ten minutes later muscle spasms, profuse sweating and the erection of hairs may be widespread. Saliva pours out and by now the patient may be quite confused and possibly lapsing into unconsciousness. The pupils may be widely dilated and the pulse rate very rapid. Copious quantities of white fluid rise from the lungs. The deeply comatose victim may die at any time from fifteen minutes to twenty-three hours after the bite. Patients who recovered did so over several weeks and usually made a complete recovery.

This is a summary of the data available in 1970. The death of a young woman that year had not produced any new information which might have benefited other victims. Therefore, I felt it was essential to study an experimental animal which would be similar to a human poisoned by this spider.

Getting Close to Monkeys

Monkeys are surprisingly strong and their bite especially painful. Early on I had observed the response of a normally quiet animal handler who was bitten on the undersurface of his forearm. As his screams died down, I promised to treat monkeys with the same caution as snakes and spiders.

My daughter Susie fondly remembers an incident when a monkey displayed intelligence and an unnerving sense of humour. The little monkey used a bent nail to imitate me smoking my pipe. Every time I turned my back he would entertain his fellow inmates with a superb rendition of me smoking my pipe and pompously striding up and down. She recalled the monkeys screeching with glee but, as soon as I faced them, all would go quiet, pretending nothing at all had happened.

In 1972, with the help of these wonderful creatures, some of the mysteries of this venom were exposed. Dr Wiener had induced Funnel-web spiders to bite several monkeys and had observed the venom-produced illness progress to death. I had studied the effects of various venom components on unmonitored monkeys. It needed a lot more organisation to prepare a monkey whose blood pressure, pulse and other vital functions could be regularly recorded. The first monitored monkey experiment was significant and memorable.

To use the appropriate equipment it was necessary to take the monkey over to the University of Melbourne six kilometres away. The monkeys were under quarantine at CSL so permission had first to be obtained from the Quarantine Department. The plan was to take a heavily drugged monkey over in the boot of my car. On the phone, the Quarantine Officer put a long series of obstacles in the path of this venture and one by one his objections had been overcome. 'All right then,' he said. 'What if you hit a tram?' I promised to take special care not to hit a Melbourne tram.

A nice quiet laboratory in the Physiology building had been booked for the experiment. It became less quiet as more and more curious people packed themselves into the small space. Each newcomer wanted an explanation of what the monkey and I were endeavouring to achieve but they were very helpful and eagerly raced off for extra monitoring devices and anything else which might prove useful.

As those who knew him might expect, the Reader in Physiology, the venerable Dr Everton Trethewie, tried to take over the experiment. Ten times I was forced to say, 'Get out of it, Treth.' Finally, the soundly sleeping monkey was fully monitored and ready to receive the venom. The recording gear was started and an accurate record of blood pressure and pulse, etc., was made on graph paper. This paper steadily extruded itself from the recorder and, like wallpaper, rolled itself up when it reached the floor.

Male Funnel-web venom was injected into a vein and within a second the monkey's blood pressure dived. It stayed low for a minute or so and then it suddenly rocketed up and nearly went off the chart. The pulse rate rose with it. Four minutes after the injection, the monkey's muscles were contracting spasmodically. Profuse saliva and tears were produced. The pupils were fixed and dilated and it was obviously profoundly comatose. Its breathing efforts were inadequate and it required assistance. For over an hour the monkey's heart raced at a breakneck speed while the muscles twitched and salivation continued.

It was horrifying and no-one had seen anything like it before. Over the next few hours the effects of the venom diminished and there were periods when the monkey would suddenly stop breathing and require ventilation. The most significant trend after two hours was a steady and remorseless fall in the monkey's blood pressure. Eventually this fall in blood pressure led to a quiet death.

Further work on monkeys during this period confirmed that there appeared to be two critical stages in the poisoning. The first involved severe changes in the circulation, loss of consciousness and a failure to breathe. The second occurred some time later as the blood pressure steadily fell. It seemed possible that children had died early due to the first effect, and adults later because of low blood pressure. In fact, it turned out that matters were far more complex.

One bright spot was the discovery that atraxotoxin could produce the full syndrome in a monitored monkey. This confirmed that we were on the right track.

Reports of the relative immunity shown by domestic animals to this spider's venom were investigated and found to be valid. One cat became a constant reminder of its superiority in this regard over man and monkey.

It was a pure white half-grown cat which tolerated the venom yield from five spiders after which it was neatly sewn up. A girl technician volunteered to take the cat home and look after it overnight rather than have it put to sleep.

My neighbour's daughter had been wanting a cat for some time and so that night I told her about this particular animal. Her parents wanted it spayed and so, when it was brought back to work next day, I decided to have a go at spaying it. I found one ovary easily on one side but had the dickens of a job locating the other. I rang one of CSL's vets and explained my problem. He was silent for a moment and then gave appropriate instructions to complete the operation. After doing this, he paused for a minute and said, 'Incidentally if you want any of your female technicians sterilised, you might care to send them down to me.' He hung up abruptly.

For the next ten years this white cat, known as Peppermint, happily prowled the neighbourhood, blissfully unaware of its uniqueness.

Complacency and No Antivenom in Sight

We made further attempts at antivenom preparation using the purified atraxotoxin but, although there was a hint of success at times, further testing indicated yet another failure. We decided to study how the venom affected nerve fibres. It appeared to produce its effects by widespread and direct stimulation of nerves, but no permanent damage was done to them. Part of the syndrome could be described by the old term 'autonomic storm' in which broad stimulation of the involuntary nervous system occurs.

Venom was sent to many overseas laboratories for special investigations aimed at finding a pharmacological antidote. But, by the mid-seventies the Funnel-web project had become relatively unimportant whilst attention was being turned to other matters.

The general medical opinion was that a patient suffering from the effects of a Funnel-web spider bite should do quite well if managed in a modern intensive care unit. Drugs in routine use could be used to help control various stages of the illness. Clinicians were warned to keep a sharp eye on the blood pressure during the second stage of the illness. I was frequently quoted as saying that the patient who got to hospital in good health soon after the bite had an excellent chance of surviving. I was proved very wrong.

The Death of Mrs Christine Sturges

This thirty-one-year-old woman was bitten on the wrist by a male Funnel-web spider when making a bed early in the morning one day in

January 1979. She was in hospital within an hour, by which stage there was clear evidence of significant poisoning. After surviving the early stages her blood pressure fell steadily and by the end of the day there was difficulty maintaining it at a reasonable level. After a long conversation with her doctor that evening I had no more suggestions to make and felt quite helpless. When the patient died six days later I was despondent and considered myself in part responsible.

The death of Mrs Sturges had a number of effects. Firstly, although the media might have amplified the response there was a real concern in Sydney regarding the danger of Funnel-web spiders and the all-too-apparent lack of an antivenom. 'The Christine Sturges Funnel-web Appeal' was launched by her husband to make funds available to various university and other groups in an attempt to solve the problem.

Secondly, doctors who had managed a number of near-fatal cases were finding problems not demonstrated in the animal studies. In particular, Dr Malcolm Fisher of the Royal North Shore Hospital was adopting a particularly aggressive management in the early stages of these cases. By thinking fast on his feet and attempting to counteract each twist and turn of the syndrome before it got too advanced, Malcolm established himself as the leading clinician in the management of Funnel-web envenomations. Obviously there would have to be more communication between the clinicians handling the cases and the so-called backroom boys. Questions were occurring to Malcolm and his colleagues which I was finding it increasingly difficult to answer.

The third effect of the death of this woman was on me, personally. For years I had had control over the whole project. I'd played down the risk in modern hospitals and, worst of all, I'd produced no antidote or antivenom. The obvious question was whether someone else would have had more success. After a while these gloomy thoughts were replaced with a positive, almost combative, attitude. This was in part a reaction to some of the ill-informed comments coming from the city of my birth. A variety of people were getting into the act and a bit of healthy competition proved to be most stimulating.

The Final Assault

1979 was a busy year which saw the study of the movement of snake venoms in monkeys completed. This provided the experimental evidence for the use of the pressure-immobilisation technique as the best first aid for snake bite. The main paper had been published in *The Lancet* in January and I spent a lot of time explaining and promoting and, indeed, at times, defending the new type of first aid. To develop this technique it had been necessary to study the movements of all the important Australian snake venoms, and this study was near completion. (See page 226)

The first venom detection kits were also assembled in the Immunology Research Department and these had been issued free for clinical trial to many hospitals by the end of the year. (See page 241)

My working conditions and relationships with the CSL management had taken a turn for the worse (see Chapter 9), so it was not a comfortable time for expanding the attack on the Funnel-web problem. It was also a peak year in regard to numbers of scientific papers written, and there was a heavy, but fulfilling, programme of lectures. Also the WHO had invited me to Zurich for a major meeting on the development of antivenoms. Life was going on at a cracking pace!

In planning the next stage of our work on Funnel-web venom, I tried to pretend to be someone else. Surely, I thought, there must be areas that this bloke Sutherland has not pursued thoroughly or some results incorrectly interpreted. Although antibodies had been made to some venom components, there seemed no immediate chance of producing anything that would neutralise the main toxin. The best approach appeared to be to look for some drug or drugs which would be especially helpful in managing patients and to investigate more deeply the effects of the venom on susceptible creatures.

On 31 October 1979 I wrote to Dr Kester Brown, Director of Anaesthesia at the Royal Children's Hospital, to seek his help. In the past Kester had taken a close interest in venom studies. With my letter, I enclosed several papers and wrote the following:

> The main reason I would like you to read this is that I could see great benefit from collaborative work between CSL and some keen anaesthetists from your Department. With this in mind, also enclosed is part of our monkey summary in which you will see we have got monitoring of conscious monkeys from a toxicological point of view down to a fairly fine art. On page 217 you will see a plea for assistance re monitoring vital signs, etc. We now have adequate Funnel-web spider venom and some monkeys are available, and I wonder if we could do some studies on them re pulmonary oedema, role of diazepam, catecholamine levels, etc. When you get a chance to glance at the enclosed give me a ring so we can get together.

At the Royal Children's Hospital on 22 November 1979 Kester introduced me to Dr Alan Duncan, Director of Intensive Care, and his Deputy Dr James Tibballs. They were very enthusiastic about the proposed studies. In no time, experiments were planned for their very specialised experimental unit. Since approval by the Hospital animal ethics committee had to be obtained, the studies were hopefully to begin early in February.

Thus 1979 closed with exciting experiments planned to help solve the Funnel-web spider problem and from the snake bite victims' point of view, 1979 had brought two new benefits—a safer type of first aid and the availability for the first time of a venom detection kit.

The Funnel-web Strikes Again

On 3 January 1980, two-year-old James Culley was bitten on his hand by a Funnel-web spider near Terrigal some 95 km north of Sydney. He had put his right arm into his tracksuit when he suddenly started to cry and yell out, 'Needle!, needle!' He rapidly became critically ill and remained in a coma until his death three days later. Not surprisingly, this tragedy received intense publicity and was coupled with the death of Mrs Sturges less than twelve months earlier. The heat was turned on CSL and on me in particular. August bodies such as the Royal Australasian College of Surgeons requested an update on research progress and a justification of the recommended management of patients. Researchers in other States stepped up their efforts, some of which were funded by the Christine Sturges Appeal.

Success at Last

The next twelve months were rather eventful and are best narrated by taking selected diary extracts with appropriate comments.

Mid-Jan. 1980 — Preliminary work in monkeys at CSL finds the pressure-immobilisation type of first aid also appears to work well with Funnel-web venom. This method will replace the previously recommended arterial tourniquet for Funnel-web spider bites. To study this technique in monkeys more closely next month at the Children's Hospital.

5 Feb. 1980 — A heated discussion with the CSL Director, Dr McCarthy, over staff cuts (see page 259 for more details). As a result, two days later I am suspended until 21 February. Legal manoeuvres etc. continue for several more weeks.

13 March 1980 — At last, the first monkey experiment conducted at Royal Children's Hospital. The restraining frames made in my back shed for the first aid experiments proved ideal for safely and comfortably transferring the monkeys from CSL to the Hospital.

It is immediately apparent that Alan

Duncan and Jim Tibballs are most skillful and knowledgeable people. Over the subsequent weeks a number of new and important findings are made. The dramatic changes in blood pressure are closely followed and some animals produce much fluid from their lungs as seen in severe human cases. All monkeys developed an acute metabolic acidosis; other gross disturbances were detected.

The most exciting finding was that the pressure inside some monkeys' skulls rose very dramatically. Such a unique effect of the venom could well explain part of the mechanism which caused the deaths of humans. A number of other findings were retrospectively discovered in human cases when reports and x-rays were re-examined. An example was the unnoticed presence of acute gastric dilatation.

Several departments at the Children's Hospital co-operated closely in this work and within six weeks we were really coming to grips with the problem. There seemed a logical sequence to the various effects of the venom, and Jim and Alan started considering various types of drugs which might be used to advantage.

17 April 1980 Samples of high potency Funnel-web venom sent off overseas to various workers.

18 April 1980 A final attempt to raise antivenom in rabbits is started. The work at the Children's Hospital is going so well, it seems that it might be useful to have available antibodies to some of the components in the Funnel-web venom. Antibodies made against spreading factors in the venom might be a useful adjunct to drug therapy, even though the main toxin could not be neutralised.

Four healthy adolescent rabbits commenced an intense immunisation course. Once a week, they would receive a milligram of the most potent male venom available, which was mixed with an adjuvant and injected with great care into four separate sites. A ten-week course was planned. As far

7 May 1980 as I was concerned this was a minor sideline. Dr Ron Atkinson of the Darling Downs Institute of Advanced Education has claimed that normal rat serum is a potential antivenom because it neutralises the venom of the Sydney Funnel-web spider. He has been funded by the Sturges Appeal and has come down to Sydney to test the rat serum component in monkeys. I don't believe the experimental model employed by Dr Atkinson is likely to solve the problem.

12 June 1980 Two monkeys have now survived massive doses of male Funnel-web venom by the use of a variety of drugs to block or dampen certain venom actions. This is quite a triumph; it required countless manipulations, intense monitoring and maximal drug dosage. Jim Tibballs, being the youngest member of the team, had the job of staying up all night with the patient!

19 June 1980 Another monkey survives with full therapy. Supplies of Funnel-web venom are now under threat. I am beginning to curse the Sturges Appeal because someone involved with the rat serum theory has approached the Federal Minister for Health asking that any venom held by CSL be released for their experiments. (Note: It was not.) I am certain that rat serum will not work and but will have a look at it in the near future.

Having just put the phone down re the rat serum matter, I am rung by a Sydney newspaper asking for a comment on Dr Merlin Howden's plans to produce a vaccine to immunise the whole population of Sydney against Funnel-web spiders. Merlin of Macquarie University is an excellent biochemist but on this occasion out of his depth. I was getting very browned off but my comments on the proposed vaccine were fortunately not quoted in full.

20 June 1980 (Friday) Serum from an immunised rabbit mixed and incubated with Funnel-web venom and then refrigerated prior to injection into newborn mice on 23 June. My ever

capable technical assistant, Erin Lovering, is asked to inject them first thing in the morning and to check them after lunch and again late that afternoon.

24 June 1980 'This is a bit funny,' mutters Erin from the other side of the laboratory. 'What's funny?' I ask. Erin replies, 'Some of the mice that were alive after lunch died later in the afternoon, and some of those that were alive later in the afternoon have died overnight.' In the past, the vast majority of newborn mice injected with Funnel-web venom were either dead in four hours or survived. They never died six hours later or overnight.

There really was something funny going on. Over the last ten years a few newborn mice would survive longer than their companions but further testing of the serum had not produced a similar finding. Looking at the table Erin had produced, there was no doubt that some had had their life prolonged. The controls had all died in the required time span. A repeat experiment was planned and in the meantime on 24 June 1980 I wrote, 'By George, looks like some neutralisation of male Funnel-web venom. A ray of light?'

26 June 1980 Dr Coles (my immediate superior) informs me that my department is no longer to be called 'Immunology Research'. He thinks 'Venom Laboratory' might be more appropriate. (Dr Coles may well have felt slightly put out by my earlier finding that the Funnel-web toxin he described in 1964 was not the main one.) Jim Tibballs and I study the effect of venom in a cat and confirm that this creature is extremely resistant to the effects of the venom. This experiment was far more sophisticated than the first cat experiment done eight years earlier.

1 July 1980 A monitored monkey is given eight lethal doses of male Funnel-web venom. Crepe bandages and immobilisation are applied and left in place for two hours and then removed. The monkey remains in perfect health. This is a most pleasing finding.

8 July 1980	A repeat experiment confirms that several of the rabbit sera appear to be temporarily prolonging the life of newborn mice. Alan Coulter and Rodney Harris commence work on these rabbit sera using small glass columns filled with a newly available substance, Protein A-Sepharose. This allows removal from the serum of everything except the antibody, which can then be used at a more concentrated level.
18 July 1980	For the first time, mice survive lethal doses of male Funnel-web venom when it is given with the antibodies that Alan and Rod have isolated. An antivenom, at least for newborn mice, appears to be at hand! Life is most exciting. At the same time we tested rat serum and found it to have no such activity. Nor did the serum of the Blue-tongue lizard which had to be investigated at the same time because it had been claimed to have healing properties.
19 July 1980	Saturday, my daughter Susie's fifteenth birthday. At CSL at 7.45 a.m. and prepare two injections for paired monkeys. One consists of venom alone and the other a mixture of venom and antivenom. Both are incubated at 30°C for twenty minutes and then, with the help of an animal technician Ian Roberts, both monkeys are injected intravenously. The monkey that receives venom alone is critically ill within ten minutes but, wonder of wonders, the second monkey that we named Susie remains free of any symptoms. An hour later Susie is still completely normal and it is proved that the antivenom will neutralise the venom when allowed to react with it in a test tube in an incubator. (This is called in vitro neutralisation because the reaction takes place 'in glass' and is then proven to have occurred by use of a test animal.)

That same morning four more rabbits start their immunisation schedule. Ian is scheduled to finish work at midday. Since he lives nearby he offers to drop in a couple of times to check monkey Susie's condition and report back to me. At daughter Susie's party I'm a bit filled

21 July 1980

with myself. Ian reports that the venom-only monkey has died but Susie is as fit as a fiddle. Ian Roberts gets a blast from his superior for coming in when off duty without getting permission. I tell him he's a good bloke. I was very impatient to get going and would have liked to have done today's monkey experiment yesterday but could not find where Rodney had put the other bottle of antivenom when I was in on the weekend.

At last, everything is ready and the two monkeys receive an injection of venom which is roughly equivalent to the amount of venom which could be collected by milking three mature male Funnel-web spiders. After ten minutes the injected legs of both monkeys are starting to show gross muscle twitching and one is given an intravenous injection of antivenom. The monkey which received the antivenom developed no signs of general poisoning and the local muscle twitching ceased abruptly twenty minutes after the injection of antivenom. The other monkey became critically ill, showing all the features of severe envenomation.

As if to prove its first class condition the monkey that had received antivenom, which we named Megan after my wife, managed to open its cage and appeared keen to join us all for morning tea. But for her sharp teeth, I would have picked this monkey up and hugged it. I was certain we now had the Funnel-web spider on the run.

22 July 1980

After discussions with Dr Schiff we agree that some antivenom will be sent to our Branch in Sydney. It was sterile-filtered by Rodney Harris and is clearly labelled an experimental antivenom. Since a child had died in Sydney from a Funnel-web bite seven months earlier in Sydney it seemed reasonable to keep half our stock rather than holding it all in Melbourne. I told Malcolm Fisher at the Royal North Shore Hospital that it would arrive later that day and he was delighted, to say the least. On the other hand, the other clinician in Sydney who was considerably experienced in managing Funnel-web spider bites, Dr Tom Torda, did not respond quite so enthusiastically. Tom had been caught up in the testing of the rat serum products in monkeys—admittedly with a more cautious attitude to the likely outcome than the co-investigators. Fortunately, Tom was won around a little later.

I made a most important call to Eric and Robyn Worrell who ran the Australian Reptile Park in Gosford. These two had provided me with Funnel-web venom at no cost for all the years the project had been running. During the difficulties of recent months they had been unfailing in their support. I told them of the recent results and put it to them that, if they could provide CSL with significantly increased quantities of male venom in the next few months, we could have an antivenom available by Christmas.

They responded magnificently and did everything possible to induce the local population to deliver spiders for milking to the Reptile Park. Lyn Abra, the biologist in charge of the expanding spider colony, coaxed them along to maintain maximum venom output. Without this help progress would have been limited and the rabbit immunisation halted. The Worrells declined cash offers for venom from other researchers and continued to send the venom to me gratis.

23 July 1980

The fact that an experimental antivenom is available in Sydney becomes known. I stress that it is experimental, only enough for perhaps one serious case, and it will be replaced as soon as possible with a formally approved product. I cited the example of Dr Wiener who sent an experimental antivenom to Sydney twenty years earlier. This Funnel-web antivenom was made in horses and was recalled to CSL a little later when further experimentation showed it to be ineffective.

24 July 1980	Claims and counterclaims are now emanating from CSL. The Acting Director of CSL, Mr Vivian Davey, plays down the importance of the discovery. In *The Age* he was quoted as saying, 'I wouldn't be happy to see it tried on the evidence we have at the moment.' In the same article I was quoted thus: 'If a child was critically ill, withholding it would put us in a very difficult position.' *The Age* Science Reporter Peter Roberts stated that Sydney doctors would be willing to use it in an emergency.
25 July 1980	A press release from the Minister for Health, Senator Guilfoyle, is published extensively. The substance of the release has been prepared by Mr Davey and as reported in *The Age*, opens with the following paragraph: 'The Federal Government yesterday denied claims that an antidote had been found for the deadly bite of the Funnel-web spider.'

While some people at CSL seemed to be tripping over themselves to issue statements such as: 'There is no way that the Federal Government would ever allow an antivenom developed in rabbits to be used in humans ...', a lot of spontaneous outside support was evident.

Dave Robson, the Chairman of the Christine Sturges Funnel-web Appeal, sent a telex on 24/7 as follows: 'Congratulations on your development of an antivenom serum for Funnel-web spiders. Would appreciate any data you can give us to help us with our efforts to have a Funnel-web antidote freely available to people who require it.' My reply dated 25/7/90 is as follows: 'Thank you for your kind message. I personally am confident that adequate supplies of an effective antivenom will be available by December if more male spider venom is obtained. Please contact Mr Eric Worrell of Gosford who is our main venom supplier. Best wishes.' I ceased cursing the Christine Sturges Funnel-web Appeal.

25 July 1980   Getting annoyed by some of the statements being issued. Davey's latest one: 'It is premature to speculate when all the necessary data required for such a new product registration will be available.' I point out that with a little bit of help with things such as typing, December seems quite feasible. (It was.)

The best way to combat these adverse and confusing reports was to get the findings to date published as soon as possible. I spoke to Dr Laurel Thomas, editor of *The Medical Journal of Australia* who was very supportive and she agreed to hold space for three articles in October provided the articles reached her within three weeks. Another deadline, but the more irons in the fire the better, and the publication of such articles would assist in the registration process.

The steps required to meet this December deadline were fairly obvious.

The bickering at CSL had to stop, at least for the time being. As much male venom as possible was urgently required, not only to continue the immunisation of current rabbits but to allow additional rabbits to join their colleagues. Extra venom was also required for the testing of serum, etc.

Our aim was to produce as large a pool of immune serum as possible. A control pool of normal rabbit serum was also required to be used in the initial testing of the equipment used to purify the antivenom. To scale up the process we had to obtain urgently large quantities of the fairly expensive material Protein A-Sepharose from Sweden.

When the antivenom had been purified and concentrated, it had to be dispensed and subjected to extensive testing in monkeys and to a wide range of standard and special Quality Assurance tests.

Meantime, every important step had to be meticulously recorded as part of the submission to Canberra required to obtain early approval for a clinical trial of the antivenom. A precise protocol for the use of the antivenom had to be prepared and a physician or physicians found who would be willing to take charge of the actual clinical trial. Further down the track, we would need to collate the findings of the clinical trial and hopefully obtain permission for unrestricted issue of the antivenom to any hospitals requesting it. Clinical aspects of the use of the antivenom would later be published in a medical journal.

27 July 1980   Sunday, in at CSL by 7.30 a.m. Spend the whole day writing up the results to date and preparing appropriate tables, etc. in response to requests by both the Health Department and Mr Davey.

28 July 1980   A sharp memo from Mr Davey which includes the statement: 'The delay in submission of this

	information is inconsistent with the spirit of your letter of 22 July 1980 and the expectations generated by your recent statements to the media.' Added to Davey's memo was a note from Dr Schiff: 'Struan, I received your draft report after this memo. Will discuss.'
	Mr Davey is remarkably quiet thereafter. A good Sunday's work.
29 July 1980	Five new rabbits commenced on immunisation. Bleeds from best rabbit, No. 907, are dropped and lost. Technician asked not to do that again.
2 Aug. 1980	Attend graduation ceremony at University of Melbourne and receive a doctorate of medicine from the Chancellor, Professor Pansy Wright. Sitting on the stage with all the nobs wearing a big floppy hat, I nervously and regularly glanced at a little card I had prepared. My personal instructions were as follows:

Usher will take you to the spot.
When name read out, lift hat to Pansy.
After citation, go and shake hands with Pansy.
Lift hat.
Go back to spot.
Lift hat to Dr Penington, and sit down.

I tried not to stamp around like a Clydesdale and the whole thing went quite smoothly. The audience clapped languidly as I approached Pansy, who grasped my hand firmly and refused to let go.

Restraining me with a surprisingly vice-like grip, Pansy leant forward and said, 'We're all very pleased with you, very pleased,' then he paused and said, 'Now you go off and study the venom of your administrators.' I nearly said, 'Thanks Pansy,' and left thinking, 'I would love to tell the audience what you had just said Pansy.'

In 1974 I was having dinner in London with the elderly Physiologist, Professor Peter Daniels. The dear old chap had never been to Australia and was racking his brain for Australian orientated topics. Suddenly his eyes lit up and he said, 'I remember a chap, a fellow called Wright was over here before the War. We worked together at Oxford.' Long pause, 'Absolutely dreadful fellow.' Somewhat flummoxed I concentrated on identifying what the lumps might have been in my soup.

5 Aug. 1980	The Funnel-web antivenom paper sent off to *The Medical Journal of Australia*.

	The routine immunisation of antivenom-producing rabbits has commenced, with the Veterinary Services Division undertaking the bulk of the work.
	1.00 p.m. Meaningful discussions with Drs McCarthy and Schiff and Mr Sullivan and Mr Davey. Described as a 'clear win to Department 523' (my department). They have agreed that the required amount of Protein A-Sepharose be ordered and airfreighted from Sweden as soon as possible.
12 Aug. 1980	Sufficient venom now available to continue dosing rabbits for a further three weeks. After that we may have a problem.
	We were all cheered up by a letter from Peter McKay of Coolaroo in the Melbourne *Sun*:
	'Who are they? Stand up you hardworking dedicated loyal Australians who discovered the antidote to the Funnel-web spider. I must be the first because no-one else seems to have thanked you. Congratulations and thank you.'
18 Aug. 1980	More venom arrives from Robyn Worrell. If she can keep this up we may be out of the woods.
19 Aug. 1980	Ye gods! I find that Dr Norman Coles, after reading a report in the Library of the mutagenic effects of Protein A has taken it into his own hands to cancel the order for Protein A-Sepharose from Sweden. I bail him up and point out to him as politely as possible that, firstly, he has no right to have done that and, secondly, there are many substances in common use which produce mutagenic effects on cells when they are grown in culture.
	A classic example is the drug colchicine, which has mutagenic effects in tissue culture but has been safely poured into patients with gout for many, many years.
	The order for the vital material is reactivated and we add another test to the antivenom, namely seeking mutagenic effects. This test will be carried out at the Walter and Eliza Hall Institute. (It was always negative.)

23 Aug.–9 Sept. 1980	Megan and I accept a free trip to Okinawa courtesy of the World Health Organisation to attend a meeting on antivenoms. Japanese do things in style, put VIP labels on our suitcases and fly the Australian flag outside the hotel. Quite a change from CSL. Am given a huge bundle of Japanese money to cover all expenses. Consider ourselves fantastically rich at last. Have a swim in the hotel pool while a Japanese follows me up and down yelling at me. I then find it costs the equivalent of 5 dollars Australian to swim in the pool. Hotel guests must pay in advance.
At the Governor's Reception ate some cute little rissoles. Turned out to be made from sea snakes. Given a bottle of snake wine to bring back to Australia. The wine is accompanied by a large data sheet describing its analysis and usefulness. It includes the statement: 'This is thoroughly tested in our modern lavatories.'
Leave Okinawa about 2 yen down. We understand now why the Japanese have an average bank savings of 60 thousand Australian dollars!
10 Sept. 1980	Back at CSL. The Veterinary Services have been doing a wonderful job and the rabbits are in excellent health. They have continued the habit of giving the rabbits fresh carrots immediately after their injections. This distracts them from any discomfort and makes them feel quite special since no other CSL rabbits get such a tasty diet.
7 Oct. 1980	The venom is now critically low and so eleven rabbits will have their dose reduced by half. The submission being prepared for Canberra is now five centimetres thick.
8 Oct. 1980	All equipment is now ready for the major isolation of the antivenom from the immune serum. Normal rabbit serum is being fractionated through the system to provide control material and to test everything. If accidents and mistakes are going to occur, better with normal rabbit serum than the precious immune serum. The Serum Fractionation staff

	are a joy to work with. The quietly spoken Charles Chatelier spends many hours rugged up in a cold room watching over the process and not making a single slip.
12 Oct. 1980	Sunday, sort out the best sera for pooling and then fractionation. Select rabbits for further big bleeds on the basis of the last week's assay results.
16 Oct. 1980	Big bleeds from all selected rabbits. From now on, immunisation will be continued as venom becomes available but further bleeds will be stored for fractionation possibly next year.

The draft protocol for the proposed Funnel-web antivenom trial is completed and goes to Drs Malcolm Fisher and Tom Torda in Sydney for comment.

18 Oct. 1980	Three major articles appear in *The Medical Journal of Australia*. One is a preliminary report on the development of the antivenom and its testing in newborn mice and monkeys up to the end of July. It also includes data indicating the ability of the antivenom to neutralise the venom of some other dangerous Funnel-web spiders such as the Northern Tree-dwelling Funnel-web and the Toowoomba Funnel-web.

Alan Duncan is first author of the second, very comprehensive article. It describes the effects of Funnel-web venom in monkeys and how many of these effects could be prevented or abolished by the use of a variety of potent drugs. The paper contains a number of effects of the venom which were totally unknown twelve months earlier. The third paper describes how the pressure-immobilisation type of first aid appeared to inactivate doses of venom as high as 2 mg in monkeys. The conclusion reached in this article was that these findings had immediate applications—both to the first aid and to the actual medical management of human victims.

In their different ways, these articles were landmarks in the management of Funnel-web spider bites.

21 Oct. 1980	Give a research seminar at CSL on the

Part of a photograph of the ship's company taken in Hong Kong. Captain Wells is in the centre of the front row and the author on the far left.

HMAS *Voyager*

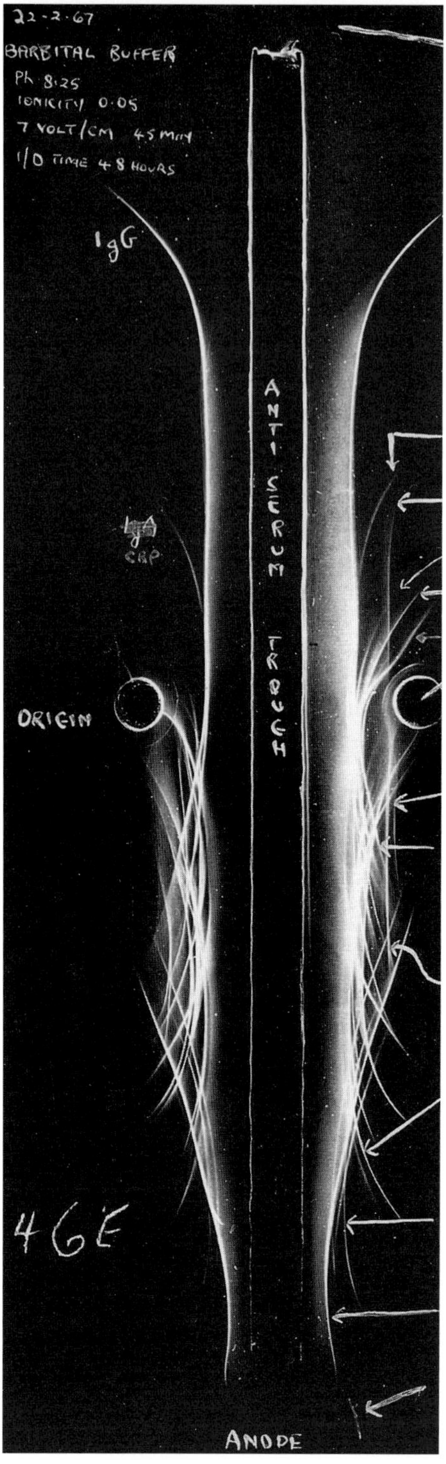

Analysis of human serum by immunoelectrophesis. Each arc represents a different component.

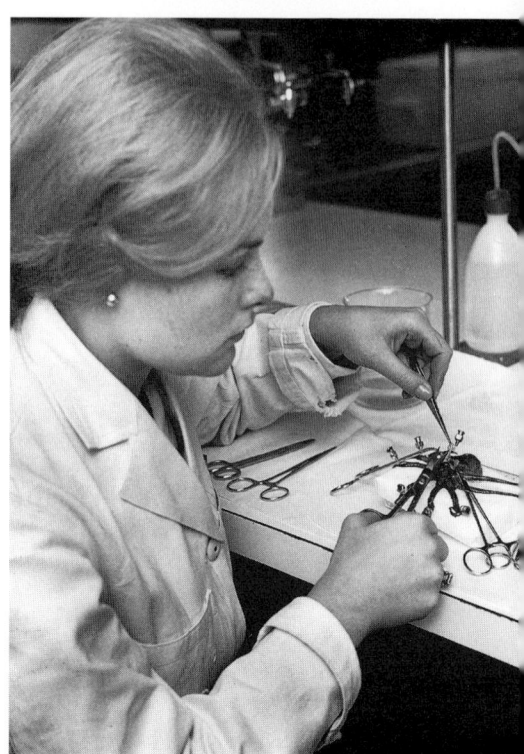

Technician Moira Whigley dissecting a Blue-ringed octopus circa 1968. (Photo F. Pietras)

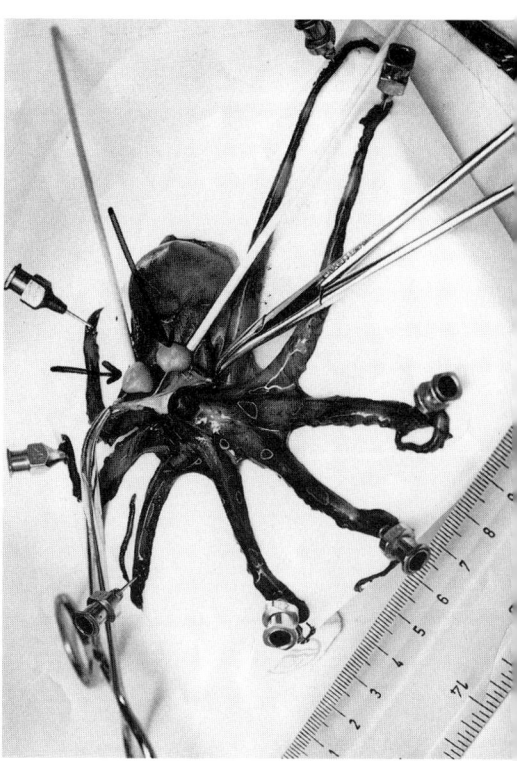

Dissection of Blue-ringed octopus. The two poison glands are arrowed. (Photo F. Pietras)

	Funnel-web spider project. This is based on the papers published several days ago and goes over reasonably well.
22 Oct. 1980	A meeting with my seniors to review progress. This is a good meeting, at last an atmosphere of co-operation prevails.
27 Oct. 1980	We now have 720 ml of serum pooled and ready for testing in newborn mice.
28 Oct. 1980	At 3.00 p.m. ordered to come to the Administration Building immediately. Face a solemn foursome, Drs McCarthy and Schiff, Mr Sullivan and Mr Davey. I am tersely reprimanded for breaking agreements not to discuss antivenom progress in public without prior approval. I am completely bewildered by this having gone to ground since early August.

Mr Davey triumphantly reads out a report in a Sydney paper dated 22 October. The other three looked suitably thin-lipped but as he continues, I realise that the material that was being quoted by *The Sydney Morning Herald* had in fact been taken from the articles published in *The Medical Journal of Australia* on 18 October. The CSL contributions to these articles had been approved by the appropriate CSL people. I was not impressed. They vaguely apologised, and if I hadn't been so busy, I would have asked each of them to put a hand out to receive a brisk smack with a ruler. |
| 29 Oct. 1980 | Funnel-web antivenom has been fractionated by Charles Chatelier and is now being dispensed into ampoules ready for tomorrow. |
| 30 Oct. 1980 | At the Children's Hospital testing the first production batch of Funnel-web antivenom, designated batch 002, in a monkey. Alan Duncan is overseas so he misses out on the excitement. Later Jim Tibballs and I would tease him, saying we would cut him out of the publications to follow this and other work.

It was almost magical to see readings of the charts return to normal within minutes of the injection of batch 002. |

31 Oct. 1980	Another monkey is successfully treated with batch 002. When it has recovered an additional larger dose of batch 002 is given. No adverse effects are noted. The experiment is summarised as 'an excellent result'. I worry that things are going too beautifully. I can sense my enemies at CSL watching, waiting and ready to pounce if the project becomes unstuck.
2 Nov. 1980	Sunday, all graphs, etc. and tables relevant to last week's monkey testing of batch 002 are prepared for submission.
3 Nov. 1980	Jim and I do another monkey, delaying the antivenom administration. Good response seen when antivenom is given.
9 Nov. 1980	Write up the complete summary of the testing of batch 002 in monkeys.
11 Nov. 1980	Most of the quality assurance studies done on batch 002 are now complete and being collated.
13 Nov. 1980	Another monkey done, this time using a large quantity of venom and a smaller quantity of antivenom. A good result obtained.
17 Nov. 1980	Disaster. One out of 16 ampoules of batch 002 may have been contaminated by bugs. A more extensive sterility test to be set up immediately.
28 Nov. 1980	Everything has now gone to Canberra except the repeated sterility test, which is still underway.
3 Dec. 1980	Group of us go up from Melbourne to a meeting at the Royal North Shore Hospital in Sydney with Drs Fisher and Torda. The indications for antivenom and other aspects of the trial are thrashed out. I decide Malcolm Fisher is a good bloke. The Sydney people, however, talked far too much and we missed our scheduled flight back to Melbourne.

Over a few drinks while waiting for a flight I found out from Jim Tibballs and Alan Duncan that they had occasionally done little experiments on the monkeys after I had left. These experiments had discounted a theory I had advanced in earlier days regarding the use of diazepam (valium) in treating Funnel-web spider bites. You can't trust anyone nowadays!

5 Dec. 1980	Batch 002 has passed all sterility tests. Canberra now has the complete submission.
10 Dec. 1980	A day trip to Canberra to chat with some Health Department officials. Several are gloomy doubting types who put up a surprising number of objections. Others were quite pleased to see the grey doubters knocked for six by a fine performance by both Alan Duncan and Malcolm Fisher. I made sure I did not miss the exit flight from Canberra.
16 Dec. 1980	The antivenom has been cleared by Canberra and will be issued immediately to hospitals participating in the clinical trial in New South Wales.

The release of the antivenom attracted conflicting statements from some of those involved. Mark Camm in *The Australian* on 22 December reported that Neville McCarthy appeared negative about this latest product from CSL. Amongst other things Dr McCarthy said, 'There was no guarantee antivenene would work. Trials were to determine whether it "changed the course of the venom" in humans.' Dr McCarthy also warned it would not be available for general use and was not to be regarded as 'a wonder serum'.

As you can imagine Dr Fisher and I weren't too impressed by such 'supportive statements'. Dr Fisher adopted a more comfortingly laconic attitude and told the public in the Melbourne *Sun*, 20 December, 'Let's put it this way. If my kid gets bitten, he's going to get it. I see no reason why it shouldn't work.' In the same article I was quoted as saying, 'We see this as a Christmas present to Sydney from the Commonwealth Serum Laboratories.'

Christmas came and went without a significant Funnel-web spider bite. It was, as Dr Fisher said, 'As though they had gone off the bite.'

27 Jan. 1981	I received a hand-written personal letter from Dr McCarthy, the contents of which I found extraordinary. Dr McCarthy amongst other things wondered whether I intended to continue at CSL. He expressed an opinion, 'that you should seek appropriate medical advice. If by chance you are already under treatment, the benefits of sick leave should perhaps be utilised.' I showed this letter to a medical colleague at CSL and he appeared as astounded as I. I

thought, two can play at this game and sent a most conciliatory letter back to McCarthy suggesting that now was an excellent time for a fresh start. I concluded, 'I therefore propose to establish more harmonious relationships with those involved and trust that such an approach will be fully reciprocated.' That night he rang my home for the first and only time and agreed with the proposal to make a fresh start. To my surprise, copies McCarthy made of his letter were produced later during my 'trial', see page 278.

A Bite to Remember

At 11.00 p.m. precisely on 31 January 1981 Gordon Wheatley, a healthy forty-nine-year-old man, was bitten on his foot by a male Funnel-web spider at Cheltenham in Sydney. He was changing a light bulb in his dining room when the spider was trodden on. Gordon arrived at the Ryde Hospital at 11.15 p.m. complaining of pain in the right foot extending up his leg. He was perspiring profusely, his pulse rate was 100 beats per minute and he complained of palpitations. An intravenous line was inserted and he was noted to have muscle twitching in both arms which progressed to severe muscle spasms. Tears and saliva were pouring out and he began to vomit.

As Gordon turned frantically from side to side, the intravenous line was pulled out. Fifty minutes after he had been bitten, his complexion was bluish, he was irritable and would not keep the oxygen mask on. He was therefore sedated and paralysed, placed on a respirator and transferred to the Royal North Shore Hospital, arriving it 12.15 a.m. on 1 February.

The phone rang at 1.30 a.m. at my home, Brighton, Melbourne. The voice of Malcolm Fisher came down the line. He gave a quick run down on the patient's history and concluded with the statement, 'I've given him 3 ampoules of antivenom, it hasn't killed him but hasn't done him any bloody good either.' I replied, 'Well I guess all you can do is give him another dose of antivenom, because we don't really know how much he's going to need.' Fisher went back to work.

Thirty minutes later Dr Fisher was on the phone again. His opening comments were, 'Struan, you've just ruined a beautiful bloody syndrome.'

The second dose of antivenom had caused the patient's blood pressure to return to normal, he regained consciousness and a medical emergency no longer existed. The patient's relieved wife was glad she had given permission for the antivenom to be used.

While having a shower at 7.00 a.m. I suddenly recalled my conversations with Malcolm Fisher. For a moment or two I quite frankly thought the whole episode must have been a dream. The scribbled notes by the telephone confirmed that the conversation had actually occurred. A little later, sitting looking at my cornflakes, I was swept by deep elation comparable to that felt after passing final year medicine. This was then followed by thoughts of gratitude to the handful of people who had unfailingly helped in their different ways to achieve this happy result.

When the patient was discharged from hospital next day, completely recovered and with no side effects evident due to the antivenom, there was great happiness amongst the little band at CSL. Not a comment was made one way or another by the CSL hierarchy. The annual report for that year gave the antivenom the briefest of mentions.

How to Confuse the People of Sydney

Hopefully the Sydney afternoon tabloids do not closely reflect the average intelligence of a Sydneysider. Incorrect stories appear to be rarely corrected and some 'unusual' approaches to the management of the Funnel-web spider problem were given front page publicity.

The proposal by Dr Atkinson that there was a component in rat serum which held promise for treating human beings bitten by Funnel-webs has been discussed earlier. By 1981 both this theory and Dr Atkinson had faded from the Sydney papers. The end of the rat serum approach may have been encouraged by the publication of information from CSL in October 1980 that no protection could be found in rat serum and, for good measure, none in rabbit, Blue-tongue lizard, chicken, dog, cat or mouse. We could have worked our way through the animal kingdom but felt that number was sufficient for the purpose. When first asked by the Sydney papers about the future of rat serum as an antivenom, I had replied that on the evidence to date, it was nil. I later wondered how much this futile exercise on rat serum cost those good people of Sydney who had contributed to the Christine Sturges Appeal.

However, as the rat serum project faded into history, a new and bizarre approach was being proposed to the problem posed by the Sydney Funnel-web spider. A problem which, in the eyes of my better informed colleagues, had basically just been solved!

The release of the CSL antivenom in early December 1980 was not helped by a statement by Dr McCarthy that the antivenom was 'not known to be safe'. This brought a promise from Dr Merlin Howden of Macquarie University that he could 'solve the problem'. In a statement in the *Daily Mirror* dated 2 December 1980 Dr Howden said that given 2 000 live Funnel-web spiders, his team would come up with an antivenom using antibodies which would be completely compatible with humans and safer to inject. This seemed a pretty remarkable goal but a lot of Funnel-webs which might have been milked on CSL's behalf were channelled to Macquarie University. Consequent shortage of venom made ongoing rabbit immunisation difficult, but not impossible, thanks to continued support from the Australian Reptile Park at Gosford.

By March 1981 Dr Howden and his group had abandoned their original approach. They then proposed once again to develop a vaccine to immunise the 2½ million people in the Sydney region. I had no problem in classifying this concept as grossly impractical and frankly ludicrous.

Dr Howden approached Councils in the areas infested with Funnel-web spiders in an attempt to raise large sums of money for this research. Statements attributed to him, possibly incorrectly, cast doubts on the effectiveness of the CSL antivenom. One way or another Merlin asked for it and got bucketed by Malcolm Fisher, myself and others. When Malcolm found that his own Council was contemplating a large contribution he waxed most eloquent and no donation was made. I reiterated my statement that Dr Howden was a brilliant biochemist, but was straying too far from his specialty.

My opposition to active immunisation of the people of Sydney, as proposed by Dr Howden, was on a number of grounds. Technical problems aside, the rarity of extreme envenomations combined with the lack of time for an anamnestic response (as seen say after tetanus vaccination) made the concept unacceptable on both economical and medical grounds. To protect people from a particular venom by immunising them against the venom is a very tricky business. The venom has to be detoxified, it must not cause local reactions and, to maintain immunity, injections possibly as frequently as six-weekly, are required. The chosen venom must also be excellent at promoting antibodies, a property definitely not seen with Funnel-web venom. Sometimes 'venom vaccines' promote a state of severe allergy rather than immunity.

The first batch of antivenom we issued was only enough to treat perhaps seven patients. It was not known how long its 'shelf life' would be and so there was great pressure to produce a larger batch, particularly after its initial success. Getting sufficient venom to maintain the immunisation of the rabbits was our great problem. As mentioned above, it was aggravated by the activities of other parties in Sydney. A number of traumatic events occurred before the release of batch 003. The hitches and problems that occurred were not due to the rabbits or the scientific staff involved in the antivenom production. (See page 295)

Out of the Woods

Sunday morning 3 May 1981 found Malcolm Fisher and myself addressing a large medical audience in Sydney. (About the same time burglars found their way into my home in Melbourne.) Aspects of the antivenom were discussed and Malcolm presented the first clinical case of antivenom usage in considerable detail. Everyone seemed to be impressed except for one senior Government cynic, who made the point that one swallow did not make a summer.

One hour later, straight after lunch, Dr Fisher got up and announced that 'Summer has arrived'. At 11.45 a.m., whilst Malcolm and I had been holding our audience spellbound, a three-year-old boy, Liam Trehy, had been bitten in his backyard at Terrigal. His mother prised the spider from his foot and he arrived at Gosford District Hospital at 12.00 noon. The child had nearly died but recovered after two ampoules of antivenom. Fisher and I were pretty pleased with this, but some people accused us of having organised the bite. Something not even Fisher could have managed!

A number of successfully treated cases followed and the second batch of antivenom was released. In 1982 the Health Department declared the clinical trial over and the Funnel-web spider antivenom became available for use at any hospital which requested it. Sixteen years later, over one hundred patients have received the antivenom—and it has been used successfully in cases of bites by Funnel-web spiders other than the Sydney species. Such bites have occurred in northern New South Wales and in Toowoomba, Queensland. In Toowoomba it was also used to treat an eighteen-month-old infant bitten by a Mouse spider, which can produce a syndrome somewhat similar to that of the Funnel-web spider.

In 1989 no-one doubted that the antivenom had saved the life of nine-month-old Julie Burnside. Her mother Wendy heard her scream and found her with the spider on the bathroom floor. Wendy, who is a trained nurse, applied the pressure-immobilisation type of first aid while her husband Cameron, a pharmacist, rang for an ambulance. Julie, the youngest recorded Funnel-web victim, completely recovered, but required eight ampoules of antivenom. The Burnsides thereafter were great supporters of antivenom research. Julie at age six is now pictured in the Australian Venom Research Unit fundraising brochure.

On a lighter note, Malcolm Fisher and I were pleased that none of the patients developed long ears, fluffy tails or showed a preference for carrots over other food.

Funnel-webs Still have Secrets

Since 1980 biochemists have done brilliant work exploring the venoms of various Funnel-web spiders. The sequence of Atraxotoxin (now renamed more specifically, robustoxin) was published by David Sheumack and his colleagues in 1985. However, apart from structural studies, the venom still holds secrets.

The finer details concerning its targets and what it does to them are unclear. No explanation has been forthcoming as to why men and monkeys have a special vulnerability.

Here I want to jump a wee bit ahead of myself to highlight some unfinished business which is worrisome.

The closure of my research department in 1981 prevented the fine tuning of the antivenom and, most importantly, the testing its effectiveness against the venom of lesser known Funnel-web spiders. This lack of information has, on a number of occasions, turned the spider bite victim into an experimental animal—with neither the doctor giving the antivenom nor I knowing if or when a response would be seen.

A doctor in this unsatisfactory position can only pour antivenom into the patient and hope for the best. A typical example occurred on 8 February 1998. A man aged thirty-eight was bitten by a male 'paper bark' Funnel-web spider (*Hadronyche cerberea*). He was desperately ill, in fact unconcious, when he reached Wyong hospital fifteen minutes later. After a record thirteen ampoules of antivenom, he commenced a slow and shaky recovery. This man nearly became the first known adult male fatality caused by a Funnel-web spider. He was the second victim in the district of this spider, a four-year-old boy having been treated successfully with antivenom in 1989.

It is a sad reflection that for some years *H. cerberea* venom, meticulously collected by Lyn Abra at the Australian Reptile Park at Gosford, has remained untested. Lyn was called in urgently to identify the spiders involved in the two cases above and so has more than a passing interest in any results. In time, funds may filter through so we can do this work; meantime things remain a bit tough for victims of this spider and some of its relatives.

Everyone seems to agree that the antivenom made by the CSL's rabbits has been useful. To date, there has been no hint of any adverse reactions at all and it appears to be the safest antivenom yet prepared. This, in turn, suggests that antibodies raised in rabbits could possibly have a wide application in other areas of human medicine, for example, in antitumour therapy. In time, cells maintained in tanks may produce a more potent antivenom and reduce the initial demand for venom.

However, man cannot live by Funnel-web alone! Let us now turn to more straightforward research. Although not attracting the public eye like the Funnel-web spider work, the next chapter relates more important findings as they benefited a far greater number of patients.

Improving the Lot of Victims of Snake Bite and Other Bites and Stings

8

'It is no help to shelter behind the statement that snake bite accidents are a rarity and that the average doctor seldom or never will treat one. For the patient who has been bitten, it is a matter of life or death and the rarity of the event is of no interest to him.'

Translated from the Editorial, *South African Medical Journal*, 12 July 1975

Prologue	220
Some Careful Planning	221
Isolation of Snake Venom Components	222
Detection of Snake Venom in Clinical and Experimental Samples	222
Finding snake venom in dead children	224
Professional obstruction?	225
Investigation of First Aid	226
A new type of first aid for snake bite	228
Early questions posed by the new first aid	231
What happens if a larger dose of venom is given?	231
Is venom destroyed whilst trapped in the tissues?	231
Would the first aid work with other snake venoms?	231
Was the first aid applicable to bites by the Sydney Funnel-web spider?	231
Would the first aid delay the movement of low molecular weight toxins?	232
How does this first aid measure work?	232
Promoting the pressure-immobilisation type of first aid	232
Reducing Reactions to Antivenoms	235
Premedication before antivenoms	239
Surveys of Antivenom Usage	239
Snake Venom Detection Kits	241
Investigating Snake Bite Cures	242
The Most Venomous Snake in the World	243
Finding a Need and Spreading the Message	244

Media	247
Epilogue	248

Prologue

Australia can be very proud of its venomous wildlife! Apart from scorpions, most of the animals are the undisputed world champions of their respective classes. This reputation may not be a selling point for our travel industry but most people, especially tourists, are fascinated by venomous creatures. While some people are terrified of snakes or spiders others take foolish risks with them; the latter are almost invariably male and stupid. It is best to admire these creatures from a safe distance and, in case of accidents, know about the correct first aid (see Appendix).

I once estimated that in Australia there are more than 365 types of venomous animals capable of causing a medically significant illness. Admittedly some on the list like midges and march flies are not truly venomous but have a highly irritating saliva. Most of the others are about as venomous as one can get. On land there are the world's twelve most venomous snakes, the deadliest spider (the Sydney Funnel-web), the most dangerous tick (the Australian paralysis tick) and the only venomous mammal (the platypus). The Red-back spider appears to cause more frequent and severer poisonings than its American cousin, the Black Widow. Residing in the sea is the world's most venomous jellyfish (the Box jellyfish), the potentially lethal Stonefish, Conus Shells and, of course, the Blue-ringed octopus.

All these venomous animals have killed people but there is another side of the coin. Their venoms, and those of many less dangerous species, contain substances of great interest to medical science and we all believe that novel new medicines will result from research into them. Thus, venom research has two objectives: it can improve the treatment of a particular poisoning and sometimes produce unexpected but much wider benefits.

In 1970, having learnt some of the tricks of the trade, my little team expanded our venom research, encouraged and supported by my immediate superiors. Allen Broad and Alan Coulter, my principal biochemists, also became close friends and stuck by me through thick and thin. Good ideas were forthcoming from other scientists who were engaged in non-venom-related activities in the Immunology Research Department. In particular, Kevin Fahey, Geoff Shellam, Howard Chandler, John Cox and Robert Premier deserve a special mention.

This chapter describes some fairly straight 'up and down the pitch' research, mainly carried out from 1970 to 1980. There is nothing brilliant about it, but it was fun and led to some fairly practical results. Much has yet to be done in all the areas looked at.

There is nothing 'cushy' about good research and my team worked their butts off. I salute them.

Some Careful Planning

Although some subjects which attracted my attention at CSL, like the octopus work, came out-of-the-blue, the studies related to snake bite certainly did not. After we had carefully reviewed many aspects of snake bite, a number of avenues were selected to be explored and then systematically pursued. From 1970 to 1980 research into snake venom and snake bite occupied far more of my time than the Funnel-web spider work. Whereas the outcome of the latter project was uncertain to the end and associated with problems of venom supply, the snake bite project offered some potentially achievable and exciting early results.

Although thousands of Australians suffered snake bite each year modern researchers seemed to have neglected it. The more I read and heard about the subject the greater my fascination became.

After many discussions with some 'old hands' such as Drs John Trinca and John Graydon I became reasonably well informed. These two doctors generously and patiently imparted their knowledge of snake bite, enabling me to develop an understanding of its complexities and manifold variations. During 1970 I spent hours with Dr Graydon, who at that time had been at CSL for forty-five years, trying to tap in and absorb his unique familiarity with unpublished CSL research (both successful and unsuccessful). I was told of the notable cases of snake bite and reactions to antivenoms which had influenced CSL's thinking over the years and John with his quiet self-effacing reminisences also brought historical figures like Dr Charles Kellaway to life.

There were so many exciting aspects to explore it was hard to know where to start! Indeed, snake bite alone seemed a veritable gold mine of fascinating challenges. Dozens of questions awaited answers and CSL was the logical organisation to start seeking them.

The main avenues of research proposed in 1970 were as follows:

* explore the composition of snake venoms
* develop methods to detect snake venoms
* investigate first aid
* try to reduce reactions to antivenoms
* analyse reports of bites and stings and publish updated advice on patient care

The cost was relatively minor considering CSL's overall expenses and, like all other venom research done over the next ten years, it was funded directly by the Federal Government under Section 19 (b) of the CSL Act, as work considered to be 'in the National Interest'.

With the fulsome support of my superiors the project got underway. Sadly, ten years later when most of the major goals had been achieved,

John Graydon was dead and most of my other sources of guidance and inspiration had also died or retired. Thereafter, I was the only doctor at CSL especially interested in the 'venom diseases'. (I stuck out like a sore thumb.)

Let us now look at some of this work. I will try to hold the reader's interest by focusing only on the highlights.

Isolation of Snake Venom Components

The toxic proteins in Australian snake venoms had barely been investigated in modern times. Since techniques were now available to isolate and characterise the main toxins, we thought such a study might improve antivenoms and further clarify the actions of the individual toxins.

Alan Coulter first tackled the isolation and characterisation of the main neurotoxin in Mainland Tiger snake venom. A skilled and meticulous biochemist, Alan soon had a small amount of the main toxin in a pure form. Just as he was finalising his paper a Swedish group led by Karlsson beat him to the post and published details of this isolated neurotoxin which they named 'notexin'. However, despite the disappointment of coming second, everything worked out for the best. We had sufficient supplies of our own notexin, which in fact proved purer than Karlsson's, for all our experimental work. The experience Alan had gained proved invaluable as he moved on to isolate the main neurotoxin from the venom of the common Eastern Brown snake. This toxin, which he called textilotoxin, was far more difficult to isolate but Alan made sure he was not beaten to publication this time! The toxin that he isolated remains the most lethal snake neurotoxin yet discovered in the world. In addition, Alan built up stocks of some lesser but still clinically very important snake neurotoxins. He also prepared in rabbits antiserum to many of the neurotoxins which proved invaluable tools in venom detection and other studies.

Detection of Snake Venom in Clinical and Experimental Samples

Our understanding of snake bite would be vastly expanded if we could accurately measure venom and its components in the serum and tissues of animals and humans. If nothing else it would assist the Coroner in the investigation of real or possible snake bite deaths. A couple of methods were in use but they were relatively insensitive. By good fortune and hard work, CSL was soon to be a leader in this field which is important as each year more than 100 000 people die from snake bite throughout the world. The evolution of the successful assay makes an interesting story.

One day in 1969 I wandered over to Prince Henry's Hospital to hear a lecture by Dr Kevin Catt. He was to talk on a method of separating and identifying proteins which I hoped to set up at CSL. I was comfortably settled in the middle of the audience when the chairman got up and apologised for circulating the wrong title for Dr Catt's talk. Instead of hearing about protein separation I would be enlightened on radioimmunoassays. This was a subject of little interest to me, but I was trapped so I resolved to spend the next hour in peaceful meditation on other matters.

Fifty minutes later I was an ardent fan of radioimmunoassays. I thought they should be set up at CSL, since they could not only detect minute quantities of antigen, but also measure very low antibody levels. Furthermore, large numbers of assays could be carried out relatively cheaply with the use of fewer animals. (Later I learned that Dr Rosalyn Yallow, who was co-discoverer of the radioimmunoassay for insulin detection, had her first major paper on the assay rejected by an important endocrine journal. Despite the evidence presented, the journal refused to believe such low levels of insulin could be detected. They must have been red-faced when later she shared the Nobel prize for this research.)

On my return to CSL I submitted an enthusiastic report outlining possible applications of radioimmunoassays (RIA). Soon there was collaboration between CSL and Dr Catt's group. Some of our staff were trained in his laboratory and on his behalf we undertook the immunisation of rabbits to raise antibodies to digoxin and other substances. Tony Collins, in particular, did some very good basic work on tetanus antitoxin measurement.

Soon more and more RIA work was being done at CSL. Fortunately at this time, a need arose to establish an assay for the measurement of anti-D immunoglobulin which was a new preparation designed to reduce the occurrence of haemolytic disease of the newborn. Since this rather complex RIA had to be carried out regularly it gave impetus to the establishment of a proper radioisotope laboratory at CSL.

By 1971 the isotope laboratory was up and running with Alan Coulter in charge. His assistant was initially Mr Bruce Roberts and, thereafter, Rodney Harris. I cannot speak too highly of the quality of work carried out by Coulter and Harris. They were a pair of quiet achievers.

Alan's initial work on venom detection used methods pretty similar to those of Dr Catt. We coated the inside of polystyrene test-tubes with various antivenoms. After the tubes had been thoroughly washed the sample to be investigated was reacted with the thin layer of antivenom attached to the inside of the tube. A little later the tube would be washed out and radioactive venom added for a time, then the tube would be washed out again.

If there was no venom in the sample there would be plenty of antivenom molecules available to react with the radioactive venom and thus a high radiation count would be found. The more venom there was in the sample, the less radioactive venom would stick to the walls of the tube and the lower the radiation count would be.

Later Alan progressed to far more sophisticated assays which had greater sensitivity. His assays became so sensitive he had no trouble detecting levels of venom as small as one thousandth of one millionth of a gram. Looking back, we were very lucky that right from the start, we obtained clear cut and encouraging results.

Finding snake venom in dead children
The first medicolegal application of the new technique occurred at the end of 1972. On 11 November at Jandacot, WA, two-year-old Belinda Smith was seen playing with what appeared to be a piece of orange rope. Soon after, she complained of having been bitten and four marks were seen on her forearm. Her mother bathed the arm and a few minutes later Belinda began to vomit. In the meantime her father searched for the 'rope' but could find no trace of it. She was taken to hospital and kept for two and a half hours. Because it was considered that she showed no evidence of envenomation she was then sent home where she was put to bed. Her mother checked her at intervals throughout the night as instructed by the hospital and at 7.30 am on 12 November she found her comatose. Belinda was immediately taken to hospital but failed to respond to treatment and died at 8.00 am. Post mortem blood and tissue samples were collected and forwarded to CSL at Melbourne for examination.

The gang in immunology research always made a big thing of the departmental Christmas lunch. We worked on the principle that if you were going out for a long lunch then the earlier you started the better, since early departure is less obvious than the late return of a whole department. Near noon on this particular day we were sorting ourselves out as to who needed a lift and who didn't.

Alan Coulter was delaying things and clucking around, muttering something along the lines, 'There's definitely venom in these samples, there's no doubt about it, most definitely.' 'Which one, Al?' I asked. 'The ones from the little girl in Perth.' Alan pressed a couple of buttons on the automatic radiation counter so the samples and controls could be recounted whilst we had lunch.

At lunch we discussed all the exciting possibilities at hand. We could perhaps prove not only that a person had died of snake bite, but also detect the species of snake and actual quantities of venom. Much could also be done with samples from snake bite victims who had been promptly treated.

By comparing the old and new counts of the dead child's samples on our return, it became clear that Alan was absolutely right and this infant's tissues were saturated with Tiger snake venom. We could tell the Coroner the exact amount of Tiger snake venom in the post mortem blood. Thus we proved death was due to snake bite, the type of snake responsible and the extent of envenomation. This was the first time it had been done anywhere in the world.

At the conclusion of his inquiry, the Coroner issued a strong warning that hospitals should take far greater care in the management of real or suspected snake bite cases. Details of Belinda's death and the Coroner's warning received nationwide publicity. From that time onwards most hospitals introduced protocols for the management of snake bite.

Many tragic snake bite deaths and near misses were investigated by radioimmunoassay. For example, when a nine-year-old boy from Moe, Victoria, who had wagged school was found the next day trapped beneath a barbed-wire fence and profoundly comatose Alan was able to prove his subsequent death was due to a Tiger snake bite.

A seven-year-old girl at Delagate, New South Wales, was bitten by a Tiger snake while she was in bed asleep. Because of the circumstances the diagnosis of suspected snake bite was made late but she survived after extensive antivenom therapy. Tiger snake venom was found in her samples when they were analysed at CSL. This case was one of the many which stimulated the development of venom detection kits for bedside use (see page 241).

For a while, the number of snake bite deaths appeared to increase, but this was a mere reflection of our ability to prove snake bite as the cause of death. Overall there was a dramatic fall in the death rate from snake bite. A combination of coronial warnings and the publication of instructive cases made doctors aware how dangerous snake bite could be. Literature on the treatment of snake bite poured out of CSL and doctors seemed to take a greater interest in the management of snake bite as well as other envenomations.

Professional obstruction?

After a year or so the assay was working well and a number of human samples had had their venom contents accurately estimated. At this time there had only been one published report about venom being detected in a human snake bite victim. In America back in 1957 a woman had been left in charge of a reptile park and told to keep away from the King Cobra. For some reason she failed to do so, and in no time at all had died from overwhelming poisoning by this giant snake. Samples taken from one of the bite sites reacted strongly with antivenom in a simple diffusion experiment. This method and others given in earlier literature had not proved suitable for the Australian situation. Professor Findlay Russell in the United States had described one method in which red cells were specially coated and gave a reaction in the presence of venom. We tried this method but found it too insensitive.

In due course Alan and I submitted a paper on the detection of snake venom using the radioimmunoassay to the journal *Science*. The paper was rejected, the editor regretfully explaining that he himself was happy with it but the journal's policy was that if either one of two referees was not strongly in favour a paper would be rejected. Copies of the reports by the two anonymous referees were enclosed.

One referee's report was most effusive, praising not only the scientific merit of the paper but the obvious applications it had to the whole field of human envenomation. The other was unjustifiably and extraordinarily critical. It had been typed on a typewriter with which we were familiar and comparisons with letters from Professor Russell left no doubt as to who the second reviewer was. Professor Russell, or 'Fin' as he is known to all and sundry in the field of venoms, had co-founded the International Society of Toxinology and the Journal *Toxicon*.

We sent the same article off to the *Journal of Immunological Methods* which accepted it immediately.

Investigation of First Aid

At this time the first aid recommended for snake bite by CSL and other organisations was unsatisfactory but it had not been the subject of research for many years. This advice was to use an arterial tourniquet, which was extremely painful, potentially dangerous and could only be used for a relatively short period.

Although I had given up Cubs early because I couldn't stand the noise (page 25), they had effectively trained me to deal with snake bite. Like practically every Australian I knew that snake bite was almost invariably fatal unless the fang marks were slashed with a razorblade, the venom sucked out, Condy's crystals rubbed into the wound and the blood supply cut off to the region by the use of an arterial tourniquet.

'Slash, suck and bind' were the Karma Sutra-like bywords. The literature abounded with variations on these techniques, including reports of the amputation of fingers, large blocks of tissue and, in one case, the victim's penis. Sometimes gunpowder was poured onto the bitten region and ignited and once a gun was used to blast the bitten region out of existence. The snake often fared little better and frequently suffered a ritual beating, its only consolation being that it would sometimes score another victim before escaping or being bludgeoned to death.

When I took the time to look at the facts about snake bite, a few things became quite clear. First, many cases of snake bite did not develop any signs of general poisoning; they were actually dry bites with insignificant amounts of venom being introduced. My conclusion was that the first aid often caused more damage than the snake! Nerves, tendons and blood vessels were often cut and the wound contaminated by bacteria.

Second, because of the method of venom delivery such invasive methods were unlikely to remove the deposited venom. Australian snakes have fine pointed fangs and the venom is released from the tips of the fangs often under considerable pressure. The fangs allow the venom to be delivered in a similar fashion to a doctor injecting, say, tetanus vaccine. Being deposited quite deeply, it is not a practical proposition to use an excision to try to extract it before its absorption. Furthermore, sucking the bite site is most unlikely to remove venom because of the fineness of the path made by the fang.

The use of Condy's crystals was really a hangover from the desperate days before antivenoms. Although the crystals can destroy snake venom when mixed with it in a test tube, in human tissues all they produce is death of tissue and large slowly healing ulcers.

Often it was recommended that the bite site should be washed thoroughly. Even this seemed pointless since snake venom could not penetrate the skin. (Nowadays washing the bite site is strongly discouraged because doing so may reduce the chances of obtaining a positive venom detection swab.)

The second stage in the old type of first aid was the application of an arterial tourniquet to 'the first single bone part of the limb between the bite and the heart'. This was usually the upper arm or the thigh and the instructions were to release the tourniquet for a few minutes every twenty minutes to prevent loss of the affected limb by gangrene.

Applying an arterial tourniquet effectively to cut the blood supply off to a limb is sometimes fairly difficult to do. This is particularly so with the male leg, as the thigh can be very large and muscular. Arterial tourniquets also produce injuries. Apart from the development of gangrenous limbs, some of which require amputation, nerves and blood vessels are often damaged.

One aspect of arterial tourniquets, which the first aiders never appeared to consider, was the extreme discomfort which may develop. When a doctor takes your blood pressure using a large, comfortable inflatable cuff, it is moderately uncomfortable after a minute or so. Imagine an uncomfortable tourniquet in position for twenty minutes. At the beginning of the project, we applied these comfortable blood pressure cuffs to ten 'volunteers' amongst the staff and inflated them sufficiently to cut the blood supply off to their arms. All but one withdrew their voluntary participation in this exercise within ten minutes. Only the stoical Allen Broad outlasted them and fifteen minutes was his limit. This little experiment provided pretty convincing evidence that an alternative to tourniquets should be sought.

Another important factor was that, since tourniquets could of course only be used for twenty minutes at a time, then we could expect a great whoosh of venom to occur when the blood was allowed suddenly to flood back into the limb. Indeed instances had been described when tourniquets had been used to manage Funnel-web spider bites, that collapse occurred within a minute of the release of the tourniquet.

It should be pointed out that in certain operations on limbs the surgeon requires a 'bloodless field'. In these cases the patient is anaesthetised and so feels no pain and the tourniquet is applied in such a fashion that blood is squeezed out of the limb prior to the start of the operation.

A new type of first aid for snake bite
Since Alan Coulter could now measure snake venom and venom components in serum samples we could start looking at how venom circulated in the body. Once we knew how it moved we could inject snake venom into the leg of an experimental animal and apply various first aid procedures to determine which was the safest and the most efficient.

Ideally, we wanted an experimental animal which, for example, would closely mimick the situation seen when a child suffered a snake bite. We wished to monitor the early stages of snake venom poisoning, collect as much information on the effects of the venom as possible and reverse the effects with antivenom early in the syndrome. No such comprehensive study had been carried out before, and there were many gaps in our knowledge about the specific effects of various venoms.

As luck would have it, there were a large number of monkeys at CSL which had been used for yellow fever testing and therefore were not suitable for other virological studies. I spent quite some time sitting in the Monkey House watching these little monkeys (average weight 2.5 kg) jumping around in their cages whilst working out the most comfortable and gentlest way of using them for our first aid studies. The main problem was that usually we could not use anaesthetics over the two-hour experiment, because a child bitten by a snake is not anaesthetised. Our model had to be wide awake and able to move any unrestrained limbs. Apart from minimising the discomfort the monkey might experience, it was important to consider the safety of the staff.

After measuring a few monkeys, one of my daughter Susie's dolls was selected as a model for the construction of a pair of heavily padded timber frames. The most important feature of these frames was the upper part which, like a mediaeval stock, gently but firmly enclosed the monkey's neck so its teeth were active only within a prescribed range. The monkey's limbs and trunk were immobilised as required by straps and buckles attached to the frame. Plenty of foam rubber packing was used between the monkey's body and the straps to minimise discomfort.

There was a good omen at the very start of the research into first aid. Late in the afternoon of 2 December 1977 I was busy at work in my back shed constructing a monkey restraining-frame. A phone call interrupted my hammering and a nice lady from the Australian Medical Association informed me I had been awarded a research prize for my work on venoms. It was called the Triennial 1935 BMA Prize. She wanted to know if I would be available to receive it the following Wednesday night. I said that would be fine and in a rather jaunty frame of mind resumed hammering and banging in the back shed. That's amazing news, I thought. It was even more amazing that a year later, almost to the day, details of a new type of first aid would be announced.

We had to take many small blood samples from each monkey during the course of the experiment, so we adopted a routine hospital practice. A fine flexible tube was introduced into a vein in the monkey's left leg and run upwards until it reached one of the larger central veins. The left leg was padded and left undisturbed during the two-hour experiment. Via this tube, samples of blood could be taken at any time without causing the monkey any distress. Antivenom was also administered through the tube when the monkey showed early signs of being affected by the snake venom.

The right leg was the site of injection of snake venom and also, in most cases, the leg subjected to first aid measures.

Initially we worked only with Tiger snake venom and injected 0.3 mg of this venom into the lower calf of the monkey's right leg. The needle included a special guard so that it penetrated to the same depth as an average Tiger snake bite. The venom dosage was selected because it would produce symptoms in the monkey by 120 minutes. This meant that antivenom could be given about this time to reverse these early symptoms and the monkey would fully recover.

Preliminary work in December 1977 showed that the model was quite promising. When venom was injected into the monkey's leg and no first aid measures were applied, we could detect venom components in the animal's circulation within three or four minutes. After sixty minutes the circulating venom reached its highest levels. We could soon produce graphs giving a clear picture of the pattern of venom levels in the absence of first aid. In addition, small blood samples allowed us to study the effects of the venom on blood clotting, red blood cells, etc. In collaboration with the Royal Melbourne Hospital, batches of monkey samples ran through all sorts of posh automatic machinery. This gave us a vast amount of information from small samples and single monkey experiments.

Since the first frame worked satisfactorily a second one was constructed and thereafter all experiments were done as paired studies. One monkey would receive some type of first aid whilst his or her companion received none. By June 1978, a variety of different first aid methods had been studied. Short term use of a 'reasonably comfortable' arterial tourniquet showed that all the venom remained in the monkey's limb until the tourniquet was released. We studied the role of anaesthesia on venom flow, and also the application of uniform external pressure to the limb. A special chamber had to be tailor-made to perform the latter experiment. We also looked at 'a loose venous tourniquet', which did not cut off the blood supply to the limb but slowed the movement of blood in the veins. None of these procedures alone effectively slowed the escape of venom from the bite site into the bloodstream.

I had 'a bee in my bonnet' about inflatable air splints because they were new and most ambulances carried them for immobilising injured limbs. They encased the damaged limb, were zippered up and inflated to the required pressure to give firm support. However, a few experiments with monkeys showed that they didn't prevent venom from getting into the circulation.

27 April 1978 was a red-letter day although we were not to know it at the time. The dictated summary notes read as follows:

> Monkey No. 11, weight unknown at this stage, female. The monkey was set up as was the previous one and received a similar dose of venom. Immediately after the injection of venom the envenomated limb was firmly bound by a crêpe bandage from the bite site upwards, it was then immobilsed with two boards packed with foam rubber and bound firmly still.
>
> Sixty minutes later further blood samples were taken and then all restrictions on the limb removed. Five minutes later another blood sample was taken for radioimmunoassay.

A week later Rodney Harris wandered into my office and said that 'Monkey Eleven's quite interesting'. And interesting she was. At fifty-eight minutes after the injection of venom there was no venom detected in the serum. However, at sixty-nine minutes the level was 172 ng/ml and it subsequently rose higher. This was quite a surprise. Firm bandaging and splinting of the limb seemed to have retarded venom movement completely.

Monkey 11 became the first snake bite 'victim' to have the movement of venom retarded by what became known as the 'pressure-immobilisation' type of first aid.

We tried another crêpe bandage and immobilisation experiment and found it worked just as well as on the first occasion. Two weeks later we repeated it, this time only bandaging below the knee and obtained another good result. It appeared that we now had a type of first aid which was effective for at least sixty minutes and caused no injury to the limb nor apparent distress to the 'patient'.

A number of obvious questions arose which required answers quite urgently and the extra experiments and hundreds of associated assays had to be fitted in with other work. All day we would hike backwards and forwards from the Monkey House at the bottom of the hill at CSL up to the main laboratory. A team member always seemed to be in transit from the two sites and morning tea and lunch became mere memories.

This work was challenging enough without a sudden bureaucratic disruption to the monkey supply which occurred in July 1978. (See page 258)

Early questions posed by the new first aid

We attempted systematically to seek answers to the following questions:
What happens if a larger dose of venom is given?
A monkey was injected with 3 mg of tiger snake venom, an amount sufficient to rapidly kill sixty monkeys and ten times the amount given in the previous experiments. The first aid measures were applied for sixty minutes, during which time the monkey remained perfectly normal. After removal of the splint and bandages the venom levels in the serum rose very rapidly, antivenom was given and the monkey survived. This experiment with a high venom dose was only done once. We followed Sir Macfarlane Burnet's adage 'never repeat a successful experiment'.
Is venom destroyed whilst trapped in the tissues?
In the case of snake venom we found the main paralysing toxin was not altered while it was trapped and it would attack nerve tissue as soon as it was released. (Funnel-web venom, however, see page 198, was a different story.)

There was some evidence that the snake venom component which caused disturbances in blood clotting lost some of its activity whilst the first aid measures were in place.
Would the first aid work with other snake venoms?
At least fifteen other types of Australian snake venoms, including sea snake venom, were studied in the monkey model and it was found the first aid measures invariably delayed their movement. Within a year extensive publications had appeared describing the results obtained with every monkey used.

We also examined some exotic snake venoms. We were able to show that this first aid was highly effective in retarding the movement of the Indian cobra venom. Since this snake killed ten thousand people in India every year we published our finding in the *Indian Journal of Medical Research*. (We were tempted to suggest turbans might be an alternative to crêpe bandages!) Our work on rattlesnake venom was published in the United States with the intention of alerting clinicians to the new technique.
Was the first aid applicable to bites by the Sydney Funnel-web spider?
In January 1980 a two-year-old boy died after being bitten by a Funnel-web spider. That month preliminary experiments with Funnel-web venom in monkeys suggested that not only did the new first aid measures delay the onset of general poisoning by this venom but the poisoning seemed to be markedly reduced by delaying venom movement (see page 198).

By May 1980 collaborative work at the Children's Hospital had clearly shown the first aid measures to be highly effective and, at least in monkeys, some local inactivation of the venom appeared to occur. On 18 October 1980, Alan Duncan, Jim Tibballs and myself published a paper in the *Medical Journal of Australia* entitled 'Local inactivation of funnel-web spider

(*Atrax robustus*) venom by first aid measures: Potentially life saving part of treatment'.

Would the first aid delay the movement of low molecular weight toxins?
The lethal components in snake venoms are fair sized proteins of molecular weight from 7000 upwards. We thought smaller molecules might pose more of a problem. It was a simple enough experiment to inject a little radioactive iodine into a monkey and monitor its absorption. We were frankly surprised to find the movement was markedly retarded. If iodine, with a molecular weight of 125, was retarded then it was most likely that the toxin of the Blue-ringed octopus (molecular weight 319) would also have its movement slowed. This finding suggested the first aid would effectively slow the movement of all toxins. This was great news indeed!

How does this first aid measure work?
When venom is injected it is a blob of liquid which had been pushed into a space made in the tissues by the sheer pressure of its delivery. This is similar to a vaccination shot which is delivered by hydraulic pressure originating from the pressure of the doctor's thumb on the syringe plunger.

The different components of snake venom have various targets around the body and to reach these effectively they have to get into the circulation. They can do this by either of two ways. One is by slowly spreading through tissue planes and the other is by entering a tube in the region through which fluid is flowing such as lymph vessels and blood capillaries.

Some evidence suggests lymph vessels are the preferred option taken by the venom. The flow of liquid through lymph vessels is very dependent on the activity of the muscles of the limb. However, after the lymph channels combine, the venom-containing lymph enters the bloodstream near the heart and is immediately spread far and wide. The venom components thus arrive at their various targets.

The pressure in the lymph channels is quite low. If muscles are restrained from moving by use of a splint and firm pressure is applied over the bitten area, then very little fluid will flow through the lymph channels. They will also be partially collapsed by the direct pressure. Local pressure and reduced limb activity also markedly reduce the flow of blood through capillaries around the deposited venom. The blob of venom will thus find itself in a situation not unlike a commuter when a major strike has stopped most forms of public transport. It won't be going anywhere fast!

Promoting the pressure-immobilisation type of first aid
By September 1978 sufficient work had been done to seek publication of the experiments and to consider promoting this new type of first aid for snake bite. A lengthy paper was sent to *The Lancet* and an abstract of

the findings submitted to the Australian Society of Clinical and Experimental Pharmacologists to be presented early December.

The publicity generated by the presentation of our work on monkeys at the December ASCEP meeting was astounding. It might have been because there was little news around at the time—or that it coincided with a number of snake bite deaths—but every paper seemed to carry reports of the new technique. On a front page spread in the Melbourne *Age* the procedure was described as 'revolutionary' and, much to my embarrassment, a statement was attributed to me as comparing the breakthrough with the discovery of gravity. My friends knew what I meant by this statement and most people got a good laugh out of it. Even Neville McCarthy rang me and sent a high pitched cackle down the line for good measure.

At the same time we were considering the best way to inform bushwalkers, boy scouts, etc., of the technique. An eight-page booklet, *First Aid for Snake Bite in Australia,* was published in February 1979 and after many revisions and reprints is still in print. (The latest edition is available from the Australian Venom Research Unit at the University of Melbourne, CSL or the World Wide Web. It is also reproduced in a modified form as the Appendix.)

The method was frequently demonstrated for television channels and the Australian Information Service despatched an illustrated summary of the procedure to many overseas outlets. It was fascinating to see how some of these were published. AIS sent me press cuttings from dozens of countries mostly in languages I could not understand. The media attention was intense and every opportunity to push the new method and condemn old procedures was grasped.

The Sunday after the ASCEP meeting, I was strolling across the back lawn past my wife Megan, who was sunbaking, when I spotted a small black dot on her leg. She said she had only noticed it for the first time a week before. Two operations were performed over the next three weeks and the poor girl was left with a dreadful scar and never sunbaked again. To our relief there was no recurrence of this particularly dangerous cancer.

This put a dampener on our annual Christmas Street Party. The previous year fancy dress had been the order of the day. I was coerced to introduce a little culture so I dressed as a ballerina clad in football boots. When I appeared as a bearded vision in a pink tutu my pirouettes drew great applause.

In January 1979 *The Lancet* published the article by Alan Coulter, Rodney Harris and myself entitled 'The rationalisation of first aid measures for elapid snake bite'. This took the wind out of the sails of the odd critic who popped up here and there and was influential in various national bodies backing the promotion of the new first aid.

On Saturday, 14 July 1979, Mr Doug Leslie, Chairman of the Australian Resuscitation Council, invited me to demonstrate the technique before his Council. This was done and subsequently the Council fully endorsed the method. As a flow-on from this and other meetings, the Australian Government banned the sale of the old style snake bite kits which contained blades, Condy's crystals, etc., in July 1980. The importation of basically useless suction devices, etc., from various overseas countries was also discouraged.

In early 1981 Geoff Maslen of *The Age* decided the Education Section should promote the new type of first aid. *The Age* held a song competition and in due course the judges assembled in a corner of *The Age* building to decide on the winner. The judges were myself, Geoff and *The Age* music critic Kenneth Hince. Mr Hince gave the impression of taking himself quite seriously. It was therefore quite a joy when Geoff and I convinced him that, since he was the music critic and we had no musical instruments available, he should whistle each tune. Admittedly he did this quite well. Judy McKinty's 'Snake bite Sam' was the winner and eventually a launching was held at the Child Accident Prevention Centre in Melbourne. Singer, Mr Henri was backed by a mob of madly enthusiastic school children and Dr Jim Tibballs gave a masterly demonstration of the First Aid.

There was increasing interest in the application of the Australian type of first aid to snake bite in overseas countries. We did not actively promote the technique for overseas snake bite, since many of their venom's actions are different to those seen in Australia. Nevertheless, after the publication of the work on rattlesnake venom and case reports of its use in humans, a number of countries have adopted the technique. In Papua New Guinea, Rotary sponsored the printing of posters in two languages demonstrating the technique. It became the recommended procedure in South Africa and has its advocates in many other countries. This type of first aid appears to be slowly taking root in the United States of America where unfortunately the management of snake bite is somewhat confused, controversial and not monitored by any national authority.

It has been grand to read case reports where the first aid measures may have been life-saving. Sometimes they have had to be kept in place for six or more hours because of the remoteness of medical care. Despite this duration, venom has remained trapped at the bite site and the patient has remained well. A number of snake bite victims became ill when they eventually reached hospital and their first aid measures were removed, but they all responded well to antivenom. In one remarkable case a young boy was bitten by a King Brown snake in a remote part of Queensland. Having no crêpe bandages, his quick-thinking mother improvised with panty hose which proved highly effective.

Some venoms cause a little local damage when trapped in the tissues for a long time by this method. This is usually only a minor problem with Australian snake venoms but it is more common with the venom of many overseas snakes. However, it is preferable for the patient to be alive next day with some local damage at the bite site than to be dead!

Reducing Reactions to Antivenoms

In their own quiet way, antivenoms can be truly miraculous. Given promptly and in sufficient quantity, they will usually prove life-saving. Antivenoms have been around for a long time—since 1894 in fact. That year the Frenchman, Calmette, immunised a horse with snake venom and later collected some of its serum. The serum was found to contain antibodies to the venom and was successfully used to treat some snake bite victims. The secret was to start the horse with a tiny dose of venom then successively increase it. For years horses were also used to raise tetanus antibodies which were often routinely given to patients with tetanus-prone wounds. Many Australians received anti-tetanus shots made in horses until they were phased out in the sixties.

Life-saving these antibodies might be, but they can be associated with some nasty reactions. Fortunately, the side effects became rarer and usually milder as manufacturers improved the methods used to extract the antibodies from the collected horse serum.

The adverse reactions were basically of two types: immediate or delayed. The immediate ones saw the patient's blood pressure plummet, often with other features of a severe allergic reaction such as breathing and swallowing difficulties, generalised rashes, etc. Death was not uncommon, even after receiving tiny amounts of the horse preparation. The slower reaction, often called a delayed serum reaction, set in about seven days after the injection. A widespread itchy rash and enlarged lymph nodes made the patient very miserable indeed. Fortunately, after a few days of extreme discomfort, complete recovery was the norm.

When I came to CSL in 1966 I was scared of antivenoms because of their reputation for side effects. Only once had I heard them mentioned at Medical School—I was told how in casualty at the Royal Melbourne Hospital a symptomless snake bite victim was given Tiger snake antivenom, immediately collapsed and remained unconscious until the next day. A far worse case occurred whilst I was living in Frankston. A fourteen-year-old boy had died within minutes of receiving Tiger snake antivenom. He was showing no evidence of poisoning but the doctor, who was one of those for whom I did locums, decided to use antivenom because he could see fang marks. (Only in about one in ten cases has the snake injected sufficient venom to cause an illness.)

To reinforce my prejudice against antivenoms, I had read that the only snake bite related death that had occurred in the United Kingdom during the previous hundred years had been a result of an anaphylactic (acute allergic) reaction to a French antivenom.

All this had tended to push me towards the body of opinion that antivenoms are a case where the 'cure is more dangerous than the disease' and to steer clear from responsibilities involving CSL's antivenoms. Little did I know what lay ahead! Within a few years I had become CSL's principal consultant on antivenoms and commenced a lifelong battle against that 'body of opinion'. In summary, my views are that when an antivenom is clearly indicated it should never be withheld. Since 1976 there have been protocols described which allow their safe administration. Let us look briefly at the evolution of these protocols. (It is described in greater detail in my book *Australian Animal Toxins.)*

During my first few weeks at CSL Dr Trinca, the Deputy Director, put me on a fast learning curve about the establishment's products. Antivenoms were high on the list since all the establishment's doctors had to be able to advise outside colleagues about these and other agents likely to be used in an emergency. (I will return to the problems I found in this self-education shortly.)

Antivenoms were employed far more often than I had imagined. For example, more than three hundred patients received Red-back spider antivenom per year. Hundreds of ampoules of mainly Brown snake or Tiger snake antivenom save the lives of pet dogs and cats every summer. Information on the illnesses caused by the venomous creatures of Australia poured into CSL. Apart from telephone calls, there were letters, press cuttings and returned questionnaires which had been sent out with each ampoule of antivenom.

For decades CSL was the only antivenom manufacturer in the world to have monitored the performance of its products by questionnaires. It became my privilege to analyse and often probe deeply into this unique collection of more than 25 000 case reports.

I was told CSL antivenoms were amongst the world's best. Although they caused a reaction of significance in about 7 per cent of patients the rate was higher than 40 per cent for some foreign antivenoms. The variation in the size of antivenoms was striking. To cater for the potential output of certain snakes, some of CSL's antivenoms were whoppers. An ampoule of Taipan antivenom held a massive 40 ml of a 17 per cent protein solution. The biggest overseas one I could find was a South African one which weighed in at 10 ml of a 10 per cent protein solution. Some commonly used Australian snake antivenoms were quite small, for example an ampoule of Brown snake antivenom contained less than 5 ml.

My predecessors at CSL had given the reaction rate considerable thought. They had decided that skin testing for allergy to horse protein before antivenom therapy was unreliable and delayed urgent specific treatment. Although this practice has been abandoned in this country for over forty years some overseas doctors still do skin testing. In the fifties CSL was recommending that an antihistamine precede the antivenom to dampen any allergic reaction. This instruction is still given today and, although the usefulness of antihistamines in this role is limited, it seems wise to continue to give them, at least for the present.

Studying the more serious reactions I was particularly struck by some that had occurred in children. There was no way they could have developed an allergy to horse protein as, unlike many adults, they had never received the tetanus shots made in horses—these had been replaced with ones made from human blood before these particular children had been born. There had to be a factor other than allergy at work in at least half of these severe reactions.

As the years went by I continued to worry about the adverse effects of antivenom. They reminded me of Russian roulette. Most problematical were the large volume snake antivenoms which were given intravenously. Time and time again I've been rung up about patients desperately sick from snake bite who had gone into a shock-like state moments after antivenom was commenced. The doctor then had to deal with two potentially fatal conditions instead of just one. Often the doctor I spoke to was working solo in a remote small rural hospital. In these circumstances, with limited facilities, the hazards of receiving antivenom seemed especially undesirable.

In 1974 I visited the late Dr Alistair Reid at the Liverpool School of Tropical Medicine. Years earlier, from Penang, Alistair had spurred CSL on to make a sea snake antivenom which he successfully tested. I had greatly looked forward to meeting this renowned English gentleman and was not disappointed. We got on like a house on fire. I was puzzled why he was rather vague about the arrangements for the evening. In mid afternoon his wife arrived, inspected me and gave him a nod. He then announced I must dine at their home that night. A sensible process I thought, but when Alistair and I got going, we made quite a night of it. We were still talking shop at 2 am when his wife went to bed.

There were two results from our meeting. We agreed to a basic agenda for a WHO meeting on antivenoms which, when it was held in Zurich in 1979, accepted our proposal regarding the term 'antivenom'. (Alistair and I had agreed it was less confusing then 'antivenene' or 'antivenin'.)

The other result was that I came away convinced that there was a mechanism other than allergy which caused reactions to antivenoms. Alistair had just returned from working in Nigeria where he had found the rate of severe reaction to be at least 6 per cent among the indigenous population. He believed it very unlikely they had any previous exposure to horse proteins. This was similar to the experience of Dr Charles Campbell in Papua New Guinea and my own observations of the younger Australians. The question posed was what were the mechanisms of these non-allergic reactions and, more importantly, how could they be reduced?

When I got back to CSL I imported samples of commercial antivenoms from nine overseas countries and along with the local preparations put them through a few hoops. They differed considerably, for example there was great variation in the purity and protein content. Particularly alarming was the discovery that some were heavily contaminated with unwanted horse proteins. The sample of the American Black Widow spider antivenom was especially poor in this regard. This surprised us because we had expected a higher standard, but it did suggest a reason for the strong reluctance of American doctors to use antivenom. Overall, this early analysis only indicated which antivenoms could be processed better. Then I got an idea.

The reader may recall that towards the beginning of my time at CSL I had been involved in the preparation of an intravenous human gamma globulin for immunodeficient children (page 171). One test done on this product measured its anti-complementary activity (ACA). The lower the ACA the better. It had been estimated by infusing the child with the preparation and measuring any fall in circulating complement. (I'll explain the term complement in a minute.) The trouble was the fall in complement was associated with a shock-like state in the child so I had had to become familiar with a laboratory method of estimating ACA.

Complement is a hideously complicated system of at least twenty plasma proteins and its description occupies a full page in my medical dictionary. It is a multifunctional affair, heavily involved in severe allergic reactions. Tampering with it can set off allergy-like reactions which are called anaphylactoid reactions as distinct from the anaphylaxis ones. Anaphylaxis, meaning 'without protection', is a purely allergic reaction requiring prior exposure, as seen with bee venom allergy (see page 310).

My idea was that, if antivenoms interacted with complement, they might set off an anaphylactoid reaction and this proved to be the case. We found almost every antivenom had a very high ACA, far higher than the most reactive human gamma globulin preparations. The highest ACA was found in the least pure of the overseas antivenoms. Other horse products like Gas Gangrene antitoxin had similar levels so it was not related to the venom used to immunise the horses.

The ACA activity of human globulin can be reduced by exposure to the enzyme, pepsin. However, snake antivenoms retained plenty of ACA despite a vigorous pepsin digestion during their manufacture. Thus, from a practical point of view, all the manufacturer can do is make the product as pure as possible and stress it be well diluted before infusion. Since 1975 CSL has recommended that antivenoms that are going to be given intravenously be first diluted 1 in 10 in an infusion bottle. This procedure has reduced but not eliminated reactions. Fortunately, greater safety was achieved by the introduction of a novel precaution before starting the administration of the diluted antivenom.

Premedication before antivenoms

A couple of horrific reactions to diluted snake antivenoms which really put the wind up me occurred in youngsters. Although children are less likely to die from severe allergic reactions than adults, these two went as close as you could get. They were only saved by prompt injections of adrenaline.

Brooding about these cases I wondered if a little adrenaline could safely be given before the antivenom to abort or lessen unpredictable reactions. Adrenaline is the drug *par excellence* for treating both anaphylactic and anaphylactoid reactions. The sooner it is given the better the chance of recovery. For many years CSL's antivenom leaflets had recommended that a syringe filled with adrenaline be on hand to treat bad reactions. Would it not be sensible/prudent to get some into the patient before a possibly highly dangerous reaction is triggered off by antivenom?

There was a precedent: premedication with certain drugs makes an anaesthetic both safer and smoother. It seemed that judicial use of adrenaline as premedication before antivenom therapy, if safe and tolerated by the patient, might have a lot going for it. I checked the idea out with several physicians and their opinion was favourable. The dose should be small and given by subcutaneous injection. The patient might experience an increase in the 'jumpiness' they already felt through the snake bite and they should be so warned.

This suggestion was accepted with the result that from 1976 CSL recommended that as well as diluting antivenoms 1 in 10, adrenaline should be given before their infusions.

Since that date, amongst thousands of patients, when the recommended premedication has been followed there have only been a couple of reports of severe reactions to antivenoms. When adrenaline is not used, catastrophic reactions occur in about 7 per cent of cases. There is currently some controversy about adrenaline premedication which concerns me but I will discuss it later (Ch. 10). Suffice to say, after reading thousands of case reports of antivenom usage, I would be frightened to give snake antivenom without premedication. It has removed my biggest worry about these 'magic bullets' and allowed me to sleep more peacefully at night. It has harmed no patient.

Surveys of Antivenom Usage

When I arrived at CSL in 1966 only about a dozen papers on snake and spider bite had emerged from that organisation since its foundation. The best of these were written by Dr Saul Wiener and included the only published survey of antivenom usage. Saul's 1961 study of 167 cases of Red-back spider bite recorded early experience with his new antivenom.

In 1978 John Trinca and I published a 1992 survey of 2144 cases of usage of Red-back antivenom from 1963 to 1976.

These two surveys provided sound information on many aspects of this important spider bite. Certain trends became evident. An example was the decline in bites to the genital area from 21 per cent to 10 per cent, presumably due to the increase of inside toilets. (In a 1992 survey the incidence was 2 per cent. Unfortunately bites due to spiders nesting in ear muffs and safety helmets had rocketed!) From these cases we could build a picture of the 'average' case and its course. Important also was the cataloguing of the rarer effects of the venom, like painful teeth and convulsions.

The conclusion was that Red-back spider antivenom performed well and was associated with few adverse reactions. Although it was made in horses, unlike snake antivenom it was a minute injection which was given by the intramuscular route. This was much safer than the mandatory intravenous route used to give snake antivenom.

I was concerned that follow-up information on other types of envenomation was lacking on almost all the cases reported to CSL. The antivenom usage form was generally filled in early and was often sparse in detail. There was no mechanism in place for monitoring long-term effects of either the various venoms or antivenoms. To shed light on the situation there seemed no option but to undertake a more thorough survey.

Merely sorting out the raw data for the Red-back survey had taken six exhausting weekends but fortunately my co-author, Erin Lovering, more than pulled her weight in our next study of antivenoms. It proved worth the effort.

Erin and I tenaciously followed every known case of all types of antivenom usage in Australia and Papua New Guinea from 1 July 1978 to 30 June 1979. The survey appeared in the *Medical Journal of Australia*, 29 December 1979. The results confirmed the relative safety of Red-back spider antivenom but drew attention to two serious disadvantages in the widely used polyvalent snake antivenom. This antivenom is used in most parts of Australia when the identity of the offending snake is unknown. One ampoule can neutralise the average output of venom from the five main species of snakes. Containing about 50 ml, it is the largest and most expensive antivenom in the world. Unfortunately, immediate reactions and delayed serum sickness were both more common with this antivenom. The rate of delayed serum sickness was an alarming 10 per cent.

Fortunately it was possible to ameliorate the situation. We recommended that, whenever polyvalent antivenom was used, it be followed by five days of steroid therapy. This short-term suppression of the immune response minimised the chances and severity of delayed serum sickness. Some doctors now consider it is sound practice to give steroids routinely except when the smallest volume antivenoms have been used. In the 1992 survey mentioned earlier, I found three cases of serum sickness amongst eighty-six treated snake bite cases. All had had large doses of antivenom—one man had received a possible record five polyvalent ampoules. None had received prophylactic steroids which seemed to indirectly support their use. The incidence of delayed serum sickness, as well as some other antivenom reactions, was further reduced by the introduction of venom detection kits.

Snake Venom Detection Kits

On 1 December 1979, the *Medical Journal of Australia* carried a letter advising readers of a free issue of venom detection kits for evaluation. These did not need the special equipment and staff required for the earlier radioimmunoassay. The kits relied upon antibodies linked to an enzyme rather than a radioactive isotope. Early on, I was of the opinion that such an enzyme-immunoassay would lack the desired sensitivity. Fortunately I was proved very wrong.

The purpose of the kits was to place into one of five groups any venom obtained by swabbing the bitten area. This would aid in the selection of the appropriate antivenom. Identification of the snake involved is often a problem in the 1 in 10 snake bite cases which need antivenom as the snake may not have been seen or the victim may be a youngster. Mistakes have often occurred, sometimes with fatal consequences when the wrong antivenom was selected.

Knowing the right antivenom to use also reduces the need for polyvalent antivenom with the danger of reactions or 'gunshot' therapy with mixtures of monovalent antivenoms. It puts the patient at less risk of side effects and saves the tax payer's money. There are other benefits: if the doctor knows the type of snake involved he or she will be better prepared to ward off particular complications. For example, Tiger snake venom can cause widespread muscle damage, a complication not reported after a Brown snake bite.

I'm still amazed that the manufacture and distribution of these first kits went so smoothly. There were two types. One was for Victoria and Tasmania. The other, more complicated one, was for the other states which are blessed with more species of snakes.

Alan Coulter did all the hard design work on the kit. Problems just seemed to melt away. Difficulties with the substrate were solved by Norman Ackland. We had a working bee assembling the 600 kits in our main lab. My son John, who was twelve at the time, was roped in to help. The printed instructions and attractive packaging arrived on time. Andrew Macintosh oversaw the national distribution and soon we were awaiting the response from the field.

Despite being a bit 'fiddly', the kit was a success. As far as I know it was the first diagnostic kit which could detect and distinguish between five groups of proteins (in this case venoms) yet could be carried out by doctors or nurses. It did not require prior laboratory training. The kit had some notable wins which quickly established it as a must for any hospital offering optimum care to real or suspected snake bite victims. International attention became focused on CSL when it was realised that the kits were unique. A side benefit was that hospitals and doctors took a renewed interest in the good management of snake bite.

The first kit was far from perfect and Alan Coulter and I were quietly exploring alternatives when the project was wrenched away from us during the 1981 upheaval described in the next chapter.

Thereafter we had little input into the design of the kits. Although those who took over from us showed considerable ingenuity, they lacked background knowledge of the potential complexity of both venoms and clinical situations. Ironically, although excluded from the ongoing research aspects, for almost twenty years any problems with their performance have landed in my basket. I had the unenviable task of finding a series of inbuilt sources of errors in the kits and, when my findings were validated at CSL, some saw my motives as suspect. There was uproar in 1992, when I stumbled on a design fault in the latest kit. I was using the kit to screen a post mortem sample with a suitable high-level venom control. The venom control confused the kit which had only been tested at low venom levels. The so called 'hook effect' was soon confirmed and the kit instructions rewritten to warn of this complication.

In recent years the kits have been more rigorously tested over a wider range of venoms. They are still unique, which is a pity as Alan Coulter and I had plans to make a cheap reliable venom detection kit suitable for use in developing countries. I'm certain it could be done and be as useful as the one built for Australian use. Hopefully someone will do it soon!

Investigating Snake Bite Cures

Since snakes often inject little or no venom it follows quack remedies are bound to have some successes. For this reason, regular as clockwork, ardent proponents of new or the same old 'cures' catch media attention. One day it may be Vitamin C or an imported suction device, the next a herbal brew. 'Magic stones' from India can be a profitable line.

From time to time when an illogical treatment is gathering support, it becomes necessary to do experiments to prove the obvious. Such an instance occurred in 1975 when some Chinese researchers declared they could cure snake bite with injections of the enzyme, trypsin. Trypsin might destroy venom in a test tube but there is no place for it in patient care. When the enthusiasts for trypsin treatment were joined by Japanese

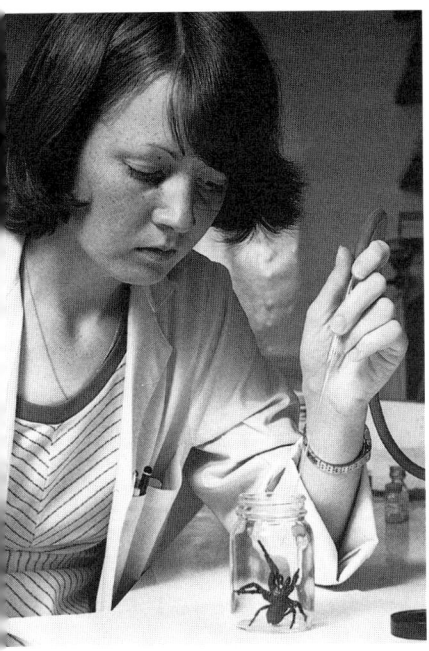
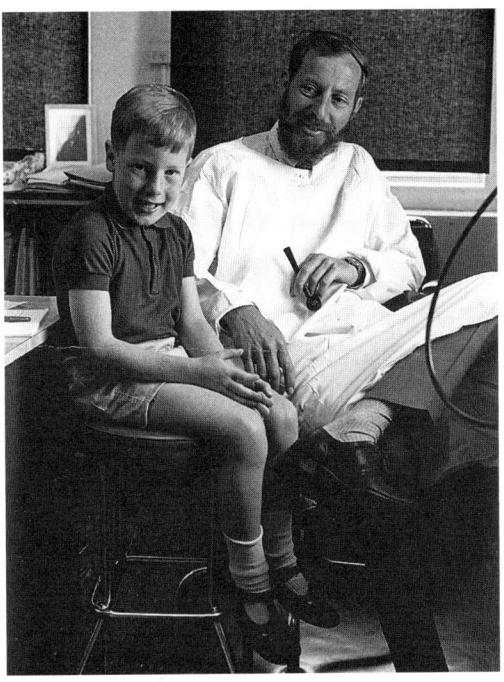

...e Miller cautiously 'milking' a female
...ey Funnel-web spider. A safety lid is
... with the potentially lethal male spider.
...tographer unknown)

John going to work with his dad, circa 1970.
(Photographer unknown)

...munology Research, May 1979. Back row: A. J. Broad, S. K. Sutherland, H. M. Chandler,
...C. Cox, A. R. Coulter, R. R. Premier and R. D. Harris.
...ont row: Erin Lovering, Rhonda Horsburgh, Mirella Barbaro, Angela Reale-Key and Jan
...urnbull

The author with the founder of the Australian Reptile Park, the late Eric Worrell, 1981. (Photo Laurie Andrews)

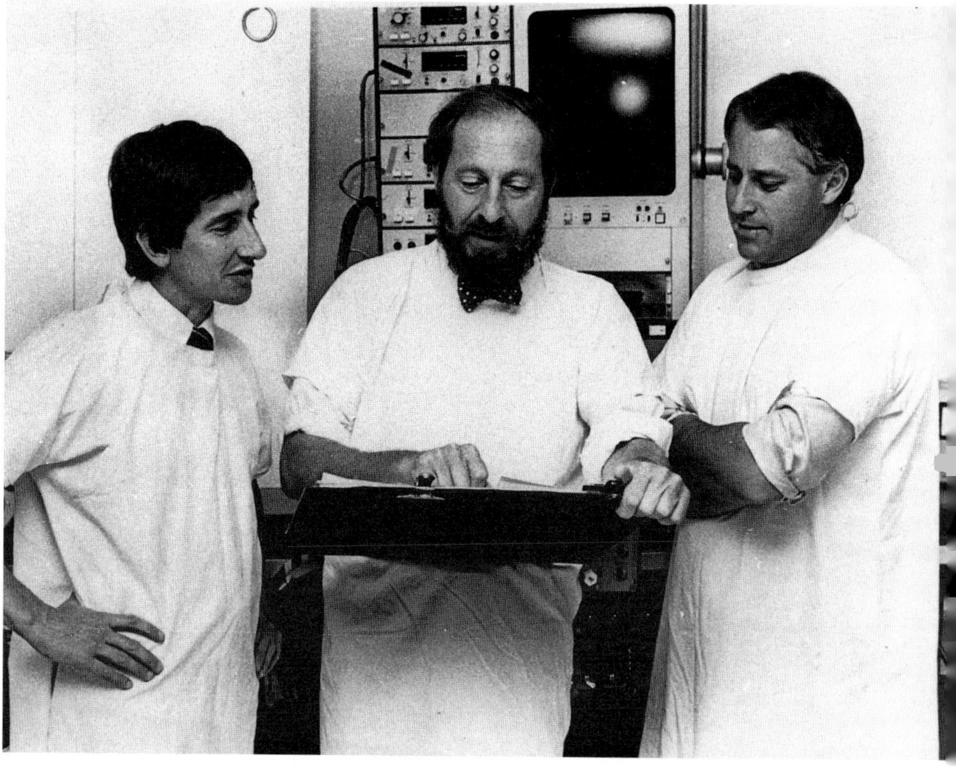

Successful collaborators: from left, Dr James Tibballs, the author and Dr Alan Duncan. (Photo the *Australian*)

advocates of tannic acid injections it was too much. In a series of neat experiments, Allen Broad showed that neither treatment was likely to benefit a patient. We got no response from the scientists in either China or Japan. However, Allen's work effectively curbed the use of either trypsin or tannic acid as a snake bite treatment.

One literally shocking treatment still surfaces every year or so. In 1986 some overseas doctors claimed to have cured snake bites and scorpion stings by giving the patients electric shocks with a modified stun gun. Experimentally, the treatment was demonstrated as a dismal failure. The only way an electric shock could prevent a man dying from snake bite was if he were to be bitten by a snake just before being strapped into an electric chair!

It is important that national health authorities should remain vigilant and react promptly to misleading media reports which might sabotage current teaching, particularly of first aid. On the other hand, venom researchers should remain receptive to new ideas and try to feign tolerance towards illogical cures. I never could!

The Most Venomous Snake in the World

In 1974 Dr John Wombey was investigating a rat plague in western Queensland when he captured a brown coloured snake. He sent it for identification to Jeanette Covacevitch at the Queensland Museum. Jeanette's discovery that it was a Small-scaled snake attracted great scientific interest. This species, which is also known as the Fierce snake or Western Taipan, was considered to be extremely rare. Its venom had never been studied in modern times and naturally Allen Broad and I were itching to get our hands on a fresh sample. We got some soon after a pocket of snakes were found and transferred to Cooktown, where the late Charles Tanner coddled, milked and successfully bred them.

The venom of the Small-scaled snake turned out to be spectacular! Allen first had to deal with a few practical matters such as whether it could be neutralised by Taipan antivenom. To test the antivenom Allen had first to calculate the precise toxicity of the venom in mice and then to find the number of units of antivenom required to protect the mice. In Allen's parlance this was going to be a 'piece of cake'. It was not. He hit a brick wall with his first experiment. His most dilute venom solutions wiped all the mice out. Thinking he had made a mistake (unlikely I thought) he repeated the whole test with the same outcome. This provoked a lot of head scratching.

Six weeks later Allen demonstrated that the Small-scaled snake had the most toxic venom of any snake in the world. Preliminary yields of the venom 'milked' by Charles suggested that the average output of this snake was sufficient to kill over 100 000 mice. One champion produced an amount which could potentially kill more than 250 000 mice. With such a potent venom and large output this snake would have no problem paralysing the odd rat!

Allen showed that Taipan antivenom would effectively treat this snake's bites. This was just as well because, as captive-bred Small-scaled snakes reached zoos and private collections, so did the snake bites occur. To date severe envenomations have occurred in places as far away from the natural habitat of this snake as Ballarat in Victoria and Tanunda in South Australia. All the customers for antivenom have been herpetologists.

There were two other important spin offs from this work. One was the finding that Small-scaled snake venom was inclined to leave some of its toxic component behind on glassware. This was reminiscent of a problem encountered with Funnel-web spider venom (page 188). The issue was solved by adding a trace of albumin to the saline used to dissolve the dried venom. This restored the venom to maximum potency.

When Allen looked at some of the common Australian snake venoms he found some of these also showed less than their maximum toxicity unless albumin was used. This cast doubt on the validity of many widely quoted figures comparing the toxicities of snake venom. The matter was resolved by an extensive survey of all the important snake venoms using only the best quality local and overseas venoms. The results made us proud to be Australians! Of the thirty most toxic snake venoms in the world, the top twelve were Australian! The thirteenth was the Indian Cobra. The poor old American Rattle-snake came last, its maximum output a potential killer of a mere 2500 mice. It is no wonder American toxinologists find this country so interesting and the local toxinologists rather smug.

Finding a Need and Spreading the Message

In 1966, although CSL had a wealth of information on antivenom performance and venomous bites and stings, it was difficult to access. It seemed to me that the only processed data was stored in the heads of some of my seniors. This resulted in an *ad hoc* learning process which made answering outside doctors' inquiries often difficult. Sometimes when left to handle an emergency call I failed to find guidance from either CSL records or the library. A number of times I had to frantically back-pedal when further hunting suggested that my proffered advice was wrong. My ignorance caused loss of face both to me and CSL. Occasionally after-hours calls utterly stumped me. If appropriate CSL colleagues were away for the weekend I could offer the doctor no alternative source of help.

After a few years at CSL I had reached several important conclusions. The first was easy. I was grossly ignorant about venomous creatures and the care of their victims and I had better systematically get my house in order. I needed to have the facts at my finger tips at all hours. Doctors in the field deserved the most up-to-date information and to this end I duplicated all important references, one copy to be filed at home.

The second conclusion, which is still valid in 1998, was that we have much to learn about all aspects of our venomous wild life. Many questions remain unanswered and there are interesting challenges awaiting the researcher. My group had a good run for a few years but to my mind did not completely solve any significant issue.

My final conclusion was that there was a real need for relevant and appropriately packaged information to be made available not only to medical care workers, but also to the general public. If the average doctor was half as ignorant as I had been, urgent action was required. (Successful moves were later made to have all medical students receive formal lectures in the treatment of envenomation.)

The leaflets that went out with CSL's antivenoms were useful but tended to be hidden away in hospital pharmacies. They were, by necessity, brief and related only to the bites of creatures treatable with that particular antivenom. In 1966 they were incorporated in CSL's medical handbook which was widely distributed. This was edited by Dr John Trinca and helped introduce me to the discipline of this writing genre.

In 1968 CSL published *Venomous Australian Animals Dangerous to Man*. This was edited and largely written by J. Ros Garnet, a CSL scientist and keen naturalist who died aged ninety-one years in January 1998. He was considered the father of the national park movement in Victoria. He had a fine sense of humour which enhanced the pleasure of working with him on his book. It was a very popular work, served its purpose well and for some years was my bible. Particularly valuable were the maps showing the distribution of the most important Australian snakes. As far as I know these were a first and represented many hours work.

My own contributions started cautiously but in a few years I was in full swing. Early on the family got a sniff of things to come when I spent the whole Christmas break writing a review on spider venoms. Review articles for medical journals are of transitory value but one I did turned out to be especially significant for me. In 1992 Dr Kester Brown, editor of the journal *Anaesthesia and Intensive Care,* invited me to submit a comprehensive review for publication. The review, entitled 'Venomous Australian Creatures: the action of their toxins and the care of the envenomed patient', appeared late 1974. It took six months to research and write. Thanks to Kester my filing system and scholarship were forced to expand, paving the way to later, larger works.

In the years that followed I published over three hundred articles. This may sound impressive but many were reviews or short communications. They did, however, successfully unbottle many of CSL's records and provide updated advice on patient management. Amongst the most useful were a pair published on snake and spider bite in April 1976 (see page 254). They appeared in the *Australian Family Physician* and were revised several times. The 1990 revisions enjoyed a particularly large print run.

Following my lightweight *Family Guide to Dangerous Plants and Animals of Australia* (Rigby, 1979), Oxford University Press approached me to write a book. After discussions, which included Al Knight and Anne Godden of Hyland House, the foundations of two OUP books were laid. Anne was to edit them and later books. The slim *Venomous Australian Animals* appeared in 1981 and has been in print ever since. Currently my son John has joined me as coauthor of the next edition.

Australian Animal Toxins: The creatures, their toxin and care of the poisoned patient was published in 1983. This 527-page baby had been a long time coming and I still feel ambivalence towards it. When people admired it I had some understanding of what it would be like to see an unwanted pregnancy through. For ten years it had sucked up my time to the proven detriment of more precious things. On the positive side the book provided easy access to much new information. I must admit to pride in the comprehensive index—I designed it with the panic-stricken in mind!

The weekend after I finished *Australian Animal Toxins* I wrote the draft of *Take Care! Poisonous Australian Animals*. This little book was aimed at young children and was fun to do. Updated versions are still circulating which pleases me because they bear the dedication: 'To all children who have died from snake bite—especially Maree V. aged 3 years.'

The death of Maree Vaccaro in 1977 deserves special mention because it hardened my heart towards colleagues who appeared not to take snake bite seriously. Maree was bitten by a Tiger snake which her father killed and later brought into hospital. When the father appeared with the dead snake as requested by a junior doctor, his senior said, 'Good. Put it in the fridge and I'll have it for breakfast.' Tragically by breakfast time the little girl was dying. Antivenom was not given until seven hours after the bite. She was rushed to Melbourne where I noted that her urine was almost solid with products of venom-affected muscle.

The subsequent coronial inquiry found me seething when the senior doctor left the witness box. Called next, after I gave details of the results of tests done at CSL the Coroner, Mr Pascoe, questioned me on the management of snake bite. One thing led to another and soon Mr Pascoe was allowing me from the witness box to question the previous witness. This turn of events rather flummoxed his lawyer who half stood up several times. I got some of the answers I wanted.

Mr Pascoe's verdict was highly critical of the senior doctor. Afterwards a reporter expressed surprise that he had seen me having a long and apparently friendly conversation with the same doctor in the car park. I explained that the air had been cleared and we now agreed we had a common enemy—snakes!

Poor Maree caused a rare blip in the good relationship I have had with colleagues outside CSL. With one rare exception I have always felt welcome when visiting hospitals to give talks. I was met by a rather grouchy doctor who explained that they rarely got snake bites and had only asked me to come because I had been making such a fuss about the need for good treatment. He kindly repeated this view when he introduced me to a surprisingly expectant-looking audience. Later, just when I got back to CSL, I received what at first I thought was a hoax call from the hospital. On the line was a casualty registrar who had a patient suffering from snake bite. And no, he had not been to my lecture ...

Talks at country hospitals can be especially entertaining because of audience participation. Fascinating cases have come to light and occasionally I have had some part in the doctor involved publishing a report.

Media

Venomous creatures are good television so I became a frequent but very mediocre performer on the media. Occasionally I found myself a national authority on some critter after a mere morning's work on its venom. The *ad hoc* research and subsequent television appearance was usually prompted by a novel case report. Sometimes, in fact often, my facts were wrong and I would be chided by *the* expert from the depths of a distant museum. In time I grew to know many such people who, without exception, gave freely of their advice. Admittedly, some of our long standing friendships got off to a rocky start. No-one likes to see a grossly distorted interpretation of their life's work on the evening news. I know the feeling. My children keenly followed this aspect of my career, but sometimes for the wrong reasons. One night I noticed my daughter Susie frantically taking notes while I spoke eloquently about the danger posed by European wasps. Later I found out she was recording my 'ums' to hopefully win a bet!

Television seemed an excellent way to make people of all ages more aware of our dangerous wildlife. I made a series entitled *Holiday Hazards* with the ABC in 1976. The series, which was conceived by Len Lamb and directed by Paul Drane, was rescreened many times and the episode dealing with snake bite was revised in 1979 to introduce the new first aid. This episode, which is still used by the ABC occasionally as a filler, captures the beauty of Megan, my second wife.

I was fairly proud of the series at the time but, seeing it recently, found it embarrassing viewing. My amateur performance aside, it does capture the 'state of the art' first aid which a few years later had radically changed. I show appalling ignorance of the dangers of certain creatures, use sexist language, even tell a mother-in-law joke! The episodes were cobbled together from generally single takes.

The episode on stinging fish was made on a trawler in Port Phillip Bay where Paul Drane became dreadfully seasick. Our problems were compounded by the netting of all sorts of fish which I had to talk about. Some I had never seen before. The ending tested everyone's patience. While a green-faced director sucked dry biscuits, we steamed in circles dodging shipping near the mouth of the Yarra. Unfortunately it was the school holidays and so I had brought my son John along for the day. Later he could recount to the family how he liked my ninth attempt best. It went: 'So, until next week, this is goodbye from Haliday Hozards.'

Epilogue

Most of my working life has been a delight. In the last few chapters I have tried to share the at times joyous experience of being involved in venom research. I was fortunate indeed to stumble into an engrossing field which suited my limited abilities. Fortunate also that I timed my arrival at CSL when it was at its peak as a caring national institute. Had I arrived a few years later, my career path would probably have done a U turn within weeks.

I have looked at some of the 'nuts and bolts' of venom research, and must now turn to the at times not so mundane atmosphere of my workplace. At CSL the work of my group was helped or hindered by particular individuals. As a general rule, the higher their salary, the greater the desire to curb our enthusiasm. If we had had a little less enthusiasm, perhaps I might not have got into such trouble. Too late now! As Frank Sinatra sang ...

Getting into Trouble at CSL

9

'The man who makes no mistakes does not usually make anything.'

E. J. Phelps (1822–1900)

Beavering Away with Enthusiasm	250
Meeting the New Director	252
Learning not to Like Neville At All	254
Naming New Buildings	255
The Director's Relations with his Medical Colleagues	256
Expensive Exams	256
Row No. 1	259
You are a swine!	259
Finding friends	261
Row No. 2	265
The background	265
McCarthy and the WHO	265
Quality of antivenoms	266
Secretarial assistance	266
Dr Norman Coles	267
Middle-aged scientist brutally sprinkled with paper clips	269
Prelude	269
Suspended again	273
The hearing	275
Round One—Thursday, 16 July 1981	275
Between hearings	282
Round Two—Wednesday, 29 July 1981	283
Round Three—Wednesday, 5 August 1981	285
Between hearings	287

Round Four—Wednesday, 12 August 1981   287
Between hearings   290
Round Five—Wednesday, 26 August 1981   291
Getting Back to 'Normal'   293
Paying the Bills   294
Disruption of the Production of Funnel-web
  Spider Antivenom   295
The Bulletin   296
Fun with '60 Minutes'   296
Going Tippy-toes with the Media   297
A Last Skirmish: The Launching of *Australian Animal Toxins*   298
Epilogue   299

Before describing several unfortunate incidents which happened to me at CSL, I'd like to present a summary of some activities there prior to 1980.

In 1966 I had found CSL research pretty laid back. There seemed little urgency, most workers were middle-aged and there were a handful of old-fashioned experts who were absolute delights. I was surprised to find that researchers only gave a talk on their work every two years at the most. This may have been because the only opportunity for such presentations was at highly formal meetings entitled 'The Scientific Discussion Group', where many of the presentations were from non-CSL people.

Beavering Away with Enthusiasm

In 1968, Dr Zana Hird and I started informal weekly meetings which we called 'research seminars'. Soon the younger scientists, in particular, were competing to perform at these meetings and it was wonderful to see how some of the shyer people blossomed. Convincing management to provide free coffee for these functions was a hard slog. One long memo listed the reasons why it was not possible and our response concluded by pointing out 'it did not hold water or, in this case, coffee'.

The problem was solved when we volunteered to set the coffee urn up ourselves. Our coffee was particularly strong and its pleasant aroma penetrated the whole building. For some of the technical staff, the only other free things at CSL were the use of toilets and lifts. The novelty of free coffee was an initial attraction for our regular audience and the popularity of these seminars was a catalyst for the construction of a lecture theatre a year later. Thirty years on, free coffee was still being served at these meetings, however, it was the instant variety.

For years Senior Medical students had come out to CSL for instruction on smallpox vaccination. When this was over they were usually given afternoon tea and a senior staff member would talk about CSL and its history. I listened in and was rather startled when one of the students put a blunt question to the venerable Dr John Graydon. When asked, 'Doctor, what world shaking discoveries have been made at CSL?,' John was absolutely flummoxed. It was a fair question and, of course, the answer was none. Most of the millions poured into research and development went into pedestrian but important activities such as fine-tuning vaccines and increasing penicillin production. Research was run on public servant lines and often scientists who had proven unable to handle the stresses, say of vaccine production, gravitated to Research. No scientist who stayed at CSL established an international reputation. But many of the scientists who left early in their careers did, for example Professor Kevin Lafferty became head of the John Curtin School of Medical Research.

In 1972 a short-lived Research Guild was formed. This Guild consisted of the more active and vocal research staff and its aims were to improve the quality and effectiveness of CSL's research. Some enthusiastic members like Kevin Fahey and Howard Chandler subsequently left CSL and their careers took off like rockets.

In 1972 the first Radioisotope Laboratory at CSL was set up in the Department and Alan Coulter started work amongst other things on a sophisticated method to detect snake venom in bodily fluids.

In 1974 a proposal of mine to establish a Pharmacology Laboratory at CSL was accepted. This was for venom research and also to develop methods to reduce the use of experimental animals in assays. The establishment of this new facility was supported by the Health Authorities in Canberra and I was nominated to be in charge. I oversaw the purchase of equipment and selection of technical staff but had no say in the appointment of a Senior Pharmacologist. In no time at all my control over the direction of this Laboratory was markedly curtailed. As far as I know, the Pharmacology Laboratory made no contribution towards reducing the numbers of animals used in assays.

By September 1978 any essential venom pharmacology could only be done in Immunology Research and, a few years later, due to lack of enthusiasm and drive, the Pharmacology Laboratory was closed down and its beautiful equipment disposed of.

Despite the time wasted over the Pharmacology Laboratory, satisfactory work was proceeding in other fields. In 1978, after exhaustive studies in monkeys, a new type of first aid for snake bite in Australia was developed (see page 228). The output of papers from the venom group in that year was twenty-seven. Weekly Department Research meetings were held and once a month a 'guest critic' would be welcomed to these Monday morning meetings. 'Guest critics' during this period included Sir Macfarlane Burnet, Sir Gustav Nossal, Dr Eric French and many others. They were most stimulating and encouraged greater output. The minutes were sent out to all relevant parties at CSL the next day and included requests for advice, access to equipment and uncommon reagents or chemicals. My group was bubbling with enthusiasm and confidence and this may have laid the seeds of the chronic professional jealousy which was later to have a drastic effect on the team.

By this time, I was averaging one outside lecture a week but had to limit them to universities, biggish hospitals or conferences. I really enjoyed pushing the principles of good patient care and talking about the group's research. The feedback from audiences was also very useful and stimulating. The same could be said of the increasing demand from doctors for advice on patient care. This kept me up-to-date with practical matters such as where and what creatures were currently poisoning the Australian population!

In September 1979 the World Health Organisation paid my expenses to go to a meeting in Zurich to discuss standardisation of antivenoms, etc. Dr Alister Reid, who was Chairman of this conference, was kind enough to point out that 50 per cent of the new data contained in the final report came from my laboratory at CSL. It was at this meeting that Alister and I managed to have the WHO adopt the clearer term 'antivenom' for official usage rather than 'antivenene or antivenin' (see page 237).

Thus at the end of 1979, although those working in venom research only numbered four out of the sixteen in the Immunology Research Department, the effectiveness and output of this group was quite significant. Considering publications alone, the three professionals accounted for over a third of the papers put out by the whole R and D group which had a total staff of about eighty.

There were a lot of things to look forward to in 1980 but, as it turned out, we were going to have problems to deal with other than those involving venoms.

Meeting the New Director

The death of Dr Bill Lane led to the appointment in 1974 of Dr Neville John McCarthy as Director of CSL. Few, if any, of the medical or scientific staff had ever heard of him. Inquiries disclosed that he had graduated in 1953 with my sister, had attempted to obtain surgical qualifications while at St Vincent's Hospital and afterwards had gone into general practice at Yarrawonga in northern Victoria for ten years until 1969. He then became Regional Medical Director for the international pharmaceutical company E. R. Squibb and Sons until his appointment to CSL. In those days, I was under the impression that the medical officer for a pharmaceutical company offered a well paid, often cushy job, which was almost opting out of medicine. CSL on the other hand was a different kettle of fish.

McCarthy's job with Squibb was possibly not especially onerous since their products were limited in number. During this period he obtained a Master's Degree in Business Administration from Monash University. He also became involved with efficiency surveys conducted at the Southern Memorial Hospital. (At that time it could hardly be considered one of the great hospitals of Victoria.) Somehow, he caught Canberra's eye and, although having no higher medical or scientific qualifications, was appointed Director of CSL with instructions to shake it up and generally make it more profitable.

In fairness, before going any further, I must confess to being stubborn, headstrong and even occasionally short-tempered. Generally when I have got into trouble it is because unnecessary obstructions have been put in the way of getting on with the job at hand. It should also be pointed out that I had a few healthy disagreements with McCarthy's predecessor but, with Dr Lane, one would win some arguments and lose some, but there was never any ill-feeling. For example in 1971 there were moves afoot to pinch one of my key biochemists, Allen Broad, who had no desire to be pinched. I declared that if he went, then I would resign as head of the department. The matter was fully and openly discussed and the proposal to move Allen was dropped. I never subsequently used the ploy of threatening to resign, which is just as well, because it would have been accepted by management within seconds!

I didn't think the appointment of Dr McCarthy would make a great deal of difference to my situation. Most of the venom work was directly funded by the Federal Government under the provisions of the CSL Act as work 'in the national interest'. It was not particularly expensive as research goes and the short-term goals were practical and easily understood by even the most stolid of visiting politicians. Moreover half the group's efforts were being directed at the development of a single-dose slow-release veterinary vaccine which was showing considerable promise. This vaccine was to immunise farm animals against five important diseases.

Dr John Trinca had been Acting Director prior to the new Director's appointment and during this time the Laboratories ticked along smoothly, everyone got on with their job and there were no crises. On Dr McCarthy's second day at CSL, John Trinca called me over to meet the new Director.

I immediately found he had two habits that irritated me. One was closing his eyes from time to time when talking to me and the other was smiling at seemingly inappropriate times. There was also a slight whine to his voice, which was related to his hearing defect. The poor chap couldn't help the whine and I thought it unfair that the wags referred to him as 'The Bionic Ear'. In the final years he was generally merely referred to as 'McCarthy'. At the conclusion of our first meeting I was left with the impression he was not particularly interested in what my group was doing.

In the months that followed, I observed a great flurry of activity with feasibility studies, reorganisation, outside experts and the commencement of the expansion of Administration. The titles of executives were changed with a bewildering frequency. One executive recently admitted to some eighteen different titles, though these were not all during the McCarthy era. At first McCarthy sat in his office with the door open but soon a second door protected his inner sanctum.

I kept out of the way and got on with my work. Various reports filtered through to me which caused concern. For instance, the head of the Bacteriology research department, Dr Gulasekharam, had a blazing row with McCarthy and resigned over what he described as an ethical matter. In contrast, certain obsequious lickspittles were promoted at a rate which positively dazzled the observer. The oddest people got company cars and it was breathtaking to see what was spent on interstate and overseas air travel.

Learning not to Like Neville At All

The Christmas break in 1975 and every weekend in January 1976 were spent writing two articles for the journal *Australian Family Medicine*. At the time these were the longest articles on snake and spider bite published in Australia. They were to be published in the April edition and, since CSL frequently advertised in this journal, I specifically asked that the CSL ad not appear in the centre of these articles, as I considered it would be unethical for CSL to advertise their products right in the middle of a scientific article written by one of their medicos.

When the articles appeared, the centre of the snake one was dominated by a full page advertisement featuring a snake and a screaming headline, 'Is this the only time you think of CSL?' It was followed with material promoting CSL antibiotics.

After discussions with colleagues, I composed a letter to the Associate Editor of the journal stating my disappointment over the placement of this advertisement, despite CSL's reassurances. In it I stated that this advertisement, 'certainly distracts the reader from the paper to berate him or her on their prescribing failures'. When I had somewhat cooled down I decided against sending the letter to the Associate Editor, instead I forwarded a copy to the CSL Marketing person who was responsible for insertion, contrary to the prior agreement. I thought 'that will fix you' and left the matter at that.

The articles themselves were highly successful and reprints ran to over 10 000 copies. It is fair to say these papers helped improve the management of snake and spider bite cases in Australia. A completely revised edition of both articles appeared in the January 1990 edition of *The Australian Family Physician*.

On 11 May 1976 McCarthy rang me at 10.00 am and told me to come over immediately. I was ushered into the Board Room where McCarthy and a handful of the executives were seated. When seated, McCarthy said, 'We have obtained a legal opinion from Madden, Butler and Co regarding your letter to the Associate Editor of *The Australian Family Physician* dated 29 April 1976. Legal opinion is that grounds for libel exist.'

When at last I got a chance to speak, I pointed out the letter had never been sent out of CSL. If it hadn't been sent to anyone other than the person it was indirectly aimed to criticise, how on earth could it be libelous? As this sunk in, McCarthy hesitated and then decided it was not relevant and went back into the attack. He ordered me to apologise to all present and retract any criticism of the Marketing Division, otherwise a writ would be slapped upon me. Somewhat shaken by this turn of events, I mouthed an apology.

I returned to my laboratory thinking what an unpleasant fellow McCarthy was to set up a medical colleague like that without any prior discussion. The fact he would seek legal opinion over such a matter without even talking to me should have warned me about his *modus operandi*. I no longer felt any loyalty or respect for him, although some people might believe that his actions were appropriate for a Chief Executive. There were, however, no CSL advertisements in the 1990 edition of the above-mentioned articles.

Naming New Buildings

In March 1976 a site was being cleared for the construction of a new library. To name the new library after the late Dr Lane seemed a nice idea and so I mentioned it to Neville McCarthy. He said he would think about it and on 1 April I sent a two line memo to him on the subject, wondered if Dr Lane's widow might be written to seeking her approval for such a proposal and concluded by saying, 'If you are agreeable, it might be nice to send her a sketch of the new building.' Next day the Director rang me and said he'd considered the idea but it was contrary to Public Service practice to do such a thing.

A little over a year later, on 20 May 1977, a series of ceremonies at CSL were presided over by the Minister for Health, Mr Hunt. They involved the naming of the library after Sir Macfarlane Burnet and a number of other buildings after previous directors. Large numbers of outside and inside people were invited to these ceremonies. I was not on the invitation list. The new Quality Control building was named after Dr Lane. Mrs Lane informed me later that she specifically inquired after me but had been informed that I had not been invited. This became common practice. I was cheered up a little when a small temporary plaque appeared on a soon to be demolished stable naming it the Neville J. McCarthy building. The tradition of unofficially naming buildings was to continue.

The gem of them all was a plaque outside the Welding shop dedicated to 'Dr J. X. Xuereb' aka 'Joe the Welder' a skilled and popular identity.

The Director's Relations with his Medical Colleagues

When I arrived at CSL there were nine medicos, but by mid-1977 there were only five. Senior positions had not been filled and increasing responsibilities had been taken on by those who remained. As a result, a meeting of the doctors and Dr McCarthy was held in his office on Friday, 14 October to discuss the urgent need to recruit additional medical staff. Present were: Drs Hartman, Schiff, Feery and myself. We were aware that the positions existed. We also knew that no matter how talented the new persons were it would be a matter of months before they learnt the ropes.

The discussions were quite amiable in spite of McCarthy's comment that he considered 'Medical Officers are in fact expensive luxuries'. As my colleagues were getting over that comment I raised the need for a replacement of a younger medical colleague who had recently moved overseas. I stressed the importance of having someone trained up in the fairly extensive field I covered. This produced another McCarthy gem, 'That's not necessary. If you drop dead we can just look up what you've written.'

Another concern was the matter of promotion. Three of us were sitting on Class 2 positions and there were two vacant Class 3 positions due to the retirement of Dr Trinca and the abrupt departure of Dr Gulasekharam. Dr Feery wanted to know exactly what requirements were necessary in order to be promoted. These were described as either recognition as an established expert in a field and/or obtaining a higher qualification by a learned college or a higher degree. (When two of us subsequently more than satisfied these requirements, particularly in regard to higher qualifications, no promotion was granted.)

Expensive Exams

The year after the above meeting I'd been admitted as a Fellow of the Royal College of Pathologists of Australia and also as a Fellow of the Royal Australian College of Physicians. The examination fees were fairly high and I'd had to fly to Sydney to face the Board of Censors of the College of Physicians in September 1978. It had always been the custom at CSL for any employee gaining qualifications highly relevant to their work to have the cost of obtaining such qualifications refunded. Only two other people at CSL had ever gained entry to these Colleges and all their expenses had been paid. Accordingly, I gathered up the receipts and submitted a request for refund of some $1200 in expenses. This was big money to me.

On 3 October 1978 I was told that McCarthy would not authorise the reimbursement. I wrote him a number of memos outlining past precedents and the fact that I had informed him in writing of my intention regarding these qualifications. His final reply included the remark, 'there is no appropriate role for CSL at this stage'.

I then faced the problem of the hefty annual subscriptions required to retain membership of these Colleges. CSL finally agreed to pay one subscription but 'rationalised' and markedly reduced payments on my behalf to other societies I belonged to.

I thought this was unfair and decided to make a submission to the Chairman of CSL.

On 18 October 1978 I submitted a ten-page submission to the then Chairman of the Commission, Mr R. T. Shelmerdine. In a respectful, almost forelock-tugging fashion, the facts were laid bare. As far as I knew no-one had made such an approach to the Commission before and I hoped for a favourable outcome. After all Mr Shelmerdine had been positively benevolent to me at the CSL Director's Christmas Party. A few weeks earlier I had won the Triennial 1935 BMA Prize and at the party he had beamed down on me and said, 'Winning that prize when you did was the greatest PR thing in the history of the CSL—especially just before the elections.'

A letter from the Commission dated 27 October 1978 disclosed that every request I made was refused. No reasons were given.

In 1978 things were becoming difficult in other areas. Before 1975 I reported directly to Dr Peter Schiff, who was Chief of Research and Development, and he reported directly to Dr McCarthy. In 1975, Mr Merv Hinton was appointed R and D Manager, this position being inserted between the department research heads and Dr Schiff. Merv was a delightful big-framed sixty-year-old who worked very long hours and did everything possible to see that research activities ran smoothly. We were good friends and drove each other into work. He was a father figure who had developed the Brown snake antivenom and retained an active interest in venom research.

By 1978, however, the organisation of CSL was getting top-heavy. Mr Vivian Davey had been appointed Technical Director over Dr Schiff and Dr Norman Coles was slipped in as Assistant R and D Manager.

Vivian Davey had been at CSL for years. As Head of Quality Control, his finicky ways admirably suited the job, but as CSL's 'scientific supremo' his unpopularity was to reach new heights. He was a master of obstructionism, and could seize on one small negative point in a large proposal and expand it so that the whole plan was condemned. He certainly made my working life far more difficult.

When I found out that Viv Davey and McCarthy attended the same Anglican Church in Kew I was flabbergasted. The thought of them both being Christians was something that had not occurred to me. However, it occurred to them when setting up the animal ethics committee to pop their vicar on the said committee.

In 1978, CSL and the Royal Melbourne Hospital were collaborating on a major study on the effects of Australian snake venoms in monkeys and investigation of first aid measures. A new type of first aid (page 226) was emerging from this work so when an outside expert offered to videorecord some of it for free it seemed an excellent idea. I sent a memo up the line to the Director requesting permission that this be discreetly done. There was little in it to upset animal lovers and, since the findings appeared of increasing importance, it was appropriate that they be recorded as comprehensively as possible. It might also reduce the need for repeat experiments. Back came a reply from Davey stating, amongst other matters, that the monkey work should cease immediately and not recommence without reference to him. On 10 August I put in a five-page summary to the Director, via Davey, outlining reasons why this ban should be lifted.

The abrupt halt to this work was very embarrassing. A great deal of organisation had gone into co-ordinating the co-workers at the Royal Melbourne Hospital who were performing a large array of tests on the monkey samples. They could not understand this disruption since the results had been regularly presented at meetings held at CSL during the year. I directed the submission to the Director as I believed that Davey, not being a medical scientist, couldn't appreciate the significance of the monkey work. Days passed and I discovered that Davey had decided to hold on to the submission.

Five weeks later, I at last managed to convene a meeting at CSL to discuss the matter. The decision was that I could recommence the monkey studies immediately without any new restrictions. The proposal to videotape was abandoned. In January 1979 a major paper describing this work was published in *The Lancet*. We didn't feel inclined to mention Davey in the acknowledgments.

In July 1979 Davey was reponsible for another irritation. Dr Frank Perkins, Head of WHO Biological Standardisation, had invited me to a meeting at Zurich to discuss antivenom standardisation. The airfare and accommodation were to be paid for by WHO. This was the first of three overseas trips I was to undertake for WHO and as far as I know at that time no-one else at CSL had been so funded. In planning this trip I decided to fly eastward and arranged to visit a number of laboratories on the way. The logic of this route was that these laboratories could provide updates useful to the Geneva meeting and it's far more comfortable flying eastward around the world than westward. All the plans were settled when Dr Perkins made a brief visit to CSL.

I was somewhat astounded when McCarthy informed me that Perkins had told Davey that he'd like Dr Sutherland to visit a certain Institute in India on the way to Geneva. This really messed up my itinerary but, everything considered, Dr Perkins had every right to request it. Later that day, by sheer luck, I came around a corner and went smack into Dr Perkins. So I asked him, 'Why do you want me to visit the Haffkine Institute on the way to Zurich?' He replied, 'I don't want you to. What do you want to do that for?' I proceeded as planned in an easterly direction.

At this stage I was getting on quite well with most of the other 1200 employees at CSL. However, I discovered Davey, Coles and McCarthy, for various reasons, were combining their talents in a plan to lighten my workload.

Row No. 1

You are a swine!
Early Tuesday morning, 6 February 1980, Dr Peter Schiff rang me to say that the Director wanted to see us both at 11.30 a.m. that morning to discuss some 'reorganisation'. I said, 'That's sounds a bit ominous,' and Dr Schiff replied, 'It's nothing to worry about mate.' As Peter Schiff and I had been close colleagues for some fourteen years, I felt somewhat reassured. Since preliminary work had been completed on a new invention, I brought the basic device with me to the Administration Building so I could discuss it with my seniors.

Some time was spent in Dr Schiff's office discussing this device. He quickly grasped the potential of the invention and showed considerable enthusiasm when a number of applications were discussed. We then went into the Director's office at 11.35 a.m. For the first ten minutes we talked about the invention but, to my disappointment, McCarthy appeared uninterested in it.

To my surprise, he asked in what way the device differed from the snake venom detection kit. As described on page 241 the Department had made 600 of these unique kits for Australia-wide distribution immediately before Christmas 1979. We wanted to obtain urgent advice in regard to patenting the new device, but no decision was made and the subject was soon dropped when the main point of the meeting was broached.

The need for a stronger immunochemistry force at CSL was discussed at length and I wholly agreed with some expansion in the field. My group had established the Radioisotope Laboratory and was regularly assisting interested parties at CSL on aspects of immunochemistry. These included the isolation and labelling of proteins and advance work on immunoassays.

McCarthy then outlined his proposal to establish a new unit of immunochemistry at CSL. He suggested that four of my five biochemists, their associated staff and laboratories would be removed from my jurisdiction in the near future. When I protested that this would leave me with one biochemist and two technicians to continue existing projects, he smiled for the first time and said, 'That's the idea.' The faster

I advanced arguments against the proposal, the more he smiled. I asked if I could receive written details of the proposal and the rationale behind it so that I could put up a case to modify it. This was refused out of hand.

As it dawned upon me that I was struggling for the survival of my research team, I became more vocal and McCarthy seemed very pleased at this turn of events. I told McCarthy I thought he was a swine and had no feeling for biological science. These were the only derogatory remarks I made directly to McCarthy in the presence of Dr Schiff. Dr Schiff, who had taken no part in the conversation to date, continued to remain totally silent. Had he, as a third party and as my divisional head, made some conciliatory noises, some of the heat might have been taken out of what was in fact a confrontation between two people who obviously disliked each other intensely. I told them both that I would appeal against the decision, and that the atmosphere in the office stank.

I headed for the door with the Director close behind me. As I stood there with the invention in my left hand and my right hand on the door knob, the Director held his face within a foot or so of mine and, raising his chin, rocked it from side to side. I took a really deep breath to control my anger, opened the door and left. The time was 12.10 p.m.

This Immunochemistry Department was subsequently headed by one of McCarthy's favourites, who spent a great deal of time travelling the globe visiting institutes and commercial establishments. He boasted that he could ring the Director any time day or night as he globetrotted. Within a couple of years both he and his offsider had left CSL and were employed by one of the establishments in Canada that they had regularly visited at CSL's expense.

My diary records two other interesting events that occurred on 6 February 1980. In their separate canvas bags, two live Brown snakes and two Tiger snakes arrived. They were to be used in a study to determine in which parts of their venom glands the different components of venom were manufactured. I had been told they would be quite happy and safe left in these bags for a number of days.

The other event was two recordings by phone with the BBC, London. One was for the science programme of the BBC World Service and the other for developing countries in the series, 'Hello Tomorrow'. I tried not to sound too ocker.

I started to prepare a case for the preservation of my staff for the Commission to consider. Previous experience suggested that this would be futile as the Commissioners didn't want to get involved in 'domestic matters'. It thought that outside influence would be the only way to reverse the decision.

On Thursday, 7 February 1980, the newspapers carried a story that the pressure-immobilisation type of first aid had been highly effective in the case of Funnel-web spider bites. The child that died last month from a Funnel-web spider bite might have survived had this first aid been used.

At 8.15 am, when I arrived at the lower entrance to CSL, the guard indicated he wanted to speak to me. He was not his normal friendly self and looked quite shaken. He told me his instructions were to direct me to proceed immediately to the Administration car park, not to stop anywhere and report to immediately to Mr Sullivan (Peter Sullivan was the CSL Commercial Director at the time).

Hello, I thought, something's up. As I approached my normal car park I decided it wouldn't do any harm to have a witness or two. I parked the car in its usual spot, went up a narrow path between two large buildings (subsequently I always called this my Via Dolorosa), and popped into the library which was empty except for Dr Norman Coles and Dr Brian Feery. They'll do, I thought. One was a medical colleague and Dr Coles had recently been made my immediate superior.

I bustled them both up the stairs to the top floor of the Administration Building, although I nearly lost Brian Feery when he decided that he didn't particularly want to get involved. I rounded him up again and in no time at all the three of us were seated in the office of Peter Sullivan.

Peter Sullivan was looking extremely serious and at the same time a little bit more jumpy than usual. He handed me a document signed by Neville McCarthy and dated 6 February. This was a notice of suspension prior to the laying of unspecified charges. I was told to leave the area immediately and no longer undertake any CSL duties. I asked if this also meant not giving advice on patient care and was told this activity was also to cease. I then asked if this also included giving advice after hours, and again was told this was to cease. (This turned out to be a fatal but unwitting mistake by CSL.)

I asked if I could collect some items from my office and was told I could do this provided I remained accompanied and did it swiftly.

With the formal aspects of the suspension over, I collected the articles I was currently working on, discreetly popped the box of live snakes into a cupboard and was on my way home shortly after 9.00 a.m. (My shortest day at CSL ever!)

Finding friends

Over the next week I was amazed at the support that came from so many directions. Some was based on personal friendships, or some came from those who considered the research I was doing was important. The first positive support came as a result of the ban on giving advice on patient care. After I had got home to an empty house on the morning of the suspension I wasn't fantastically happy. However, after a cup of coffee I decided to make the best use of the time on my hands. I rang the Australian Medical Association and asked them to relay the fact of my suspension to their Federal Industrial Department. I included the information that I had been instructed not to give doctors any advice on managing venomous bites and stings.

This restriction was picked up within a matter of hours by the Victorian State President of the AMA, Dr Bryce Phillips. Bryce had been in my year in medicine and did not mince words when he said what he thought of this restriction. By Tuesday, 12 February, every paper was carrying the story and the relevant unions were producing strings of pithy quotes. CSL decided to lift the ban, but by that stage the damage had been done.

Letters and phone calls poured in. On one day I received more that fifty phone calls and it was quite enlightening to hear what certain distinguished medicos had to say about my Director.

The Professional Officers' Association swung into action early in the piece and by 8 February they had addressed a letter to 'Mr N.J. McCarthy' which was full of legalese and seemed to tear apart every aspect of the suspension. On 13 February there was a meeting of a hundred members of the POA at CSL. Two motions to be conveyed to the CSL Commission were passed. The first, which was passed by a narrow majority, expressed the wish that the problem be solved by conciliation and compromise rather than confrontation. The second motion unanimously expressed full support for the removal of my suspension.

It was agreed that another meeting of the POA would be called if the situation was not resolved within one week. By 14 February the matter had become one of daily updates in the press.

Meantime the AMA had been busy. Mr John St Vincent Welch, its Federal Industrial Officer, was dealing directly with the Serum Laboratories. John had a certain direct and effective approach and he arranged an after-hours meeting with Dr McCarthy and the Commission Secretary Alf Brogan on 15 February. I was rather astounded to find that CSL had hired a posh room at the Royal Park Motel in Parkville for this meeting. I picked John Welch up at Tullamarine and the four of us then started polite discussions at about 5.00 p.m. John Welch was very impressive in the way he laid into McCarthy and he explained in considerable detail what the AMA thought about the whole matter. John was aware of the Medical Officers' earlier discontent at CSL and had been convinced that in time something would, as he said, 'blow up'.

Like the POA, the AMA's attitude was that the Commission itself should not hear the case but it should go to some independent tribunal.

It really was an extraordinary meeting. Each time McCarthy started backing away from a near-agreement, John Welch would be after him like a terrier. McCarthy's attempts at his usual long monologues were cut short. For the first time in six years I was witnessing someone standing up to McCarthy (I had missed Dr Gulasekharam's outburst some years earlier). I was pleased John was on my side as he was more than a match for Alf Brogan, who, to his credit, broke down some of McCarthy's stubbornness.

By 8.00 p.m. it had been agreed that early the following week the Commission would announce 'the whole affair had been satisfactorily resolved', and that would be the end of the matter. McCarthy looked very tired and subdued. When he and Brogan left, John got on the telephone to report to his boss. His description of what he thought of McCarthy would have made even a bullocky stop and listen in awe.

When I got home late Friday night, I hoped that the matter was over and that possibly things would improve at CSL. The next day the papers carried accounts of the affair. *The National Times* had a major article entitled, 'A research career is not meant to be easy'. This article concluded with an interesting quote from McCarthy: 'I don't know research careers were ever meant to be easy,' McCarthy replied. 'It's difficult when people want to stay in one place. There was a time when people didn't expect to stay in research all their lives.' Apparently the Director felt that scientific creativity often passed at a certain stage in life, say in the forties. 'After that,' he said, 'people go into something like administration ...'

On Sunday, 17 February, I tried to keep myself busy. I spent four hours writing a paper entitled, 'First Aid for Funnel-web spider bites' and then another two hours on a paper for the journal *Pathology*. On Monday, 18 February, I was well into the article for *Pathology* when Alf Brogan rang me at home. He informed me that the Commission had just discussed the matter and would be meeting again in two days' time. I was invited to attend this meeting, although it was not compulsory. Next day, a CSL driver delivered the invitation to my home. It was a hot day so I invited him into the kitchen for a glass of cordial. He volunteered the information that the workers were on my side. He kindly took the draft articles back to give to my secretary Marjory Davey.

In his official history of CSL, Brogan stated that the Friday night meeting was unproductive. Brogan also reported that at its 19 February meeting the CSL Commission spent *three hours* discussing the case and the various reports and documents concerning it. Obviously Neville had got his energy back and was not going to give an inch.

On Wednesday, 20 February, I went on my lonesome to face the Commission. They were stony-faced, hostile and unhelpful. Hardly a question was asked.

Later that day another CSL driver came to my home with a letter from the Chairman, Dr Forbes. Dr Forbes was a former Federal Minister and at that stage Federal President of the Liberal Party. The letter informed me that the Commission had considered the reports of the incidents in the Director's office on 5 February (I of course had not seen these reports, so I couldn't challenge their contents). Dr Forbes informed me that the Commission took an extremely serious view of the matter and found me guilty of disgraceful and/or improper conduct. The letter contained a reprimand but lifted the suspension and generously informed me 'you are free to return to work'.

Thus the agreement to drop the whole matter worked out on the previous Friday night had been ignored. Furthermore, I had been lured into a kangaroo court without representation and without even being given access to the details of the charges.

Next day I returned to CSL and was pleased to find that the snakes were still happily meditating in their individual bags. In hindsight, I should have taken them to the Commission meeting the day before!

Back at work the immediate problem was rescheduling the work which had been so abruptly stopped. In particular, the new push towards solving the Funnel-web spider problem (page 198) had to be restarted and it wasn't until 13 March that the intensive studies on monkeys could start at the Children's Hospital. The preparation of venom for rabbit immunisation was likewise several weeks behind.

From the start of this episode, I had been selfish in neglecting to realise the effect all the drama had on my staff. I had not appreciated how dreadful it was for them to be in limbo and facing uncertain futures because of their boss's action. They had little idea what was going on and were constantly quizzed by other CSL staff. Some were mentally and emotionally highly involved with the projects. The recent death of a young boy after a Funnel-web spider bite had particularly affected those involved in that project.

In the meantime, dissatisfaction with CSL's handling of the matter rumbled on. The unions maintained that I had been treated unfairly, denied natural justice and left with a slur on my name. The matter went before Deputy President Isaacs at the Arbitration Court on 28 February but was negated by legal technicalities since I was now back at work. Many attempts by the unions were made to have the issue re-opened and inquired into independently. There was concern that the accuser was also the jury and the sentencing judge, and that other staff members could easily find themselves trapped in a similar situation.

The Staff Rules needed to be clarified and some equitable system set up for investigating possible infringements of the rules. Most approaches were directed to the newly-appointed Minister for Health, Michael MacKellar. MacKellar did a great job of washing his hands and stating he couldn't interfere in the internal matters of CSL. When MacKellar resigned as a result of the failure to declare a colour television set I was not madly upset.

My local MP, Ian McPhee, had known about my situation at CSL for several years. Ian had lunched on a regular basis with the former Minister for Health, Ralph Hunt, and could pass over the odd bit of information to the Minister. However, Ian could do nothing with Mr MacKellar.

Ian McPhee at times demonstrated a fine sense of humour. Once when he, as Minister for Productivity, was opening a symposium Neville McCarthy came up afterwards and introduced himself. When Neville had finished his spiel, Ian cocked his head to one side and said, 'Ah, CSL.

I do admire your Dr Sutherland.' He said that McCarthy's face was fascinating to watch.

I continued to receive support from the general public, the medical profession and former and current employees of CSL. Here are some examples. From Mr Lindsay Money, a retired CSL biochemist, dated 15 February 1980:

> My assessment of Dr Sutherland is that he is a particularly dedicated and conscientious scientist, possessed of exceptional ability. I cannot under any circumstances accept that a charge of 'disgraceful conduct' could be sustained against him. This case indicates to me that the situation which I, along with other senior officers, feared might develop, has indeed done so.

From Mr Jan Birner, a recently retired CSL biochemist, dated 14 February 1980:

> I read with astonishment the news about your temporary suspension in scientific work at CSL. During my work as a research scientist at CSL I had the opportunity and privilege to work with you as a biochemist, mostly in research on Funnel-web spider venom. I still admire your ability, devotion, organisation, and successes in your work, and mostly your attitude to your staff and co-workers.

Things at CSL continued in a fairly uneasy fashion until mid-1981. When Row No. 2 occurred the combatants went into action taking into account the lessons learnt from Row No. 1.

Row No. 2

The background
Row No. 1 was a mere grassfire compared with Row No. 2. In my opinion the bushfire of Row No. 2 could have been prevented by sensible 'people management' as I consider it sprang from the unresolved factors arising from my first suspension, plus a number of other reasons. These may appear trifling but when combined they got the better of me.
McCarthy and the WHO
1981 was a busy year right from the start. The first successful use of the Funnel-web spider antivenom (page 214) counterbalanced the letter I received from McCarthy suggesting that perhaps I should go off and rest my brain for a while (page 213). In February, I had to go to Fiji for a WHO meeting on seafood poisoning with particular emphasis on ciguatera. This was my third WHO meeting in as many years. I passed this all-expenses-paid request from Dr Nakajima, the Regional Director of

WHO, up the line to McCarthy for approval, which was grudgingly given provided I went there and back as fast as possible. I subsequently spent five days from eight in the morning to six at night in the conference room at Suva and flew back home anything but refreshed.

I did, however, find out in Fiji that McCarthy had written a blistering letter to Dr Nakajima on 18 December 1980 criticising him for approaching me direct. The letter, of which I have a copy, included the sentence, 'I believe it appropriate that we advise you on whether we are able to assist and whether we have the relevant skills and expertise for specific projects' (i.e. the chances of recommending Sutherland are zilch). Dr Hiroshi Nakajima went on to become WHO's Director-General. Over the next ten years McCarthy never made my services available to WHO.

Quality of antivenoms
Improving the quality of antivenoms was something dear to my heart. Comparative studies of antivenoms from different countries and experience with the WHO committees, had clarified my views on certain practices. In 1979, I finally convinced CSL to stop the procedure of reprocessing out-of-date antivenoms. Many of these had been returned time expired and for all we knew might have been kept in the glovebox of a veterinarian's truck. The recycling of out-of-date biologicals by incorporating them into new batches seemed a very bad and possibly dangerous practice.

On 12 May 1981 I was moved to fire a bobbydazzler of a memo up the line. I discovered that CSL had manufactured a cobra antivenom for sale in Malaysia, which I believed was downright dangerous. It was unpurified, freeze-dried whole horse serum. Such a crude product had not been used in Australia for over fifty years because it carried a high risk of very serious reactions. My three-page memo reviewed the literature on such crude preparations and claimed that its export would damage the reputation of CSL, particularly in the eyes of WHO. The new product release form was signed by Charles Guthrie, Vivian Davey, Peter Sullivan and Neville McCarthy.

There was no response to this memo, but plans for further batches were abandoned. It is hard to understand the motive behind the venture since even the profitability was marginal. The return on factory costs was estimated at less than $1 Australian per vial. I was annoyed about the secrecy of this project, a situation which was later corrected in Robert Drewe's article in *The Bulletin* (see page 296).

Secretarial assistance
For years I'd battled to get regular secretarial assistance. After I'd moved from the Administration Building to the Laboratories proper I had to tramp from place to place to get any typing done. I liked to reply promptly to requests and, despite short cuts, the workload increased as I became better known. This correspondence combined with scientific

papers and reports were sufficient to keep nearly one and a half secretaries busy full-time.

A number of people at CSL had secretaries with little work to do, who were basically status symbols. Some of these girls knitted during the day and I had observed Mr Davey's secretary reading novels. These secretaries helped me enormously over the years, but often the dog-in-the-manger attitude of their masters necessitated a stealthy approach.

As time went by and the workload increased, the burden became almost intolerable. I was the only person at CSL actively doing laboratory work and, at the same time, receiving dozens of phone calls from doctors, a great deal of mail and lecturing. Without some screening of phone calls it became impossible to complete experiments. From early 1978 I had submitted regular requests for secretarial assistance but they were either ignored or knocked back.

By 1979, the situation had become quite ludicrous. I collected all the previous memos on the subject and, with a covering note, bundled them up the line. Success at last was achieved. It was agreed that the R and D Manager, Merv Hinton, would get a new secretary who would be a stenographer and I would share her with him. She would be situated on the floor above me, but was free to come down to take dictation, and do filing. Most important of all, my telephone calls would be channelled through her and she could take messages. In January 1980, Mrs Marjory Davey commenced work as our joint secretary. The effect was quite dramatic. She rapidly learnt to sort out urgent from non-urgent calls and became pretty good at furnishing the requested information herself. Photocopying I still had to do myself and much typing still went to other secretaries.

This arrangement worked most satisfactorily until the benign Merv Hinton retired on 15 December 1980. A dozen of us took the dear chap out for a counter lunch. As the group were getting seated two of us were collecting drinks at the bar. I turned to my companion and said, 'You're about to see a dastardly act performed.' He said, 'How's that?' As I moved off with my tray I looked back at him and said, 'You're going to have to sit next to Colesy.' 'You bastard!' he said. Dr Coles was to step into Merv Hinton's shoes the following Monday.

Dr Norman Coles

At the outset I never took Dr Coles particularly seriously nor did I harbour any hostility towards him. Professionally, he was a competent enough scientist. He had a PhD in Biochemistry and had arrived at CSL some six years before me. As some PhD's are wont to do, he appeared to have had a fairly narrow outlook on medical science and possibly, too, on the world in general.

In 1981, Norman was a bachelor in his mid-forties, near bald and with a pinkish complexion. He gave the impression of being both finicky and secretive. In 1969 one of his biochemists, Allen Broad, was transferred to me because he was desperate for a change. A comment he made a few weeks later always stuck in my mind. He was comparing his changed working conditions: 'This Lab is much better. Dr Coles wouldn't even issue me a box of matches unless I returned the empty box.'

When Dr Coles was promoted to the chair vacated by Mr Hinton, he soon started to throw his weight about. Within two weeks I was certain it was going to prove intolerable to be controlled by two non-medical people. This may appear arrogant but to my mind neither Coles or Davey were qualified to either evaluate or appreciate the work being done by my now considerably diminished group. Dr Schiff, the third member in the line-up between me and the Director, may not have given tangible support in the recent past, but at least I respected his ability to assess the work.

McCarthy refused to alter the arrangements when I complained, and pointed out in a lengthy memo that my three immediate superiors had, 'Years of experience and considerable administrative ability'.

A few months into 1981 Dr Coles seemed to me to be plotting to make my busy life even more difficult. In March, Dr Chris Tyson, who was the only other medical graduate engaged in research, resigned. He said he had been told by Dr Coles that he was no longer to be associated with me and he was to come under a non-medical supervisor. This decision resulted in the termination of the important bee venom project (page 310).

The day this doctor quit I asked Coles to get out of my laboratory. Three months earlier, I had put in a carefully composed request for Rodney Harris to be considered for promotion to the level of Technical Officer. We had all waited patiently for a response from the Administration but none had come. When I asked Coles whether he would chase it up on Rod's behalf he replied that it was still on his desk. I was astounded by this and asked why he had delayed it. He expressed the view that it was not likely to succeed. I expressed the view that it could never succeed if no-one could read it and that, if my memo wasn't in the Administration building by next day, an evil fate would befall him. He was then instructed to clear out, which he swiftly did. Next day as part of the downgrading process Coles arranged for the sign 'Immunology Research' to be removed from the entrance to my Department. It was left dumped on my desk, but I did not react.

Meanwhile Coles was making Marjory Davey's life more and more difficult. He gave her little work and he tapped away on a small portable typewriter in his adjacent office. He may have suspected that, because of Majory's loyalty to me, she might have told me the contents of some of his correspondence. If this was his attitude he misjudged Marjory, who as a true professional would not divulge confidences.

On 25 May 1981 Dr Coles had a major triumph. I received a memo signed by Dr Schiff stating that Marjory could no longer come down to my Department to take dictation but I was at liberty, if it was convenient to Dr Coles, to come upstairs and give her dictation. Furthermore, unlike any other secretary in the Serum Laboratories, she was not to be asked to leave her office to perform tasks anywhere else. I unsuccessfully appealed against this ruling to McCarthy. When dictating, I often go from file to file, reference to reference, and to do my dictation away from source material was almost impractical. Marj and I tried doing it by phone but that blocked incoming phone calls and had to be abandoned. Nobody gave a hoot about Marjory's feelings either; she had been hired to do shorthand, she enjoyed doing shorthand and I was the only one that gave her shorthand. Neither of us was pleased.

On other fronts, a few things had happened. A second case had been treated successfully with Funnel-web spider antivenom (page 217), but unfortunately whilst I was attending a meeting in Sydney a burglar or burglars had ransacked my home in Melbourne. Anyone who has been through such an experience realises that it can leave one somewhat perturbed and edgy for weeks.

Middle-aged scientist brutally sprinkled with paper clips
Prelude

On 17 June 1981 I turned forty-five years old. To celebrate this I chaired a meeting which decided to freeze-dry the largest batch of Funnel-web spider antivenom yet prepared. After a family dinner it became apparent there might be a problem with the woman I loved. My wife Megan, exasperated by my working habits and single-mindedness, hinted at an impending change in our relationship.

The following day I was at work early because I had to give two major lectures at the Canberra Hospital next day. One was arranged by the College of Surgeons and the other by the College of Physicians. I wanted my slides sorted and packed as soon as possible in order to get other things done before flying out later in the day.

I was just finishing with the slides when Marjory Davey rang with the information that Coles wanted to see me some time in his office to discuss the budgets for next year. I said to Marj, 'Look, I'll come up right away because I've got lots of thing to do down here.' So up I went, saying hello to Marj as I went past her and into Coles' office. I sat down and for a few minutes we just chatted in general about the budget. He said that there were going to be some cuts since the money allocated by the Director in real terms was a reduction of 8.7 per cent. I asked him how this was going to affect me and he replied that certain groups would be cut more than others.

I asked him to get to the point, and he said, 'Well, you will lose certain projects to start off with, such as allergens.' He passed over my budget proposals and I noticed there was a fair amount of scribbling on them in his handwriting. He said, 'In two weeks you'll be left with only Allen

Broad and a Technical Assistant.' This was a shocking revelation and I responded, 'Well how long will I have Allen and the Technical Assistant?' Coles said, 'It will be reviewed in September.' 'What's that mean?' I asked. 'It means you'll lose them all,' he replied.

I jumped to my feet and emitted a four letter word which I very rarely use and then only in times of sudden extreme pain or terrifying surprise, 'Colesy, this makes me very angry.' His response was, '... the more angry you become the happier I will be.' I slammed the budget papers down on his desk which made other papers take off in different directions and called him 'a bloody little worm'. This description, in hindsight, was rather generous.

He grinned happily and made some smart comment along the lines of: 'This is what we're recommending to the Commission, and a fat chance you'll have of getting them to change this decision.' I picked up a bowl of paper clips from his desk and tossed them at him. The empty bowl was slammed on his desk. He was lucky, the alternative was to pick him up and throw him out of the third storey window. Instead I picked up a bowl of pins and added them to the paper clips.

As I came out of the office I said to Marj, 'The bastards have wiped out my Department.' She looked shocked and, as Coles came round the corner and said, 'Marjory, you're a witness to this, come and pick up these clips.' I turned round and said, 'Colesy, don't you dare pick on Marj,' and to Marj, 'Don't you let him annoy you, Marj.'

The whole business was over within five minutes and I returned to my Department on the floor below. I'd had rows with Coles before and I didn't give the matter much further thought since I was flat out doing other things.

In passing this spontaneous action may have put me in the good company of Gough Whitlam.

During the 1966 Budget sittings, Whitlam made a wounding allegation against Paul Hasluck, who responded by interjecting: 'You are one of the filthiest objects ever to come into this chamber.' According to parliamentary historian Gavin Souter, Whitlam was in the act of raising a glass of water to his lips: 'As quick as a flash, Whitlam flung the water at Hasluck's face. "I was provoked," he said.'

The Canberra Hospital had asked me to fly up the night before the talk, as quite often the notorious Canberra fog disrupted the arrival of lecturers. They preferred to have their speaker comfortably tucked up in bed in Canberra the night before.

As I drove out to the airport, Coles was completing a mastery summary of the paper clip incident on his little portable typewriter. His written summary was surprisingly like mine and included my description of him as 'you little worm' which someone else later described as 'almost an endearment'. That afternoon McCarthy and Viv Davey went into a huddle and compared notes. Viv apparently volunteered the information that I had insulted him by calling him a little twit. This had been done in the privacy of his office after which I had related to an amused Peter Schiff the reason for this unseemly outburst. This occurred five years earlier when I was a single parent. (Viv said this was no excuse for not being able, hypothetically, to go interstate 'at a moment's notice if CSL so directed'.) By the time I reached Canberra the trio had probably collated my individual descriptions of them, namely, 'a swine', 'a little twit' and 'a little worm'. Years later these still seem fair descriptions.

The talks in Canberra were fun and the organisers made generous comments. Some of the top Health Department people were at the second talk and I was driven over to have afternoon tea in the Department.

I managed to get out of Canberra and at 10.00 pm that night Marjory rang me at home to say that she had been asked to sign a statement regarding 'the paper clips incident'. 'Hello,' I thought, 'something's on the move.'

On Tuesday, 23 June, I was called over after lunch by Alf Brogan to see McCarthy. I was ushered into McCarthy's office and the three of us sat down. McCarthy then announced that we would wait for Mr Davey. We sat there for some three minutes in silence until Viv Davey arrived. Then McCarthy with his nose in the air and eyes half-closed said, 'I just want to say to you that I intend to act on the report I received from Dr Coles, and I want you to be aware of the fact. That is all.' The searing hatred I felt radiating from McCarthy and Davey was extraordinary. I merely said slowly, 'Good heavens,' and returned to my Laboratory.

Next day I rang the AMA's Director of Industrial Relations, Malcolm Brown, who gave me instructions and undertook to contact McCarthy to find out what he was up to. (John Welch was no longer with the AMA.) Later Alf Brogan wandered into the Laboratory and handed me some documents. They indicated that I was going to be charged with an offence under the Staff Rules. Alf and I had quite an objective discussion about the whole matter. He gave me the impression that he thought it was a great pity that this was happening again. In return, I gave him the distinct impression that this time I wanted the whole business to be more open. After we had parted amiably I notified my various unions. I was not as yet suspended and McCarthy appeared to be planning carefully, taking his time.

Friday, 26 June, I was at work at 7.00 am drawing up contingency plans re animal immunisations, etc., in the light of my impending suspension. I duplicated all articles in preparation so one set was at work and the other was in the boot of my car. The AMA recommended that I see Mr David Wells of the legal firm Mallesons and an appointment was made at 3.30 pm. Mr Wells was described as a whiz at handling industrial disputes. At lunchtime I visited the Executive of the Commonwealth Medical Officers Association (CMOA) at the Health Department in

Spring Street. They offered firm support in this latest episode of what they believed was a chronic and unsatisfactory situation.

All but the wealthiest of clients has a sinking feeling when entering a large legal firm. Who pays for these posh offices and spectacular views across the city? He or she does. At Mallesons I met David Wells and Bruce Moore. Over the next two months or so I was to get to know these two chaps very well. Bruce, the more junior of the two, was quiet and studious.

David Wells on the other hand was distinctly aggressive and short-tempered. He was all for a full frontal attack in the Supreme Court and for slapping writs on McCarthy. At the same time he was probing what I owed on my house, what my annual salary was, and if there were any other resources that could be drawn on for the approaching legal battle. I merely wanted sound legal support at any hearing before the Commission. I also wanted any such hearing to be fully recorded. David Wells was not kindly disposed to having his advice rejected, at least in part. Later, as he and Bruce became more aware of the background to the dispute, some sound plans emerged.

Something about the navy came up and then the penny dropped. David Wells was the son of my former Captain. To some extent this accounted for the briskness of his approach. The revelation emerged in the midst of a disagreement on tactics. I told him what a great bloke his father was and virtually implied that the same did not apply to him. After a bit of a chat about his father, the fact emerged that young David had lived for some time at Flinders Naval Depot; a complete change in our relationship developed.

David threw tremendous energy into the case and certainly did his best to keep the legal costs to the minimum.

At 6.30 pm on that cold winter's night of our first legal meeting my car broke down on the way home. When I finally got home two hours later, I was feeling quite sorry for myself, particularly as getting a third mortgage on the said home seemed a distinct possibility.

For the next week or so it was a phoney war. Hours were spent documenting the background to the current dispute and planning the best way this material could be presented should the opportunity arise. A nicely complicated legal letter was sent by Mallesons to CSL's solicitors. Apparently acting on Davey's suggestion, Dr Coles had got to the POA fast to complain about me before I could complain to the POA about him! This put the POA in a slightly difficult position but, because of the background, they offered firm support.

On 1 July Malleson's were informed by CSL's solicitors that the Commissioners (excluding Dr McCarthy) were going to decide on the charge against me on Thursday, 16 July. We responded by requesting that the hearing be a public one and that I be able to be represented by my legal advisers.

They replied that a public hearing would not be appropriate and 'would not be in the interests of your client, the Commission, or the public'. They had no objections to legal representation and I would be able to call evidence or make any oral or written submissions that were considered appropriate.

We really had little idea what form the hearing before the Commission would take and I would not be able to subpoena witnesses which made things difficult, particularly since we wished to have witnesses from the CSL commercial division give evidence as to the PR value to CSL of the television series and lectures, etc., that I had been involved in. Several key marketing personnel who were approached by my solicitors declined in horror. One frankly explained that to volunteer 'would finish his career at CSL'.

Suspended again

At 4.50 pm on 6 July 1981, Viv Davey surprised my poor laboratory staff by appearing for the first time ever in my laboratory. In my absence he left an envelope marked 'Private and Confidential' on my desk. That evening, what turned out to be an identical letter was delivered to my home. Neville had obviously been very busy signing duplicates of my Orders of Suspension under the Staff Rules. This time the suspension took effect from close of work the next day and so I was free to go and say goodbye to my mates. Neville had obviously learnt something from the previous episode because this time the suspension notes included the following, 'We do not impugn your discretionary right to consult in a private capacity with medical colleagues, and to undertake patient management in areas of your expertise.'

Things then hotted up in preparation for the one-day hearing on 16 July, which was set to resolve the matter one way or the other. The AMA sent a magnificent three-page letter off to the then Minister for Health, Mr MacKellar. (It pointed out the near inevitability of the current situation, and recalled for the benefit of the Minister items from previous appeals.)

David Wells said we would need a barrister for the sixteenth and he had an appropriate one in mind. I asked him to request CSL to pay for an exact transcript of the hearing, preferably by an outside agent. I said, if necessary, I would pay half. This request was reluctantly agreed to by CSL. (The estimate for the hearing on the sixteenth was some six hundred dollars. In fact, as the hearing dragged on the total cost of transcripts was $3740. It was money well spent.)

At this time I was a little over halfway through writing *Australian Animal Toxins* for Oxford University Press, bringing together some fifteen years of collected material. In retrospect, this was good therapy. I could forget about Neville and his mates and emerge refreshed to face each new aspect of problems at CSL.

As a quick lead up to 16 July, the following are selected extracts from my diary notes of this period.

7 July 1981 — Alan Coulter rings. Following my suspension he's been called over to the Administration building with Rodney Harris. They face Viv Davey and Dr Coles and are told they are no longer part of my Department but that they may sit tight for the moment and 'we will put you somewhere'. Alan asks about his Master's thesis which I am co-supervising and which is nearing completion. He gets no firm answer.

2.10 pm   I visit Professor Bryan Hudson at the Howard Florey Institute. He is very keen to help and his only engagement on the sixteenth is a meeting at 2.00 pm for an hour. Bryan thinks the whole thing is madness. He informs me that Pansy (Professor Wright) advises to admit nothing because that is where Bazeley (a former Director of CSL) went wrong. Pansy requests copies of my recent publications and has offered to prepare a written submission.

Evening   I contact the distinguished Melbourne physicians Sir Ian Wood and Sir John Frew requesting support. Both will come on 16 July. Ian Wood wants further details of the case and Jock (Sir John) indicates that he dislikes Dr Forbes, the CSL Chairman.

8 July 1981 — I ring Mr Doug Leslie, the Chairman of the Australian Resuscitation Council, who is also keen to help.

9.08 am   I speak to Professor Attwood of the Pathology Department at the University of Melbourne. He's initially hesitant but becomes quite keen.

9.20 am   Dr Malcolm Fisher returns my call. He is forwarding a letter signed by his colleagues at the Royal North Shore Hospital. Some twenty calls are made by 1.00 pm. I then get the last of the money out of Statewide, my building society, to pay for legal costs to date.

Two dedicated right-hand men. Alan Coulter and Rodney Harris in 1978. (Photo Melbourne *Herald*)

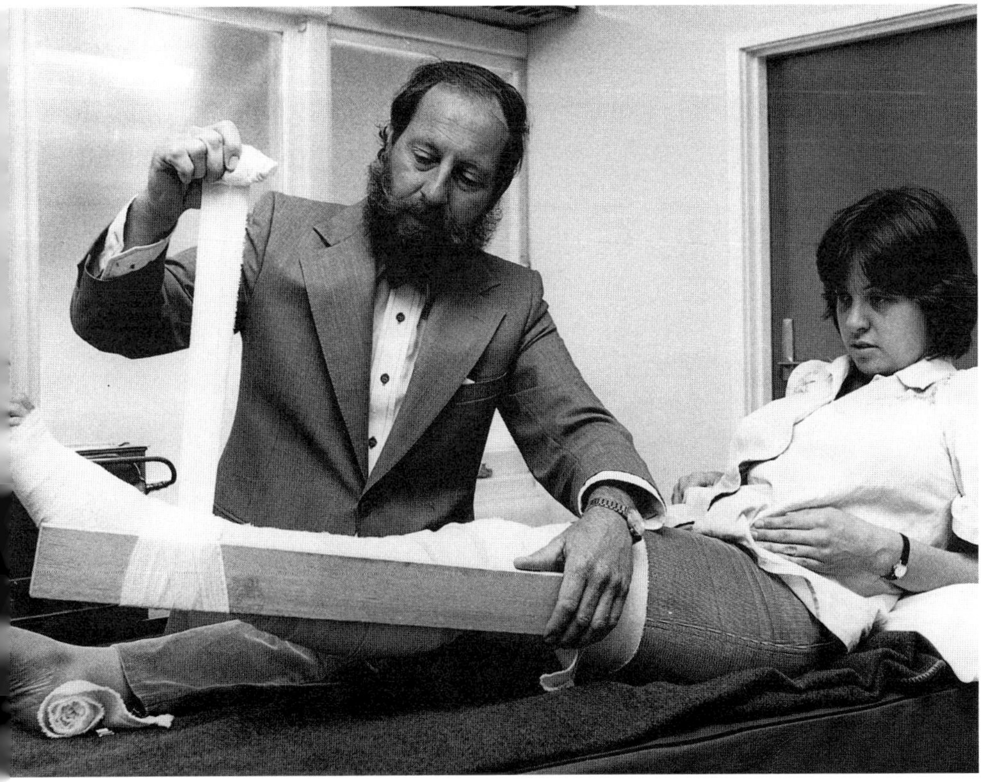

The author demonstrating the pressure-immobilisation type of first aid for snake bite, circa 1979. The rarely-subdued Erin Lovering is the 'patient'. (Photographer unknown)

Dr Bill Lane, Director of CSL from 1966 until his untimely death in 1974

2.00 pm Meeting at Mallesons. The opposition barrister is to be Mr Richard Alston. Everyone seems to know him as he was a recent President of the Victorian Liberal Party. That's a coincidence—the Chairman of the Commission, Dr Forbes, is president of the Federal Liberal Party and another Commissioner, Mrs Hardy, has Liberal Party connections. Mallesons say the barrister we hope to get should run rings round Alston because he's a wizard on administration and the technicalities involved. However, when the fellow we want is contacted, he has to admit that Richard Alston's already contacted him for his opinion on how he should handle the case and therefore he is out. Their second choice is Mr Ross Robson, who turns out to be a most fortunate selection.

Mallesons are going to contact all of the outside witnesses as well as those that have volunteered from CSL.

On Wednesday, 15 July 1981, I met my Barrister Ross Robson for the first time. He's an impressive and friendly man who has an excellent grasp of the situation. We have sorted out which witnesses are the best to use and I am appalled to find that some of the CSL people have had their pay docked for the time they spent making statements (this petty decision was later reversed).

The hearing
Round One—Thursday, 16 July 1981
It was a clear, sunny winter's day as I sat beside the guard at the main entrance to CSL. I was awaiting the arrival of my legal team while he was checking the passes of the incoming workers. A number of workmates reached up and shook my hand as they squeezed past the Guard House window. This was the first time such an inquiry had been held at CSL and it was obviously of some interest to the workers.

Just when I was getting worried, my team arrived in a Mercedes. They were allowed to drive in, but parked near the gate ready for a quick getaway. David Wells looked particularly aggressive, Bruce Moore was more scholarly than usual, whilst Ross Robson appeared most distinguished in a very English overcoat and gloves. We then waited in a small room adjacent to the Guard House.

Soon we were led up through the Administration building, furtively observed by dozens of office workers and entered the innermost of inner sanctums, the boardroom. It was quite crowded and extra tables had to be brought in and allotted to the various parties involved. I said 'Good morning Neville,' to McCarthy and we exchanged pleasantries. A minute later he seemed to disappear into a wall outside one of the boardroom doors. When he emerged, I investigated and found there was a secret toilet in the passageway. The entrance was very hard to detect but I nipped in and used it before proceedings began. It had an admirable collection of hairbrushes and I was tempted to use one to shine my shoes.

The CSL Commissioners were seated around the main table. The Chairman, Dr Jim Forbes, was in charge of the proceedings. He was a former army officer who did a PhD and then had various ministerial jobs in the Federal Parliament from 1963 to 1972. He had been minister at times for the army, navy, immigration and then finally health. About this last portfolio he admitted he 'seldom gave a thought to CSL', presumably leaving it to its own devices relatively free from direct ministerial scrutiny. He had been appointed CSL Chairman in August 1979. Dr Forbes, like his fellow Commissioners, had no legal training, but he conducted the whole hearing in a very fair and intelligent manner.

Other Commissioners present were Dr Tom Hurley, a Melbourne physician, Mrs Barbara Hardy, and Mr H. D. Huyer, about whom I knew little other than that he was a big knob in Philips. He did not attend later hearings. I didn't like the look of Mr Huyer and was pleased when he found more important things to do than to sit around pondering my fate. The last Commissioner was Mr McMullan, who took up a great deal of room. Also seated at the main table was Mr Alf Brogan, Secretary of the Commission. All these distinguished people had very large name plates in front of them which had been especially prepared.

I was pleased to see two efficient looking stenographers with fingers poised over shorthand machines. In the middle of the room was a chair and small table reserved for the witnesses and I saw with satisfaction Alf Brogan placing a Bible on this table. It was an indication that CSL did recognise the existence of a higher authority than McCarthy! At the other end of the room there were two tables reserved for the defence and prosecution. This was a fairly crowded area and McCarthy and I were to be seated only a few feet apart. He was to be chief informant and instructor for the prosecution during the proceedings. From the moment we all sat down, my left ear was dedicated to collecting audiosignals from the opposition table.

At 9.50 am we got underway. Dr Forbes explained this was a special meeting to hear disciplinary charges. He stated that it would not be open to the public and, although Dr McCarthy was an appointed member of the Commission, he would not sit on this hearing. He then read out the details of the charges, which included the claim that I had used abusive language and gave one example—'you little worm'.

Mr Richard Alston, Counsel assisting the Commission, then got up and made his opening remarks. I soon found comfort in the fact that he was no Perry Mason and awaited events with interest. He called Dr Coles in as the first witness. When poor Coles came in, he, like most of the other witnesses, looked taken aback when he saw the size of the assembled company. He took the usual oath on the (CSL) Bible. Alston then took him through a string of predictable questions which merely required yes or no answers.

Coles agreed with Alston that the whole episode had probably lasted only three minutes. Furthermore Coles testified that he had said to me, '... the more angry you become the happier I will be.' On hearing this statement, the Commissioners turned around almost in unison, and peered a little more closely at Coles.

Mr Alston next took Coles through his recollections of flying paper clips and pins. Reading the transcript years later, I am just as unenlightened as to what point he was trying to make.

My Barrister, Ross Robson, then got to his feet and gave Coles an hour-long interrogation he would never forget. First of all he had Coles confirm as correct: a long series of statements regarding the output of my research group; my standing in the scientific community; and my dedication to my work at CSL. He then probed aspects of Coles's research career and established that he had not done very much for some years. Coles agreed with Mr Robson that I did not respect him either as a person or a scientist. Furthermore, he agreed that the relationship was complicated because he was not a medical graduate and I saw him as just a bureaucrat.

These admissions seemed harmless enough but then Coles came out with a remarkable statement. He said he had made a report early that year which 'recommended that management should give every assistance to Dr Sutherland to find employment elsewhere'. This made me sit bolt upright in my chair. I was flabbergasted. When Robson said, 'You didn't want him at CSL, did you?' Coles replied, 'No.'

Robson then tackled Coles about his statement '... the more angry you become the happier I will be'. Coles explained that he adopted this attitude so that I would not have the gratification of seeing him upset. 'It is exactly the same system that Telecom give their people, their victims, who receive obscene phone calls. That is the line I took.'

Jaws dropped when this explanation was given. Later Dr Forbes gave Coles the opportunity of 'commenting on the emphasis the words appear to have'. Coles stuck to his belief that his extraordinary statement was appropriate to the situation.

By the time morning tea came I felt quite sorry for Norman Coles. He was left sitting behind the witness desk and was a bit of an embarrassment to the prosecution. My party was a little bit surprised when I took my cup of tea and sat on the desk and talked to him.

Shortly after we resumed, Coles withdrew. Mr Alston then startled me by producing a photocopy of the handwritten letter that Dr McCarthy had written to me in January of that year. (See page 214) In this letter, which was personal and confidential, he appeared to express concern about my mental health. The Chairman ruled that the letter was not relevant for the moment.

It was now our turn to wheel in a few witnesses and almost to a man or woman they were magnificent. First came Dr Alan Duncan from the Children's Hospital. Alan gave a great rundown on the collaborative Funnel-web spider work and pointed out a number of useful projects which had been temporarily interrupted. Mr Alston only asked Alan three questions, all of which seemed quite pointless. For example he asked what the incidence of Funnel-web bites had been in Victoria for the last twelve months. The answer was none. After all, there are believed to be no dangerous Funnel-web spiders in Victoria!

Next off the rank was Professor Bryan Hudson. He had been the foundation professor of medicine at Monash University and at the time of the inquiry was associate director of the Howard Florey Institute for Medical Research. Bryan, who was in a brisk and somewhat crusty mood, summarised my published works and stated that he was amazed at how much research had been done. He explained how my group had helped his team with technical advice on the production of antibodies to difficult antigens and he said a few other useful things. He also presented a short and apt letter from Professor Wright to the Chairman.

Neville had drawn up a few questions for Mr Alston to bowl down to Professor Hudson. These were smacked back so quickly he almost got his head knocked off. After a quick glare round at the assembled company, and at Mr Alston in particular, Professor Hudson went off to do something more useful.

Next came Mrs Marjory Davey. Marjory later admitted she was quite terrified of this ordeal and, like some other CSL witnesses, felt intimidated by the presence of McCarthy. Requests that he 'go for a bit of a walk' were refused.

Marjory confirmed she had heard a disturbance but she had not seen anything, since she was seated facing the opposite direction. It was also established that the bulk of her work came from me, that she had been hired because of her shorthand skills and that part of her job was to take dictation from me. Since her employment no-one else had given her dictation.

At 1.00 pm the whistle blew for lunch. No provision had been made for my band and so we meandered into the canteen which was just closing. A nice little lunch was turned on for the other mob in a room just off the Director's office. I noticed with some satisfaction as we were departing that Mr Alston had lit up a large and particularly foul smelling pipe right next to McCarthy, who loathed smoking.

I later had a word with Alf Brogan about these unsatisfactory catering arrangements. As a result, when required, a beautiful lunch was turned on in the Conference room right near my Laboratory. I made sure uneaten cakes, etc., were whisked down the corridor for the mob to have for afternoon tea.

When we resumed after lunch, Mr Alston said he had no further witnesses other than McCarthy, whom he was going to call at a later date. Mr Robson advocated that McCarthy be called immediately since he was the only other prosecution witness. This disturbed the 'planned order of things' but Neville John McCarthy had to take the oath and sit in the witness chair, well away from his legal advisers.

Mr Alston and McCarthy had obviously decided that their approach would be to dig out as much background dirt on me as possible. Despite Mr Robson's objections that this was completely out of keeping with normal legal proceedings he was overruled and the two settled down (they hoped) to totally besmirch my character.

The first thing Mr Alston winkled out of McCarthy was the matter of CSL placing an advertisement in the midst of my snakebite article in the *Australian Family Physician*, which is described on page 254. We now had every right to bring the same subject up when our chance came.

Next, McCarthy was asked about any adverse remarks I might have made about Mr Davey. He said that two or three years ago (as the transcript says—but it was really five years) I had called Mr Davey either a 'little twerp' or a 'little twit'. I noticed one Commissioner trying desperately not to smile when this bit of information was released.

Next was McCarthy's version of my suspension the previous year. He produced a stack of photocopies of the personal and confidential note he had handwritten a week or so before that particular event. He also passed out copies of my brief but carefully balanced reply (see page 214).

Things got a bit boring during the remainder of Mr Alston's questioning. McCarthy took pains to stress the insignificance of venom work and its limited place in the great plans he had for CSL. He surprised me by making the comment that the number of people affected by venom in any one year 'you could count on the fingers of both hands'.

When Ross Robson got up to question McCarthy, the comfortable relationship between questioner and witness evaporated. No longer were long meandering replies tolerated. I found the whole business most exciting, especially since it was only the second time I had seen someone giving McCarthy a hard time. David Wells communicated most efficiently with Ross. He either smoothly passed a note or whispered a suggestion out of the corner of his mouth.

Ross first established that the work I was doing relating to antivenoms, etc., was fair and square in the CSL charter. Next McCarthy had to admit that the advice I gave to doctors, etc., was 'good for CSL'. He also admitted that he was aware that Coles had made the recommendation that 'management should give every assistance to Dr Sutherland to find employment elsewhere'.

Ross Robson then went into attack, using the advertisement in the *Australian Family Physician* for ammunition. McCarthy, having played down the importance of venom research, was read selections from this full-page advertisement which included, 'CSL is the only organisation in Australia that is continually doing research into venomous bites and stings and the antivenoms to counteract them.' Then a little later, 'Write a CSL brand like Moxacin or Optipen and your prescription will help research programmes which will help benefit you and every Australian. Don't wait for a snake to bite before you think of CSL.'

At this stage McCarthy's answers were becoming increasingly monosyllabic. When Robson asked, 'Isn't it the case that CSL used the success of Dr Sutherland in the field of antivenoms to promote and sell the other products?' He could only answer, 'Yes.'

McCarthy was clearly disconcerted by this vigorous interrogation. In fact his initial expression of disbelief was similar to that seen some years later on the face of Romanian Leader Ceausescu when he heard the first howls of revolution from the crowd below the balcony.

Soon after, Dr Forbes interrupted the proceedings because he had been told that there were 'Four eminent gentlemen waiting outside'. Dr Forbes wanted to know what role these gentlemen were to play. Mr Robson said he would like them to give evidence on record 'of the overwhelming recognition of Dr Sutherland as a scientist'. Dr Forbes replied, 'I don't think anyone here is questioning Dr Sutherland's eminence and standing as a scientist. It has never crossed the Commission's mind that that particular matter was in question.' The questioning of McCarthy continued briefly. He admitted that he did not consider the statement, '... the more angry you become the happier I will be', to be a good one to use under the circumstances.

After giving his evidence, McCarthy popped out for a minute. When he returned he seemed a little distraught and on reaching his seat whispered, 'What's Doug Leslie doing out there?' To make matters worse, seated beside Doug Leslie were a couple of knights of the realm. (I had incorrectly assumed details of all our witnesses had been passed to the opposition.)

I had admired Doug Leslie for years. As a Colonel in the medical corps he had saved many lives working in the grim conditions of the Kokoda Trail. My lasting memory of this man is his extraordinarily penetrating gaze. His eyes were a lustrous flinty blue-grey which gave the impression of recording forever all that was scanned. Despite this memorable characteristic I was not unduly nervous of him as a medical student.

In 1959 he had operated on my mother after she had been rushed down from Bendigo critically ill with a bowel obstruction. When he emerged from the two-hour operation just before dawn, he took time to sit down with me and a fellow student, John Ipsen. For the next thirty minutes he drew diagrams and explained the plumbing job he had performed on my dear mum. The obstruction had been due to a gallstone and the outlook was grim but mother made a near-complete recovery and lived for another thirty years.

As Chairman of the Australian Resuscitation Council, Doug had taken a great interest in the newly developed pressure-immobilisation type of first aid. One Saturday morning in 1979, I was given the opportunity to demonstrate this technique to his committee—and thereafter it was given full national backing.

As a witness Doug Leslie assessed the value of the work I had done to date. He considered that the new first aid management of snake bite was of immense benefit not only to Australians but to other countries as well. He believed that venom research had produced enormous kudos for CSL and he said that he had always found me easy and co-operative to deal with. Prior to Mr Leslie withdrawing, the Chairman asked him if he had received particulars of the charges made against me. He had, and the same question was subsequently asked of all other outside witnesses. Earlier, I had hand delivered copies of the charges to them so that they knew the details before volunteering to give evidence.

The next witness was Harold Attwood, Professor of Pathology at the University of Melbourne. This very tall and extremely thin Scottish pathologist was not especially keen on coming along until he saw the list of charges. Harold gave a great three-minute burst on what he thought of my research activities and the benefits that had flowed from them. He even said that he considered the first aid studies to be 'so full of common sense that it thrills me'.

The next witness was Sir Ian Wood, the former Head of the Clinical Research Unit at the Walter and Eliza Hall Institute. This distinguished and self-effacing gentleman was admired by generations of doctors and medical students. His autobiography *Discovery and Healing in Peace and War*, privately published in 1984, is the diametric opposite of mine. Self promotion was not in his nature. Initially I hesitated to ask this fine person for help but self-interest won the day.

Sir Ian made some nice observations about my development over the years and sounded as though he meant every word of it. He considered CSL was the ideal place for conducting venom research and when Mr Alston asked, 'Even if it is to be carried out to the detriment of others?' Sir Ian snapped back, 'I'm not going to answer that question.'

Sir John Frew was the last witness for the day. Tall and in his late sixties, silver haired, with sharp blue eyes, he had a powerful reputation for not mucking around. As Censor in Chief of the College of Physicians, candidates had dubbed him 'The Grinning Death'. I'd been his Resident at the Royal Melbourne Hospital. You either got on well with Jock Frew, or you didn't. As a student I visited my allotted patients every day including weekends. Jock as it happened did the same, including his public patients. He was a powerful figure and had been a Commissioner at CSL starting at the same time as I did.

After saying a few kind words, Jock volunteered that I had a 'certain prima donna quality', which was something he came across very often at the Royal Melbourne Hospital. 'Handling such people,' he said, 'requires skill, which not every administrator has.' He gave the assembled company a look which implied he wasn't too impressed by the goings-on. Jock said the only time he had had to take major action was when an offence was particularly serious, such as the 'constant raping of the nursing staff', or 'uninhibited drunkenness'. He stated he would not worry a great deal about being called 'a worm'. 'In fact', he said. 'Many times I've been called much worse than that.'

After Sir John left, Mr Robson raised the point that the Public Service Board may not have approved the conditions of service of officers appointed under Section 24 of the CSL Act. I didn't understand his argument, but liked his conclusion: 'In my submission the rule is bad and therefore there's no case to answer.'

Between hearings

Thus, what was to be a one-day hearing was destined to take at least another day. In fact it was spread over weeks. The reader might think that I was having a nice little holiday on full pay between hearings, but this was far from the case. Hours were spent digging out information, planning tactics and co-ordinating the various people who wanted to help. On top of this there were financial worries and a few personal problems. I found comfort by writing furiously and having frequent tramps along Brighton Beach.

A couple of interesting things happened on Monday, 20 July. Malcolm Brown of the AMA contacted me to say he had just been rung by McCarthy and he had the distinct impression that the Commission might well back down in the near future. This seemed most encouraging.

At 2.00 pm Dr Geoffrey Metz of the Alfred Hospital rang and said his brother-in-law, Senator Gareth Evans, had been stung by some marine creature near Cairns the day before. He was most concerned by the condition of the Senator who was, as he spoke, on the way down by plane. When I saw Senator Evans at 4.15 pm at the Alfred Hospital I concluded that he might have been stung by a large black sea urchin and would survive. The Senator, who was Shadow Attorney-General at the time, said he was most interested in what was currently going on at CSL. So am I, I told him. He asked me to keep in touch while I told him to slow down, otherwise he might drop dead as Senator Greenwood had done at the same age, a few years earlier. He looked thoughtful at this suggestion and slept for eleven hours that night.

A few of my diary entries were highlighted:

Friday, 24 July 1981   Dr Coles has won a Safeway trip to Fiji and will take his mother.

A copy of the transcript of Day One picked up and my suggested tactics for the next hearing delivered to David Wells and Ross Robson.

Monday, 27 July 1981   Attend a meeting from 9.30 am to 3.30 pm with Ross Robson and Bruce Moore. I picked up my mail from CSL, all of which had to be opened in front of a Security Officer. I don't think either of us knew what he was looking for.

At 5.00 pm, Ian Gust from Fairfield Hospital rang to check on the state of play. I was cheered up when he told me that when they were trying to get rid of him from Fairfield the expression used was 'assisted to find greener pastures'.

Round Two—Wednesday, 29 July 1981
When we all assembled again nearly a fortnight later, Dr Forbes announced that because of various people's commitments the next time we would meet would be 26 August. Ross Robson objected to this and stated that this broke the original agreement.

Then we learned that one of the Commissioners, Mr Huyer, had a meeting in Sydney later that afternoon and so the hearing could not continue after lunch. This meant my evidence was to be spread over two or possibly three days, weeks apart, which Ross most correctly objected to. After about thirty minutes of discussion there was a short adjournment and then Dr Forbes announced that the best they could do was to complete the morning's sitting and then sit again on Wednesday, 5 August, for the day, and then adjourn if necessary until Wednesday the following week, 12 August. After another break, I agreed through Ross to accept this stop-start arrangement.

The hearing then started up again and Mr Alston took up the matter of validity of the staff rules and their relationship to the Public Service Board. I found the alternating submissions by the barristers confusing, but I gathered that whenever the CSL Act was amended, then the conditions of employment (including the staff rules) had to be approved by the Public Service Board, for the conditions of employment to have a continuing effect. If the Staff Rules were not valid then there would be no case to answer.

After a short adjournment, the Chairman announced that the

Commissioners were not accepting Ross's challenge to the validity of the Staff Rules and therefore to the current proceedings. Ross then told the inquiry that the Chairman's ruling should be challenged in another place. Further legal debate followed, in the midst of which the Chairman said, 'Excuse me, I am just a layman, as we all are'. Ross then explained in simple terms why we should all go to the Federal Court to get a ruling on the validity of the Staff Rules. He and Mr Alston agreed that this ruling could probably be obtained before the next scheduled meeting on 5 August.

The Federal Court eh? I thought. It seemed as good a way of spending money as any other. The legal parking meters were ticking away at a great rate but I felt I was getting value for money. The Federal Court offered a slim chance. I was advised to take it since, if the finding was favourable, it could truncate the whole proceedings.

Next day at 10.00 am Bruce Moore rang to say that we would appear in the Federal Court at 10.15 am the next day to obtain a stay of proceedings for 5 August. I told a disbelieving Bruce that it was a safe bet that McCarthy would obtain the services of a QC for the Federal Court. He reported back soon after, 'You're right, they've got Graeme Frickie, QC, assisted by Mr P. Galbally.'

That afternoon when I gave a lecture to the final year medical students at the Royal Melbourne Hospital I overheard one say to another, 'This is the bloke that keeps getting sacked from CSL.' Such is notoriety, I thought.

When I got home from the lecture I had several calls from *The Age* medical reporter, Mark Metherell. He was writing an article on spider bite which attracted more media attention when it was published the next day. Thus on Friday, 31 July, prior to leaving for the Federal Court, I had almost non-stop radio interviews from 7.00 am. (I made sure not to say anything about the disciplinary hearing.)

The Federal Court buildings were attractive and I found the case before mine quite interesting. It involved a pretty plump fellow who was fighting for the right to wear a tent-like caftan while being employed by Telecom. I had seen him on television the week before, which only added to my absorption of the case.

Our case was about to start at 10.00 am when, for some reason, we were all temporarily kicked out. McCarthy and Brogan went into a huddle in the corridor with their new legal eminence. My team entertained me with legal jokes that even I could understand. Shortly after, when I was having a quiet wee in the toilet, McCarthy joined me at the urinal. (I was very tempted to give him a bit of a push as I left!) When I emerged, one of the reporters in the court requested an interview. My legal team allowed me to discuss spider bites with her. When at last the case started at 11.55 am I noticed with satisfaction she was scribbling frantically.

Frankly, I didn't understand much of what then followed. The introductory bits and pieces were fairly straightforward, but thereafter each party took to reading excerpts from what they thought were appropriate precedents. Occasionally there was a five-minute silence while both barristers and the judge, Mr Justice Northrop, read a particular submission. Once the judge sneezed three times in a row, winced and blew his nose with great vigour. On several occasions he made a joke which had the barristers dutifully falling about, but left Alf Brogan, McCarthy and myself totally in the dark.

About an hour after the lunch break, the judge announced that he had had difficulty in reaching a decision but found in favour of CSL. His written judgement would follow in due course. Like a flash, McCarthy leant over and spoke to his QC, who shot to his feet asking for costs to be awarded against me. Mr Justice Northrop cocked his head to one side and slowly looked around while I held my breath. After a few long seconds he quietly announced that he would make no such order. Emerging from the Federal Court onto the street, there was a battery of television cameras waiting. In unison, my legal team all said, 'Say nothing', and the three of them threw out their chests and marched past the cameras looking distinguished and determined. At least they didn't tell me to pull my coat over my head and run across the road with them before being driven off at high speed!

Round Three—Wednesday, 5 August 1981
When we resumed at CSL at 9.30 am the opposition team had changed. Mr Alston was unable to attend and had been replaced by Messrs Frickie, QC, and Galbally. Mr Huyer was not attending because, although he had got up very early, the plane strike had left him stranded in Sydney. (This was the great pilot strike lead by another McCarthy which had drastic consequences for the striking pilots.) Ross Robson proposed that since Mr Huyer could not attend the day's hearing, he should drop out of the proceedings thereafter.

After a short introduction, it was my turn to sit in the witness chair. Ross steadily took me through all the events which had led to the paper clip incident. Precise details of workload, publications, petty frustrations, blocking of staff promotions and other little aggravations were relentlessly presented to the Commissioners. Each time the Chairman asked him to speed it up, Ross Robson agreed to oblige and then continued exactly as he had planned.

By the time he got to the paper clip incident I was nicely warmed up and almost enjoying myself. I even scored the occasional chuckle from everyone except McCarthy.

When we resumed after lunch I was surprised to see Mr Galbally undertake the examination while Mr Frickie, QC, stayed in the background listening to McCarthy's whispers. Mr Galbally was not the well known criminal barrister and there was nothing especially memorable about him. He did a workmanlike job—accusing me of behaving like a petulant, spoilt child—which only drew the response that I was acting under extreme provocation. He considered that the expression 'nothing but a little worm' had a touch of arrogance about it, which I denied. He then asked me if I had used what 'you politely term a four letter word' and I replied that I had.

Mr Galbally proceeded with the next line of attack. He came up with two dates of Research and Development meetings during the previous eighteen months, which I had failed to attend. I could recall a reason for missing one of these but not the other. (That night when I checked my diaries I found the reason—I missed the 7 February 1980 meeting because McCarthy had suspended me thirty minutes before the start of the meeting!)

A number of times Mr Galbally dropped a particular line of questioning when the answer coming back was clearly not the one he had expected.

The transcript bristled with 'snappy' but at times unclear banter such as:

Mr Galbally: 'Let me put it to you another abuse of respect, I suggest, to Mr Davey in January of this year.'
Response: 'I could put it to you if you go back ten years you will be battling to find more that two and a half occasions I have been frank with Mr Davey, so I assume this will be the last one.'
Galbally: 'Is that supposed to be some cutting, smart remark, Doctor?'
Response: 'No, I am just making the point that you would have just about run out of suggestions in that direction.'

A little later I expressed the opinion that Mr Galbally was doing the legal profession a disservice by one of his questions. Further on, I requested for him not to mutter to himself after hearing my reply.

We then had a debate about secretarial service and the advantages of direct shorthand versus use of recorders. Fortunately, I could use the example of the stenographers present. If they did not catch a particular word they could immediately indicate this fact to the speaker. I could also punch home that at no stage had I asked for my own secretary, only direct access to secretarial services.

Mr Galbally's final questions revolved around my statement that none of my staff had received promotion during the previous five years. He produced a statement from Mr Cox who had recently been promoted on merit. His final challenge was:

Mr Galbally: 'But the reality is that you refused to tell Mr Cox that he had been promoted.'

Response: 'That is utterly untrue—that is utterly untrue and I really take offence at that. He received notification and as I say so help me I am under oath, I found out about it several weeks later.'

To the best of my knowledge Mr Cox was the only research worker at CSL who assisted the prosecution.

During the remaining minutes, various people asked a range of useful questions. Dr Hurley asked me if I would have gone up to Dr Coles's office had I known the subject to be discussed. I said I wouldn't have gone anywhere near him. I'd have been sitting outside Dr Schiff's office since Coles was merely the middleman. The questions from the other Commissioners were sensible and even sympathetic in tone.

The day concluded with Dr Brian Feery giving evidence. Brian had been brought in for two purposes. One to give insight into a medical officer's role at CSL. The other, as a former departmental research head and my colleague, to make observations about Dr Coles. In prior discussions with my legal team he had impressed them with the clarity and strength of his views. Unfortunately these were somewhat watered down in the witness box. The only bright spot was his statement that he could think of no-one else more dedicated to his work at CSL than myself.

Before we shut down, Mr Frickie, QC, announced that he was not able to attend the following Wednesday but his learned junior would be in attendance.

Between hearings
The following are some relevant diary excerpts.

Thursday, 6 August 1981   Bruce Moore tells me that the senior CSL people we want as witnesses have all declined. Furthermore, my staff members who volunteered to be interviewed by the solicitors have had their pay docked.

Thursday, 6 August 1981   At 5.00 pm David Wells rings and suggests that proceedings be truncated and legal costs shared. I won't have a bar of this, because CSL's legal costs will be much higher than ours and I'd like the whole hearing properly completed.

Round Four—Wednesday, 12 August 1981
By now I was getting used to this style of life. I won't say it was enjoyable, but it was becoming productive. In between times I was getting much writing done and *Australian Animal Toxins* was back on schedule. The Chairman announced that the Commission had decided that Mr Huyer should take no further part in the proceedings (David Wells leant across to me and muttered 'they're learning').

Ross Robson kicked off with a request that McCarthy be asked to leave the room since all the next witnesses were CSL employees. Thus, they would be able to give their evidence more freely. Mr Galbally objected to this, saying 'Indeed, Dr McCarthy was somewhat distressed that staff members were being called "on issues which apparently are not simply issues of character".' The Commission adjourned to consider this request and decided it was inappropriate for Dr McCarthy to leave.

The first witness was Allen Broad, a good friend and one of my right-hand biochemists for some twelve years. He was a lean, athletic type and took the oath in a very positive fashion. Allen was absolutely great. He stated that my relationship with my staff was excellent, 'I felt he had concern for his staff both in terms of work and in terms of well-being. People are able to approach him on a number of subjects whether it be work problems or peripheral problems and he was only too willing to help and his staff respected him for it.' In regard to bad language, he said, 'In the thirteen or so years that I have been working with Dr Sutherland I cannot recall any time he has used bad language to his staff. I feel he has more respect for his staff than that.' Allen, who had worked with Coles for three years before joining me, said of Coles, 'He seems to be a distant sort of person and very hard to relate with. He doesn't seem to have the rapport other people may have.'

I heard a whispered instruction from the opposition table on my left 'Get rid of him.' Mr Galbally did not ask Allen Broad any questions.

Next came Alan Coulter who, in the ten years we had worked together, had had a big hand in my so-called 'achievements'. Alan was almost as good as Allen Broad. He said he was shocked when he heard I was to be stripped of my staff and stated that he thought I would need at least one full-time secretary. He said the staff cuts meant I would no longer be able to carry out any effective research. There were no questions from Mr Galbally or the Commission.

Next came my friend and colleague Dr Len Hartman. Like most other witnesses, Len looked startled when he saw the size of the assembled company and my notes say, 'Len Hartman gave me a Jesus look.' Years later, I finally interpreted this as a look of compassion on his handsome features.

Len considered that I was recognised worldwide and CSL had derived benefits from the publicity I had generated. He thought I was loyal to CSL and imagined that I would have been shocked by the announcement of the staff cuts. He confirmed that a reduction in the status of medical officers had occurred. Mr Galbally suddenly came to life to earn his fee but nothing much came from his questions.

Norman Ackland, who ran the Bioengineering group, was next. His appearance was brief because most of the proposed questions had already been done to death. The last witness was Angela Reale-Key. Angela was a lady of about forty whom I'd rescued from an unpleasant job in the

Animal House some four years earlier. She was my general laboratory assistant, who happily fetched and carried, made cups of tea for visitors and generally scuttled all over the place. Although very nervous, she was a delight as a witness. She described me as very fair and very busy. When asked if she had ever heard me swear or abuse members of the staff her response was, 'Oh no.' When asked if she would elaborate on my relationship with the staff she said, 'He treats his staff like human beings, not like work machines. I have never heard him swearing or abusing anyone.' Ross then brought up the matter of Angela having to use part of her annual leave when visiting my solicitors to make a statement. I observed Mrs Hardy looking quite thoughtful after this statement.

Mr Galbally got up in triumph and asked Angela if she had been promoted during the three years she had been with me. 'Yes,' said Angela brightly, at once seeming to knock holes in my statement that none of my staff had been promoted in five years. A quick note to Ross Robson from me had him announce that in fact Angela had not been promoted but was being paid at a higher rate because of the advanced duties she was performing. With further questioning she announced that, 'I am being paid Grade 2 wages but I am not permanently Grade 2.' She almost said, 'So there.'

Now that we had disposed of all the witnesses, Ross came to a little point he'd been itching to put for some time. He asked if the Commission had discussed the actual case formally or informally with McCarthy prior to the laying of charges on 25 June. I noticed Dr Hurley quietly nodding his head. The Chairman, Dr Forbes, said he was pretty certain that there had been some discussions. He said they would talk over the matter in our absence to decide whether they were likely to be prejudiced in regard to their ultimate decision.

After a break, during which the scene of the altercation was inspected, the Chairman announced that the minutes of the Commissioner's meeting of 24 June included a report by the Director that he was seeking legal advice about what action could be taken against Dr Sutherland. It was stated that thereafter there was no exchange of actual details about the case between the Director and his fellow Commissioners.

A somewhat heated debate then took place as to who should sum up first. Ross maintained that it was more appropriate that the prosecution had this honour. In the event Ross led off. Over an hour or so, interrupted by lunch, he beautifully built up three overlapping and interlocking themes. One was basically the contribution I had made to CSL; the second was the petty and obstructive treatment I had had from CSL; and the third was how this treatment, combined with Coles' extraordinarily provocative and gloating performance, had led to the scene in his office. Many delightfully appropriate quotes were drawn from the transcript. I thoroughly agreed with his ringing conclusion to the effect that my behaviour was in the circumstances of such a trivial nature that the charges should be dismissed. In response, Mr Galbally's approach was a lengthy lecture on natural justice and the role of tribunals, etc., which quite frankly bored me silly. He used terms such as 'of probative value' to which the Chairman inquired 'of what value'?

He considered that the charge of abusive language was proven since I had admitted to using the words 'you little worm' and also a four letter word. When it was clear that he had finished, Ross Robson rose to his feet to correct Mr Galbally on a number of errors of fact. When that was over, it was agreed that we would all meet again two weeks hence on Wednesday, 26 August, and that hopefully the matter would be concluded that day. The meeting adjourned at 3.45 pm.

Between hearings

Thursday, 13 August 1981   Dr Ron Lucas of Fairfield Hospital rings me and says, 'It's started, Stru.' 'What's started?' I ask. Ron explains that an appeal towards my legal costs has been started by hospital staff. They aim to raise at least a thousand dollars. 'Good heavens,' is all I could say. However, once I recover I ask him to keep a precise list of donors so refunds can be made if it's not needed. I will never forget that conversation.

Later, a pathologist who ran a very large private practice rings and offers me double my current salary plus attractive perks. He asks if we could meet to have a chat. I reply, 'Let's see whether or not I get sacked.' Nice to be wanted somewhere.

Tuesday, 18 August 1981   Marjory Davey tells me that the plans I made on 17 June for the next batch of Funnel-web spider antivenom production have been disregarded, upsetting some people. I learn that last week orders came through to bottle the first small batch immediately rather than waiting a few weeks to combine it with the larger second batch. I ring Len Hartman who confirms that this has happened. He agrees it was a stupid thing to do but does not know who ordered it. There will now be two quite different preparations, both of which will have to be tested in monkeys.

It seems likely that the batch was rushed through so that, if I am sacked, it can be said that CSL has plenty of antivenom ready for issue. If I'm not sacked, an explanation will be sought as to why we have much less of the stable freeze-dried antivenom.

It is hard to convey the feelings of frustration I experienced at this time. For the record I was vindicated later when the batch that was rushed through showed signs of instability and had to be replaced by the more stable freeze-dried antivenom.

Friday, 21 August 1981   The Minister for Health has received impressive letters from many professional organisations, learned societies and special interest groups (copies had been sent to me). Even the Country Women's Association, who appreciate the snake bite work, have made a submission. Some letters written by individuals are absolute rippers.

Round Five—Wednesday, 26 August 1981
Hey-ho the final round. This time Mr Frickie, QC, was back and was aided by Mr A. J. Myers.

Things got off to a brisk start: Dr Forbes announced that the Commission had found me guilty of improper conduct.

We all knew that was coming. My thoughts were, let's get the reprimand over and let me get back to doing something useful. However, McCarthy had other ideas. Mr Frickie, who had been hired for the day, announced the prosecution was going to bring in unknown witnesses 'to produce further material with respect to penalty'.

My team went through the roof. Ross Robson, in a series of exchanges with Mr Frickie, pointed out that the prosecution had had plenty of opportunity in the past to produce witnesses. Furthermore, he stated, it was universally observed procedure that the prosecution, having secured a conviction, did not take any part in the proceedings other than to cross examine any witnesses called on behalf of the defendant.

There was a good debate on whether Mr Frickie was Counsel assisting the Commission or in fact an orthodox prosecutor. Ross maintained that his learned friends had in fact 'prosecuted this matter with extreme vigour'. He firmly pointed out that not only should the prosecution not be producing witnesses at this stage but as in established legal practice all the names of prosecution witnesses should be given to the defence council beforehand.

Dr Forbes asked Ross whether he could give assurances that should I remain part of the organisation, I'd be a good boy thereafter. There was a short adjournment while my team chewed over this request. At the same time, the Commission considered whether it would be appropriate to bring in Dr Schiff and Mr Davey at this stage.

On recommencement, Ross gave an assurance on my behalf that I would maintain proper standards of behaviour and language and write all the reports, etc., as required. There was no apology to either Dr Coles or Dr McCarthy. I was adamant that this was not to be included.

After a couple more questions the Chairman said that he didn't think it necessary for Mr Frickie to bring in new witnesses. We then adjourned for lunch and I led my legal team on a quick tour of the nearby Melbourne Zoo.

The walk in the zoo obviously did Ross Robson the world of good and he was in great form when we restarted. He submitted that Coles had suffered no physical injury and that I had had no warning of the staff cuts and was surprised and shocked to hear the news. Furthermore, Coles had admitted that he was not supporting me. Ross said I had undergone the shame of being suspended, had suffered a great deal of expense and my work had been severely disrupted. He pointed out that I had apologised to the Commission and given appropriate undertakings. He pleaded that the penalty be merely a reprimand.

Mr Frickie then went into the attack, homing in on my previous dispute with McCarthy. He started reading details of some of Dr McCarthy's charges regarding that event, some of which I still strongly deny. Despite Ross's protestations that such evidence would not be allowed at this time in any court proceedings, Mr Frickie was given permission to continue.

Mr Frickie then read carefully selected parts of the handwritten letter that McCarthy had sent me in January of 1981 and the whole of my carefully constructed response to him. For some reason Mr Frickie considered that this letter (see page 213) was a clear indication that a reprimand would not be satisfactory and exhorted the Commission to grasp the nettle 'of dismissal'. When Mr Frickie had run down, Ross had a quick final go: 'He is an honest man. If he disagrees with something, he says it straight to the face. He does not go behind one's back scheming and plotting. He is an open, loyal and honest sort of chap. And in my submission he has shown that loyalty to CSL and I think that CSL should repay part of that loyalty.'

The Commission then went off into a huddle to consider the submission.

Forty-five minutes later we all reassembled and Dr Forbes spoke to me directly. He gave a very fair summary of the whole business. He told me that I had challenged the general authority of my superiors and, although the Commission did not ask for blind unthinking obedience, there were proper channels for questioning and review. He reprimanded me and said that was the end of the proceedings, i.e. no penalty!

Ross got to his feet and queried whether the Commission could in time consider: that in future I should report to a medical officer as I had previously; that my staffing position be considered; and whether the

Commission might consider making some contribution towards my costs. Dr Forbes said these matters would be considered in time, but the Commission had no power to order the payment of any costs to any party by any party. Ross then submitted the proposed press release we had knocked up which said, 'After a private hearing at CSL, the Commission has decided to lift the suspension of Dr Sutherland upon certain undertakings being given by him.' It was proposed that no other details be given and this was agreed to.

At 3.53 pm I was at last free to wander back into my Laboratory. Thus ended an eight-week farcical episode which was a horrendous waste of time, money and resources.

Fifteen minutes later, having waved my legal team goodbye, I was called back into the boardroom to discuss the press release with Alf Brogan, McCarthy and Mr Frickie. It was all very amicable. The modified press release seemed okay and so I bade them all a cheery farewell. The reports in the papers next day gave a balanced account of the whole episode.

Getting Back to 'Normal'

Next day, Friday, 28 August, at 10.20 am I was called over to the Administration Building to confront McCarthy who was dressed in 'holiday rig'. Also present were Viv Davey, Alf Brogan and Peter Schiff. McCarthy wanted to know why the press was writing up so much about wolf spiders, etc. He proffered his opinion that the AMA could consider my talking to the press was unethical behaviour. I had no clue what all this was about. He warned me that I must attend a meeting with Viv Davey the following Monday at 10.00 am. I said of course I would attend. His attack then became quite vitriolic and after five minutes I asked to leave. I was surprised and shaken by his performance.

On the Monday morning Viv Davey did his best to get under my skin. However, by this stage I was almost immune to his tactics and adopted an attitude of benign absentmindedness. I found if I repeated to myself 'Little Viv is a dickhead, Little Viv is a dickhead' nothing he said could annoy me. However, he and McCarthy were apparently not going to let up, because a week later Coles phoned to tell me that the Director had decided that Angela Reale-Key would remain in the research and development group but I would have to share her with another Department. She'd only be available either in the mornings or the afternoons.

I now only had one staff member and she was half-time. This led to an hilarious uproar in the Senate some months later. Senator Evans had taken a close interest in the whole business. He asked for a copy of the transcript and after reading it made approaches both to the Attorney-General and the Minister for Health. When he informed the Senate that I was now reduced to the services of half a cleaning-up lady, an unnamed Senator interjected wanting to know, 'Which half does he have!'

Paying the Bills

Early in October 1981 I was informed that CSL would make no contribution towards my legal bills, which had amounted to over $12 000. Without doubt my legal team could have charged me double this amount. (For example, their bill for transporting themselves over the six weeks involved was $12.90 in total.)

I will be forever indebted to David Wells. He died from cancer aged fifty-four in December 1996. His lengthy obituary in *The Age* of 14 January 1997 was appropriately headed 'Fighter for justice earnt respect'. It read in part:

> On Monday 16 December, a remarkable event took place in Melbourne. A barely advertised memorial service was held at St Paul's Cathedral for a relatively young, private individual, with little public profile.
>
> Almost 1,000 people attended, and stayed for the whole extended and deeply moving service, producing the most unusual spectacle of a nearly full cathedral in the middle of a business day.
>
> This overflow of respect and sadness was a vivid reflection of the enormous influence David had on the lives of so many people.
>
> At the St Paul's service, Peter Kelly, a partner and friend from Mallesons, summed up David's approach to the law: 'David had a deep and instinctive sense of justice, a firm belief that everyone in the community has a right of access to justice, a total commitment to the cause of his many clients and total integrity in the way he practised. David said "Law is not about making money but about justice and the proper order of things."'

Individuals and Organisations continued to write to Mr MacKellar, the Minister for Health, and also to the Attorney-General. MacKellar responded by saying the matter was merely one of an 'internal staffing issue'. The fact that I had been found guilty of improper conduct and not disgraceful conduct was not taken into consideration.

I was very lucky to receive donations towards my costs. My friends at the Fairfield Hospital gave donations and Dr Alan Duncan organised an appeal which he publicised in both *The Medical Journal* and the press. Donations were often accompanied by most moving letters. A particularly delightful one was received from Mrs Lesley Lane, the widow of CSL's previous Director.

One day I was called to the front gate of CSL to see a man who insisted upon talking to me. He was a tough looking fellow who produced a roll of $2 notes from his pocket which he pushed into my hand. 'That's for your appeal, Doc,' he said. Looking at his pretty ragged appearance, I said, 'I can't accept that.' He replied, 'I'm a contract painter, Doc, and I make plenty. Besides,' he said, 'they say you saved my granddaughter's life.' Without giving his name and without further ado he hopped into his battered truck, whose engine had been noisily clunking away in the background, and roared off down the hill. When I released the rubber band there was only a $2 note on the outside—the rest were all $20s. I held over $300 in my hand. I had tears in my eyes.

Overall nearly half of the twelve thousand dollars had been raised by donations. This saved me from taking a third mortgage out on my house!

Disruption of the Production of Funnel-web Spider Antivenom

As mentioned earlier, plans for the second batch of Funnel-web spider antivenom were finalised on 17 June 1981, the day before the paper clip incident. It was decided that the next batch would be a freeze-dried preparation. Immunisation of rabbits had to continue to achieve the large volume required and it was planned that, by the end of September 1981, the new batch would be fully tested and hopefully permission would be obtained for its release by October 1981. While I was suspended rabbit immunisation ceased and, contrary to the proposed plan, the bulk of the material available was ampouled in a liquid form.

When I returned to work on 27 August 1981 there was insufficient bulk antivenom for freeze-drying. Rabbit immunisation was restarted but it was December before sufficient serum was ready for processing. It was not until March 1982 that the freeze-dried batch was fully tested and ready for issue.

This delay caused some difficulties in meeting clinical demand. Dr Malcolm Fisher, after a hair-raising helicopter ride taking antivenom to a hospital on the north coast, was rightly vocal about the apparent delay in adequate supplies. He rang and described to McCarthy in vivid terms how the shortage of antivenom had put him at a personal risk which McCarthy was welcome to try. Malcolm later asked me if I knew how to tell if your helicopter was flying upside down, in the dark, in a storm. This stumped me. The answer: 'By the diarrhoea running down from your collar!'

I maintained that the late release of the larger freeze-dried preparation was closely related to my suspension from CSL. Whether or not this was a fact became a matter of dispute between McCarthy and *The Bulletin*. This is a good time to deal with *The Bulletin* article.

The Bulletin

In December 1981 the Sydney-based *Bulletin* sent the distinguished writer Robert Drewe to interview me as they say 'in depth'. All's fair in love and war and I gave Robert Drewe access to all the relevant documents I held. I did not give him anything which could damage CSL's commercial plans, nor of course anything that breached patient confidentiality. None of the documents were classified by CSL as confidential.

Robert Drewe concluded his piece with a quote from Professor Bryan Hudson:'There is a degree of pettiness here which is unusual in my experience. In medicine it is very unusual. I mean, what is Struan Sutherland supposed to do now? There aren't many places in Australia with venom research departments because he's it. He can hardly put his shingle in Collins Street and say "Here I am—I treat snake bite".'

Robert Drewe made a feature of the representations I had made against the release of a crude antivenom preparation for use in Malaysia (page 266).

His seven-page article appeared on 12 January 1982 and it was admittedly very pro-Sutherland. McCarthy, Davey and Coles came out of it rather poorly.

Fun with '60 Minutes'

In July 1982 60 Minutes decided to do a programme loosely based on *The Bulletin* article. I pottered into town on Monday, 26 July, to have lunch with Bruce Stannard and Jana Wendt. This was prior to them going out to CSL to interview McCarthy. They wanted a general chat and to discuss plans for the next day when I would take leave and be interviewed at home. It was a very interesting lunch and we got on like a house on fire.

That night, Bruce Stannard rang and said McCarthy had insisted that the filming be stopped at least three times.'Most extraordinary,' Bruce said. Apparently after they had left CSL McCarthy rang Gerald Stone, the producer of 60 Minutes, to complain bitterly about Jana Wendt's questions.

Next morning Jana did an interview with us both sitting in the back garden. As the cameras were being hooked up and sound checks made I was supremely confident and looking forward to being especially eloquent. Bruce Stannard's last minute advice was 'Now's your big chance Stru. Over two million people watch this programme including the Prime Minister.' Bang went my confidence and it took three takes to get anything vaguely coherent. A strange sound problem developed which was finally traced to one of my cats playing with the ice cubes that were in a glass of water under Jana's chair.

60 Minutes went to air on Sunday, 1 August. McCarthy's performance was excellent. It showed him talking with his eyes closed and popping out wonderful statements such as 'Dr Sutherland is no more important than the people who guard the gate or feed the animals.' He was almost apoplectic when Jana asked him why the successful development of the Funnel-web spider antivenom only got one line in the annual report. (Antivenom in fact got one small paragraph rather than one line.) I came over as battered, but reasonably amiable. Gareth Evans gave a brisk presentation and suggested that CSL might have lost its soul. They also interviewed the mother of the little boy who had received Funnel-web spider antivenom in May and this was a most moving performance.

Going Tippy-toes with the Media

I had always thought that giving advice in the media on first aid for snake bite and how to avoid various bites and stings, etc., was a useful public service. Special care had to be taken not to criticise CSL's policies directly. Usually the interviewer got the appropriate message after I declined to comment on such 'a controversial issue'. Most radio interviews were done at short notice and I preferred to do them by phone. A month or so after my second reprimand I went into the ABC studios to be interviewed by Andrew Potter. Coles and his mates sat down and recorded the live broadcast in case something of special interest was said. Had I known this, I would have sent them a cheerio. Believe me I would have!

When interviews were being booked some time ahead I would usually ask the party involved to ring McCarthy to get his clearance. Sometimes odd things happened. For example, on Friday, 17 September 1982, Robyn Williams rang to ask if he could interview me some time for his Science show. I told him that in fact I'd be in Sydney the following Monday because I'd been invited up to talk to the 'Friends of the Museum' at 10.00 am. He said the studio was right near the Museum and, since I was flying up on Sunday night, he would get me picked up at 8.30 so he could do a quick interview by 10. Fine I said. Please ring McCarthy to clear it.

On Saturday morning Robyn rang me from Sydney and said McCarthy had refused permission for me to be interviewed. 'I can't have my scientists being interviewed all the time', said Neville. Robyn said, 'What an extraordinary person he is.' I said, 'Sorry but that's how it is.'

On Monday morning I was having a leisurely cup of coffee in my motel room watching the traffic tearing down Parramatta Road when the receptionist rang and informed me that a car had arrived to take me to the ABC. Twenty minutes later I wandered into the Science show's tiny department. Robyn Williams looked quite startled and said, 'What are you doing here?' He had forgotten to cancel the car which he had organised on the Friday. We looked at one another, looked at the empty recording studio and popped in and did the interview. The end result was quite a chirpy interview and didn't draw any flack from CSL.

Over the weekends of 27 March and 6 April 1982 *The Australian* published a fairly gripping summary of the Funnel-web spider antivenom story by Adrian MacGregor. This kept clear of controversy and led to the making of a documentary called 'The Funnel-web—The search for an antidote' by the South Australian Film Corporation. It was subsequently shown in many parts of the world, which in turn led to a number of overseas film makers visiting Australia to do documentaries on our venomous wildlife.

A Last Skirmish: The Launching of *Australian Animal Toxins*

Australian Animal Toxins which took years to write was launched on 7 April 1983 at the Royal Melbourne Hospital by Sir Roy Wright, Chancellor of the University of Melbourne, with some additional comments being made by Dr Peter Bush (the Chief Police Surgeon of Victoria amongst other things). Invitations were sent out by Oxford University Press to a number of people, including Dr Neville McCarthy who, I believe, declined to respond to the invitation. The book was dedicated to the staff of CSL and singled out four staff members for special mention. These staff members, who applied for special leave to attend the launching, five minutes' drive away, later found their leave had been refused and their pay docked. (*The Age,* Tuesday, 12 April.)

Although CSL was not represented at the launching of this book somone high up at CSL closely read the 527 pages thereof. In June, both Oxford University Press and myself received a letter from the CSL Commission Secretary, Alf Brogan, drawing our attention to the fact that on pages 47 and 48 of this book there were verbatim extracts from a CSL published product leaflet. Furthermore, there was no reference to a paper by Chandler and Hurrell of CSL that had been published in 1982. Alf went on to say that he had received legal advice and had been told that both Oxford University Press and I were in breach of copyright and that actions by CSL, Mr Chandler and Dr Hurrell would be successful. This letter I found fascinating on two counts. Firstly, I had written the CSL leaflet under question and it had been quoted extensively by other authors without drawing any flack from CSL. Secondly, the paper by Chandler and Hurrell could not be included in the references because I only saw it after the printing of the book had commenced. If I'd been shown a draft or a preprint I could have included it as 'In Press'.

Fortunately the Oxford University Press solicitors were crash hot on copyright, etc., and sent back a gorgeous letter to CSL. After firmly demolishing CSL's arguments, CSL was advised that no public acknowledgement was considered appropriate. The letter concluded by sweetly saying, 'Thank you for drawing these matters to our attention.'

Epilogue

By 1990 McCarthy, Coles and Davey had all left CSL. Davey left the world in 1996. In 1990 Mrs Marjory Davey at last became my full-time secretary.

In 1986 Mr Richard Alston filled a casual Liberal Senate vacancy. In December 1988 Mr Ross Robson became a QC.

I can report that eventually my situation at CSL improved. However, the rehabilitation was brief because, as we shall see, the winds of change caught up and blew me fair out the front gate!

Into the Wilderness then Fighting the Odds   10

'Of course, Sutherland is his own worst enemy.'
Sometimes murmured by the CSL Management.

Battler: One who struggles continually and persistently against heavy odds
(Macquarie Dictionary).

A Bit of Isolation	302
Turning to Other Interests	304
Prison	304
Tom	305
Dick	306
Harry	308
Hydroponics	309
Allergy to Insect Venom	310
Bees in yer bloomers	310
Some unfinished research	313
Ants in yer pants	313
The White-tailed and Other House and Garden Spiders	317
Departure of Dr Neville McCarthy	319
Neville McCarthy's farewell dinners	320
Dr Brian McNamee	322
Dr Ian Gust	324
Performance Management System Agreements	326
The Beginnings of the Australian Venom Research Unit (AVRU)	337
The Sale of CSL Ltd	346
Leaving CSL	347

A Bit of Isolation

From 1980 onwards I lay low and kept my nose clean. My research proposals did not get far and few people of any standing at CSL were prepared to risk their own careers and lend support. In hindsight I should have left CSL after the disciplinary hearing in 1981 instead of spending time and energy in generally futile efforts to defeat old and powerful antagonists at CSL.

In January 1984 I was moved out of my office laboratory complex and installed in splendid isolation on the top floor of the second oldest building at CSL. I was told that the area I was using was needed urgently by Serum Fractionation. In fact, it stayed empty for two years and was subsequently, prior to demolition, used for a year or so as a lunch room for contractors. I was very sad to see the old laboratory with all its memories destroyed. Almost as bad, was seeing some of my precious equipment sitting in the rain outside the salvage store.

I no longer had direct access to equipment and I had no scientific staff. I did, however, manage to get quite a bit of research done here and there by stealth.

Some outside recognition buoyed me up from time to time. In 1983 *Australian Animal Toxins* won the Whitley Award. The next year the Royal Society of NSW presented me with their James Cook Medal. This last award stunned me because previous recipients included people like Albert Schweitzer. One ray of light was a nice letter from CSL's Chairman, Neil Batt, when I received a doctorate of science in 1985. I never met him during his two-year term.

After I had exposed the hierarchy for approving the sale of an inferior antivenom to an Asian country (see page 266), they in return kept the pressure on me until they successively retired. Few opportunities were missed to obstruct my endeavours, even though they were beneficial to CSL. My media comments and publications were routinely scrutinised for 'actionable contents', but sudden butterings-up occurred whenever there was possible contact with VIP's. Occasionally such visits raised false hopes.

In April 1984 there was great excitement at CSL, for Prime Minister Bob Hawke was to open a new Human Vaccine building, dedicated to a former director Dr Val Bazeley. The only other time a Prime Minister had come to CSL had been when the then Mr Robert Menzies visited CSL twenty-four years earlier. (A year later he sacked Dr Bazeley for publicly criticising the Government's proposal that CSL be run by a Commission.) The whole business was a great PR exercise, no expense was spared and I was about the last person at CSL to get an invitation.

Lo and behold on the morning of the Prime Minister's visit Mr Viv Davey visited me in my eyrie for the first time ever. He was accompanied by Dr Peter Schiff. All was sunshine and light: they were quite certain that in the near future I would have professional and technical staff; I could count on it from the beginning of the next financial year. Since my invitation had only arrived the night before, I stuck to the plans made and did not attend the opening. No staff materialised.

In 1985 there was another burst of excitement with the impending visit by the Minister for Health on 12 February. The order went out for everyone to have their areas neat and tidy. The week before this momentous occasion I was visited again by Peter Schiff. He said I could plan for a Biochemist Class 2 and a Technical Assistant from 1 July and that Neville did not object. He also mentioned that a fair bit of the area I was currently occupying would be taken over by the Blood Research Group. This I didn't object to since I would still have enough room left. Two years later the Blood Research Group was fully installed and I hadn't seen hide nor hair of my own staff.

When Viv Davey retired in 1985 Neville placed me under Charles Guthrie's supervision. Charles had had even less research experience than Viv Davey and had also spent his total working life at CSL. He was not the slightest bit interested in venom or antivenom research. When I asked him point blank whether he was supportive of antivenom research, he said no he wasn't. At least he didn't mince words!

In 1988, when an anonymous donor offered me $30 000 for venom research Charles' reaction was quite negative. It said a lot about the establishment when it discouraged what turned out to be a no-strings-attached grant. (Every year since, the same donor has supported my research initially at CSL and now at the Australian Venom Research Unit.)

Nevertheless, on 18 June 1987 on the sixth anniversary of the paper clip incident, I was told in a discussion with Mr Charles Guthrie and Dr George Harris that if funds arrived from Canberra I could start up an ant venom project. It was another two years before staff became available for this project. (Page 313) Even then I was denied CSL staff and had to make do with hiring scientists from an agency.

During the eighties the atmosphere of the meetings I attended would probably have astounded outsiders. Viv Davey chaired the smaller meetings involving the seven research heads. Such was the feeling of hostility and futility felt by most of those present that these compulsory meetings would often peter out after some ten minutes.

The six monthly reports to the senior staff had a similar atmosphere on a much larger scale. The eighty most senior of the twelve hundred staff would gather in the lecture theatre to be addressed by Neville and some of his favourites. The presentations were variable in quality and rarely inspiring. One of the depressing factors was the almost total lack of audience participation.

When questions were called the same old toadies would ask a couple of Dorothy Dix's and, apart from probing questions from the idiosyncratic Dr Alan Blaskett, the rest of us would sit dumb and still. Most felt it was best for their self interest not to draw attention to themselves.

A survey conducted by the Professional Officers' Association during this period found that CSL's scientists were a worried and unhappy lot. However, the official line as expressed in Brogan's book *Committed to Saving Lives* was that: 'In addition to the other benefits to CSL, the widened vision and improved management capabilities engendered in excellent and loyal senior staff represent one of the real achievements of Neville McCarthy during his term as Chief Executive' (page 247).

Turning to Other Interests

I would like to leave CSL for a few pages and muse on my outside life at this time. It was proving rather a lonely decade. By 1983 Megan had left me, CSL and Victoria. This meant, amongst other things, the end of our lunchtime walks which had been special. John and later Susie were in University colleges. Since Marjory Davey had not moved when I had, sometimes I could go twenty-four hours or more without any direct human contact. Nonetheless, my working hours sped by sometimes with non-stop calls from doctors and vets, etc., seeking advice from all corners of Australia and elsewhere.

One advantage of my solitary existence was that the family escaped the inevitable night calls. For years they had been awakened by the phone and sometimes kept in that state as I droned on.

However, I was better off than many other doctors—I was very rarely called out at night to see patients. On the other hand, with one exception, I have never charged a patient in thirty years. Doctors have sometimes urged me to send an account for telephone advice but I wouldn't because the patient could not claim a rebate and, more importantly, it might impede the collection of useful data. All in all, my contact with medical colleagues and their patients has been an uplifting and rewarding experience. In return, I have tried to keep tabs on the outcome of my advice—and learnt a few lessons in the process!

Without a research team my output of papers tapered off and, once the relevant books had been written, I channelled some of my energies into non-venomous activities. Never one for spectator sport, I packed my weekends with oil painting expeditions, bush walking and photography. I dived and photographed every coral reef in the offing. This brought me full circle back to venomous life again!

Two of my activities during this period were rather absorbing. I would like briefly to relate them.

Prison

I first went to prison in 1984. It was a business-like one too; in fact it was high security. Known as Jika Jika, it was a dreadful complex of interconnected concrete cubes set up in the 'grounds' of Pentridge Prison,

Coburg. Its design successfully incorporated the worst features of an early submarine and a malfunction-prone space station. The fog of cigarette smoke and stale cooking smells enhanced the impression that there was not too much oxygen about. Apart from the presumably proven inclinations of the residents, the atmosphere was sufficient reason for a new arrival to mentally do an about turn.

Right from the start I was treated nicely. The Governor had invited me to interact with 'clients' interested in science. I agreed, after cautious inquiries, which amongst other things cleared my superiors at CSL from any subtle plan to have me discreetly incarcerated.

My involvement with a group of the inmates lasted a number of years. Generally, I spent a morning a month at Jika Jika and an hour or so between visits talking on the telephone to my new acquaintances. (Jika Jika was renamed K Division after a fire on 29 October 1987 in which five prisoners died. It was closed down some years later.)

Curiosity, which was probably my strongest initial reaction, was replaced by commitment when I realised that there was a genuine desire among the prisoners for my participation. Overall, it was a rewarding experience and was certainly an eye opener. I'll resist the urge to describe some of the scenarios witnessed during my visits, but once I felt a nice mental glow when a female prisoner called out to the warder as I passed by, 'Who's the spunk with the bow tie?'.

I would like to summarise my dealings with three particular prisoners. I will call them Tom, Dick and Harry, and not describe their physical appearance in an attempt to veil their identity. However, for very sound reasons, Harry's identity will be revealed later.

Tom, Dick and Harry were interested in studying marine biology as best they could in their confined circumstances. They had a few well-thumbed books, a dissection kit and a magnifying glass. In a glass tank lived a variety of freshwater mussels. Six months after we got together they had better equipment including a second-hand microscope and a new computer. Funded by outside sources, this mini-lab was sited in the recreation room used by the wing's residents. Most took little interest in scientific pursuits but I was very aware of one big fellow. From first sighting he impressed me as a person not to antagonise. He tended to be constantly on the move around the perimeter of the recreation room, like a circling shark. Occasionally he abruptly stopped pacing and joined the tutorial, freely offering his opinion. Tom, Dick and Harry tolerated these interruptions. I listened with rapt attention. God, he was big!

Tom

Tom was young, very intelligent and had a good family background. His long sentence reflected the seriousness of a drug-related crime committed some years earlier. He was a voracious reader of what I would call heavy literature and he was a natural linguist. Tom made the most of the fact that the prison's polyglot population had plenty of time to expound on the niceties of their various mother tongues. When I last saw him he was mastering Arabic.

The arrangement was that, from time-to-time, I would buy fresh specimens like an octopus or large crab for the group to dissect. The Victoria Market was the usual source and I would deliver them as soon as possible, especially in summer. The quantity and, to a lesser extent, the choice of creatures were selected so as not to pose a temptation to the palate. Pickling in a dilute solution of formalin was later thought to reduce culinary interest in the specimens. Samples from far afield, like box jellyfish, arrived preserved.

Tom proved to be brilliant as a pen and ink artist, his drawings breathtaking in their detail. In particular, a frontal view of a crab with its claws wide open struck me as exquisite. It seemed to hover in space. We nearly succeeded in having these drawings published (see below). I lost contact with Tom about the time he was paroled but I hope he and his very caring mother are doing well. An update would be nice.

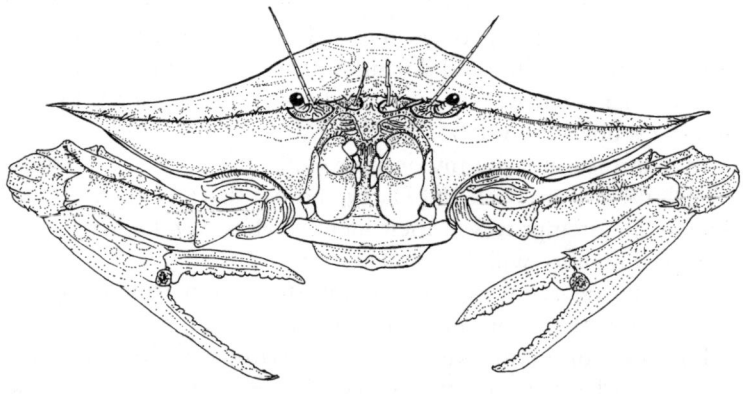

Dick

Dick was a convicted murderer and one of the oldest prisoners in Victoria. Years ago he had spent months in one of the two condemned cells directly below the gallows. During this time the gallows was tested with sandbags which had been filled to approximate Dick's weight. He estimated he was 'hanged' about fifty times. His death sentence for the shooting of an elderly farmer whilst on the run was commuted to thirty-five years in prison. Dick appeared to be respected by both the prisoners and warders, being a resilient father figure to many. He had a heart condition that caused chronic insomnia but, as we shall see, he put the extra time to good use.

Dick was the driving force behind the 'science' group. For some years he had written requesting information from environmental and other scientists. The response was patchy, but I had treated his letter as part of

Publisher's Note:

The author would have liked to have photographs of more people from his C.S.L. days, but the management of C.S.L. declined permission.

Mr Viv Davey (From the author's collection)

Dr John Trinca, who guided the author's early steps at CSL. (From the author's collection)

Dr Norman Coles (From the author's collection)

Mr Merv Hinton (From the author's collection)

Professor James Angus, Department Head
and chief mentor of AVRU

Ayse Berke

my job and pleased Dick by sending a package of literature. Months later, out of the blue, the Governor's invitation arrived.

Although a trifle wary of him, I got on well with Dick both before and after his eventual release. He was a keen student and nearly pulled off a considerable scholastic success. He decided to compile a dictionary of marine terminology and there seemed a need for one, especially with an emphasis on Australian marine life. Dick beavered away for months and months increasing my understanding of marine terminology no end. When the first draft was completed I sent it to a senior CSIRO marine biologist for his opinion. He was really fulsome in his praise and his report convinced a publisher to commence negotiations. We envisioned using Tom's striking illustrations in 'Professor Dick's' opus, but were disappointed when an overseas consultant to the publisher produced a damning critique. I agreed with some of the criticisms but felt with help they could be overcome. Fortunately by this time Dick was busily occupied in writing, with a colleague, the history of Pentridge. Entitled *Pentridge the Change 1887–1987* this was a fascinating read.

The tales of accident-prone warders would no doubt delight your average crim. One warder who liked playing with his gun, shot himself accidentally on two separate occasions. Another was practising 'twirling' his rifle in his guard tower when it slipped and fell to the outside world. He lowered a rope and two passing schoolgirls helped to rearm the warder. Serious injuries were accidentally inflicted to the guard towers themselves: from time to time roofs blasted away, even a tower dunny was once destroyed by a volley. Dick's account, which I assume remains unpublished, details some little known but ingenious escape attempts. Dick 'escaped' legally from prison in 1988.

A week before he was released on parole Dick was given a half-day of freedom. He was accompanied by a parole officer who also came in handy as a chauffeur. I took Dick to lunch in Lygon Street where he puzzled over the menu. He found having a choice a novelty in itself. Before returning to gaol Dick had to open a bank account. As we queued in the bank, Dick looked around at all the security warnings and paraphernalia. 'Things have changed,' he said rather wistfully. Last time he had been in a bank, years ago, he had held it up.

When Dick got out, his ambition was to get a Combi van and 'go bush' for a while to work on freshwater mussels. I offered to lend him half the money required. We duly arrived at the car yard that Dick had insisted we patronise. I didn't fancy the area or the proprietor. The only van in our price range looked definitely seedy. To my surprise, without even kicking the tyres, Dick said 'We'll take it.' I was stunned and tried to lead him away for a little chat. He whispered, 'Wait till he finds out who he's sold it to.'

The pair sat down to do the paperwork. I hung back at the office door. When Dick was asked his full name the answer froze the dealer's hand. He perhaps recalled that Dick was someone not to be messed about and/or he had influential friends. Taking the initiative, Dick started enumerating the things he wanted checked. New battery and tyres were mentioned. More technical matters were addressed. Indeed, I got the impression that Dick had had the van fully 'sussed out'.

When next I saw the van, on Christmas Day 1988, it looked very smart. Dick was still getting used to driving it which partly explained its collision with our front gate post. This was after a fine Christmas lunch which included crayfish. He left a very happy man.

I had two memorable conversations with him later. He rang one night to report winning some $300 000 in Tatts. He returned my loan without interest and a week later he rang me up to ask me if I would mind driving across Melbourne to immunise a cat which had to go into a cattery next day. I explained he should ask a vet. I think he might have been drinking. I never heard from him again.

Harry

Harry deserves to be dealt with brusquely. His name was Rodney Francis Mallard. He wasn't too bright but Tom and Dick showed tolerance and patience towards his illiteracy and odd behaviour. Several times he made himself sick by surreptitiously sipping the dilute formalin preservative solution that I had provided.

There was something about Mallard which led me to decline to participate in his 1989 application for parole the next year. There are some lessons to be learnt from what happened ...

Rodney Mallard, aged thirty-nine, was released on parole in March 1990. He had served sixteen years for murdering a nurse at Katoomba in 1974 and a male bank employee at Mallacoota in the same year.

On 26 May 1990 Mallard, who had changed his surname to Cameron, gave highly selected details about himself on air to a Melbourne matchmaking radio programme. He described himself as a marine biologist which sounds a fairly innocuous, even gentle, occupation. After all, the late emperor of Japan had found it a nice peaceful hobby to help pass the time during World War Two.

Maria Goellner heard him and made contact. On 22 June, Ms Goellner's body was found in room 46 in the Sky Rider Motor Inn at Katoomba where they had both been staying. Her skull had been bashed in and she had drowned in her own blood. A bunch of carnations had been placed on her chest.

On 16 October 1992 our 'marine biologist' was sentenced to be jailed for the rest of his life by Mr Justice Newman in the NSW Supreme Court. He became only the third person to receive such a sentence. Hopefully the three of them may share the same cell. I do not believe in capital punishment, having experienced an overwhelming feeling of powerlessness in the crowd outside Pentridge in 1967 when Ronald Ryan was hanged. On the other hand, I would be in a quandary if Mallard sent me a request for a stronger solution of formalin. No doubt the good people of Katoomba would have no hesitation!

My last prison visits were in 1990 after I had become a self-taught 'expert' in the gentle art of hydroponics. Helping various groups embark on this type of gardening, inside and outside prison, was highly satisfactory. At my last demonstration of hydroponics at Pentridge there was standing room only. Some in the crowd had had first-hand experience of raising hydroponic crops, particularly marijuana. Further tutorials appeared superfluous and even unwise from my point of view. So prison links were severed, as my laboratory was being re-established at CSL and I became busier. I still made time for hydroponics which deserves some further mention.

Hydroponics

I am not fanatical about hydroponics. It's just a hobby which seems to make my gardening more rewarding for the time spent. I also like playing with water.

After television, gardening is apparently the second commonest form of recreation so any significant development in the field is worthy of some attention.

One fine spring day in 1983 I was sitting in my little glasshouse armed with pencil and paper. My task was to flesh-out an editorial for a medical journal. My heart had not been in it and so, hoping for some inspiration, I had moved from my study into the midst of my hydroponic garden. I was definitely in a creative mood but not for the editorial. It would be more fun to write about hydroponics than this dreary stuff, I mused. I finally did a deal with myself. I would spend one hour recording my highly favourable impressions of hydroponics and then get back to the editorial.

I sent the result off to the editor of the *Weekend Australian* and was quite elated when it was accepted. Some of my drawings were included in the article which was headed 'How the spider man made his magic garden'. The article attracted a lot of attention. Various hydroponic societies I had never heard of absolutely loved it. Three publishers made offers for a book on the subject, despite my protestations that I had no horticultural training. Dr Brian Hanger and Fred Funnell gave me much excellent advice. Hyland House first published *Hydroponics for Everyone* in 1986. It has been in print ever since having had a major revision in 1996. My daughter Susie's reaction to the first copy she saw rather set me aback. After a lengthy perusal, she looked me straight in the eye and said, 'Dad, you always like telling people what to do!'

Apart from stimulating some writing, hydroponics led to me doing a bit of inventing. My experiences left me with a profound respect for the tenacity and single-mindnesses of successful inventors.

I had fun developing a self-contained hydroponic unit which was operated by a tyre pump. About two hundred were sold with a profit which just covered the very reasonable fees of my patent attorney. My two other ventures were technically fine but considered barely economically viable. Working with Dr Graeme Blackman of the Institute of Drug Technology we produced two types of tablets. One, intended for hydroponics, was ideal for making small volumes of nutrients. My late mother swore by them and used them to great effect. The other tablet was designed to prolong the life of cut flowers. The performance of what we simply named 'Vase Tablets' was harder to assess because of the myriad of factors involved. After a great deal of work all I could say was that our formula was not inferior to others and a clean vase was the best starting point.

I enjoyed these ventures into different aspects of horticulture. They seemed nice, peaceful pursuits compared with medical research, and it was a comfort to know they might perhaps be returned to one day.

Hydroponics was and still is very kind to me. My hydroponic garden has allowed me to continue to produce flowers and vegetables well past my physical 'use by' date. The advantages of the method to the handicapped was not lost on Don Burke who invited me onto his show in June 1998.

Allergy to Insect Venom

Bees in yer bloomers

Severe allergic reactions to insect venoms occur in many parts of the world. In the United States of America, it is claimed that bee and wasp allergy causes more deaths than all venomous spiders and snakes combined. Perhaps some 10 per cent of the Australian population will develop a significant allergy to an insect sting if they are sufficiently exposed.

In Australia the two main offenders are the imported honey bee and certain common ants. Allergy to the European wasp is also becoming more frequent. A look at insect allergy in general and bees in particular will help explain the unique therapeutic challenge posed by our ants (see page 313).

Usually a number of bee stings are required before the allergy develops. After perhaps the second sting, the reaction is greater than experienced by the 'normal patient'. Instead of mild local swelling and redness, a large itchy area may develop at the site of the sting. Later stings may result in the development of general effects such as a widespread rash, difficulty in breathing, swelling of the membranes at the back of the mouth, a severe fall in blood pressure, and sometimes death. A critical illness (anaphylaxis, see page 238) can develop within a minute or so of the allergic patient being stung. It is estimated that one in every two hundred persons stung by bees will suffer such a reaction.

For many years the management of these allergies was far from satisfactory. Obviously the patient would take care to avoid contact with the insect involved and usually emergency drugs such as adrenaline for injection would be prescribed for the patient to have on hand at all times. Many patients underwent immunotherapy or desensitisation to render them less allergic to the particular insect sting. Whole body extracts were prepared and the patient would receive increasing doses of these extracts over a period of some months. When they had reached the maximum dose they would often be kept on a maintenance dose, perhaps monthly, for many years. Until about 1980, CSL Ltd marketed a number of such whole body insect extracts for immunotherapy.

There were no properly controlled clinical trials with any of these whole body extracts. In 1974, Lichtenstein of Baltimore showed that purified bee venom was effective both from a clinical and laboratory point of view in the management of bee venom allergy. Patients were initially injected with minute amounts of bee venom. If they showed no reaction, the next week's injection would be slightly increased and the process continued until the venom given was equivalent to one bee sting. Dr Lichtenstein also demonstrated whole body extracts were of no value. Thereafter further evidence accumulated as to the effectiveness of pure venom immunotherapy and whole body extracts fell into disrepute.

Some allergists found difficulty in accepting that whole bee-body extracts were useless. I remembered how my Uncle Charles had frequently used such extracts, some of which as a boy I had helped process (page 61). Charles once had a patient who queried the effectiveness of his just completed whole-body therapy. By mutual agreement a bee sting was arranged and the patient gave Charles a hell of a fright by nearly dying! He never tried a similar experiment. On the other hand, Lichtenstein's patients rarely reacted to end-of-therapy bee stings.

In 1975 the Australian College of Allergists asked CSL to replace their whole body bee extracts with pure venom. I volunteered to take this work on board, partly in remembrance of Charles who I felt was looking over my shoulder. Funds were not a problem as the product the venom was to replace was a listed pharmaceutical benefit and, as such, was a nice little earner. An enormous world market awaited the successful manufacturer.

Allen Broad and I took up the challenge of safely collecting the venom from twenty thousand honey bees.

In 1963 an ingenious device for extracting venom from bees had been described by overseas workers as an alternative to the dreadfully slow one-by-one dissection. The device consisted of a platform with fifty fine wires on its upper surface. Under the wires was stretched a fabric with minute holes in it and below this a sheet of glass.

The apparatus was reported to work as follows: it was gently introduced into a beehive and, when a bee landed on two wires, it received a small electric shock. In a reflex action, the surprised bee drove the tip of the sting down so that it penetrated a tiny hole in the membrane. The hole was too small for anything but the tip to penetrate, so the barb did not get caught. The venom was squirted onto the glass plate from where it was later collected. The startled bee released a pheromone which quickly attracted other bees. (A pheromone is a compound an animal secretes as a form of communication.) In no time, hundreds of bees were landing, reacting and taking off, leaving copious quantities of venom behind.

We had no plans of this apparatus but we decided to build a frame from scratch. My first effort was a strange, zither-like creation which collapsed in my garage with a loud crack as I was tightening the last wire. A definitive frame, after further experimentation, was designed and built by CSL engineers. The system for delivering electric shocks had to be self-contained and was sorted out by CSL's Norman Ackland. Norm provided a series of black boxes which converted the juice from a car battery to the required 8 volts of alternating current. Sourcing the right material for the membrane was a real headache. Finally I tracked down an obscure Swiss firm which made taffeta with the right-sized holes.

During the interval Allen and I had cautiously been looking for suitable hives and an obliging bee keeper. All paths led to Russell Goodman who ran the Victorian Apiculture research station at Scoresby and soon we were regularly being stung. Allen proved to be an individual highly attractive to bees—they would be in his hair or crawling down his neck as he got out of the car. My stings were due more to carelessness, with so many bees about it was hard to avoid them.

The frame worked well from the start and we had a second one made to speed up the process. The enraged bees took some time to get clear of the frame and we found it wise to let them depart at their own pace.

Eventually Allen and I resolved to supervise and encourage Russell from the safety of his laboratory. Dressed in full protective clothing he would remove the frame from a hive and, usually too quickly, open the door to the lab and pass it to us to change the glass sheet which was still accompanied by a platoon of angry bees. While Russell placed the frame in the next beehive, we would process the venom on the glass, watching out for the bees.

After about twenty collecting days spread over two summers there was sufficient bee venom for the treatment of thousands of patients. The biochemistry and other properties of bee venom had been well known for some years and this information allowed us to demonstrate the various batches were of the highest potency.

The dried venom was to be dissolved in a special solution, sterile filtered, dispensed in ampoules and then freeze-dried. Each ampoule would contain the venom from ten bees, sufficient for one patient's desensitisation course. To obtain product registration CSL would have to provide detailed documentation on the stability and properties of two identical batches. These requirements did not appear a problem and, as part of the registration process, part of one batch was to be issued for a clinical trial organised by the College of Allergists.

Neither Allen nor I had the time for or experience preparing the pre-registration documents and we were both relieved when the task was given to Dr Chris Tyson. At the time I was struggling with a similar job with the Funnel-web antivenom. Over the months I observed Chris attending to various criticisms made by people who had changed their earlier advice on mutually agreed testing protocols. The path to medical registration seemed an arduous mutual learning experience and an obstructionist's dream. Each bee-sting death highlighted the slowness of progress. Then a bureaucratic decision effectively finished the project.

As mentioned in Chapter 9 (page 268), Chris Tyson abruptly resigned in March 1981. Chris was a medical graduate, who also had a PhD, and left after being told he was to be answerable to a young, up-and-coming PhD rather than me or Dr Hartman. The registration process ground almost to a halt with the departure from CSL of what is sometimes called in regulatory circles, the 'project champion'.

A year later CSL's marketing division had the bright idea of selling what Allen and I thought of as 'our venom' overseas. I had to point out that, although it was a stable venom, it was getting a bit long in the tooth (or sting!). If a manufacturer dispenses a venom this signals the start of its mandatory three-year shelf life and, in principle, new venom should be preferred to older material.

In 1983, Bayer (Australia) began marketing attractively packaged Albay bee venom. It soon became a pharmaceutical benefit where it remains, unopposed, at a price of $85 (May 1998). I don't know where the bee venom ended up but some years later I spotted one of the two bee frames crushed amongst a load of junk on a truck exiting CSL.

Some unfinished research
Two particular projects, which gathered tempo from 1990, are still on my 'unfinished business' list.

The first, dealing with ant venom allergy, is a neat example of my inability to learn from past experiences. The other, the White-tailed spider project, highlights a possibly widespread and knotty problem. Both projects are topical, and a matter of concern to many people.

Ants in yer pants
After seeing the bee venom project fizzle out collecting venom from other stinging insects was not on my list of priorities, yet a few years later I was back into it with a vengeance. In retrospect, a degree of support and enthusiasm from CSL for this project was manifest from the onset.

Dr George Harris threw two odd but apparently related questions at me one day in 1989 as he passed me in a corridor at CSL. The rather abrasive George, who was then assistant R and D Director, said 'How's the ant venom work going? Have you pulled your finger out yet?' I intimated that he had baffled me and invited enlightenment. It transpired that three months earlier CSL had received some millions of dollars from the Federal Government for two projects. One was to improve the whooping cough vaccine, the other was to prepare Jumper ant venom for the desensitisation of allergic patients. I was to be offered this latter project but George and the other seniors had not got around to telling me.

With no staff and precious little equipment there seemed little point in 'pulling my finger out'. When I suggested a staff member who would be ideal to help, George said I was to rely entirely on agency staff and continued on his way, never to discuss the matter with me again.

Although fairly pedestrian, the ant venom project could rightly be described as in the 'national interest'. Ant venom for allergic patients would not be provided by some international drug company, as was the case with bee venom allergy. Jumper (or Jack-jumper) and Bull ants were unique to Australia and thus there was no potentially lucrative world market for their venoms. On the other hand, the lack of a specific treatment was of great concern to patients, their families and the medical profession.

I was cautious about taking up this project for sound reasons, other than the fate of the bee venom work and the lack of CSL scientists. Firstly, I knew collecting the venom posed a problem. In 1977 I had failed to 'milk' ant nests using electrified platforms, because the ants refused to swarm over the wires like the bees did. One small group of guard ants relentlessly attacked the platform while the main body radiated out from the nest intent on stinging the dedicated scientist.

Secondly, unlike bee venom, nothing was known about the stability of these ant venoms and precious little known of their biochemistry and pharmacology.

I must admit to being influenced by the desperate mother of a seven-year-old boy, who had suffered a second near fatal ant sting the day before. She told me she would prefer a Tiger snake in her backyard than a Jumper ant nest. She exclaimed, 'They can treat snake bite! They can't do anything for my son's allergy. I'm so scared.' This story, and many like it, helped along my decision. Thus, despite unease about obstacles, known and unknown, I threw myself optimistically into the ant venom project.

I hoped that down the track, if the project went well, essential CSL expertise would be forthcoming. In the meantime, because it could not be carried out in winter, the project would allow me to conduct some long-postponed research into other venoms. Another factor in my positive attitude was the impending departure of McCarthy.

I doubt if I had ever worked harder at CSL. In 1989 the attic area I had occupied for the last five years, was declared a fire-trap. Some Commonwealth funds for re-locating people in potential fire-traps were used to set up a decent office and laboratory in the old Penicillin block.

I moved into this good accommodation in late 1990 and after ten years had my secretary, Marjory Davey, close at hand.

I at last had a laboratory again, albeit staffed by inexperienced non-CSL staff. Funding for my research was largely from a Federal Government grant and later some private donations. Each day was a blur with a stream of phone call inquiries from doctors on envenomations and to a lesser extent inquiries about vaccines and other CSL products. Marjory Davey and I were meticulously careful to maintain the Advisory Service to doctors and with over 2500 calls logged each year this service was extremely time-absorbing. We had some quaint rules. For example, it was considered a crime to go to the toilet without leaving the answering machine on! Important details of each inquiry were dictated immediately afterwards and subsequently typed up by Marjory. These diaries proved invaluable, particularly for medico-legal matters.

The ant project was as well planned as the Manhattan Project but on a smaller scale! It was six months before enough ground work had been done to warrant a hired assistant.

The aim was to make available by 1994 routine products of both standardised Jumper ant and Bull ant venom sac extract for immunotherapy of venom allergic patients. Precise standardisation and monitoring of every stage was essential to ensure all the following ones were as near identical as possible.

The project was a collaborative one, involving CSL and Dr Brian Baldo of the Kolling Institute, Royal North Shore Hospital, Sydney. The third collaborator, who was to be in charge of the clinical trials of the preparations, was Dr John Weiner of the Alfred Hospital, the representative of the Australian Society of Clinical Immunology and Allergy.

The first thing Marjory Davey and I did was establish a registry of patients awaiting venom therapy. Within a year it contained details of several hundred patients who had returned a questionnaire designed by Dr Weiner. This data showed the problem was more common and over a wider area than formerly believed.

Leaving Marjory with the paper work, I went hunting ants and their venom. Jumper ants inhabit southern and eastern parts of Australia and are particularly common in Tasmania. Bull ants, which are bigger, are limited to the south-eastern mainland and Tasmania. They are both amongst the oldest species of ants in the world, a fact reflected in their complicated venom.

Patients guided me to nests around Melbourne. Samples of ants and their nests were shovelled into plastic bins specially coated so the ants usually could not escape. Back at CSL various methods were tried to extract venom. These ranged from electrically stimulating anaesthetised ants to enticing ants to sting balloons filled with distilled water. It soon became apparent that the only way sufficient venom could be obtained was by dissecting out, then processing, hundreds of venom glands.

Always go to the expert is my motto. The CSIRO ant man took thirty minutes to show me his delicate technique which exposed the tiny venom gland. His second demonstration on another dead ant was timed at twenty-five minutes. I felt rather flat as I walked back to my lab after thanking him and seeing him safely off. That dissection rate was impossibly slow.

In the lab I put another Jumper ant under the dissecting microscope. The tip of its sting was just visible, protruding from the tip of its abdomen. I grasped the ant's waist with one pair of jeweller's forceps and the tip of the sting with another. I gave a tug and the sting came away, bringing with it the glistening pearly venom sac! Another was tried with the same result. The next three attempts by their failure showed there was a definite knack to smoothly collecting the sacs. Over the next four years at least 100 000 ants were relieved of their venom by this painstaking method. I could never do more than thirty an hour whereas Mary Papadopoulos once scored 400 in one day. Bull ants were easy because of their bigger size and tougher tissue; even I could be trusted to successfully demonstrate the process to visitors!

After extracting the venom from several batches of sacs we could determine the maximum amount of venom an ant might inject. The yield of venom sac extract from Jumper ants averaged some 40 micrograms per ant and since, theoretically, twenty such sacs would be required for immunotherapy of one patient, dissection proved a realistic approach. Bull ants yielded 150 microgram per ant and posed less of a problem.

Early in the piece, it was essential to find out whether Bull and Jumper ants collected in Victoria would produce material suitable for treatment of patients in other states. Sera were collected from over fifty patients who were highly allergic to Jumper ants and this was tested by a number of immunological procedures against Victorian ant venoms. These patients lived in a variety of places in Victoria, Tasmania and New South Wales. Preliminary work suggested that ant venom from Victoria contained all the major allergens. Arrangements were also made for samples of ants from other states to be snap-frozen so that their venom sac extracts could be studied.

Since Victorian ants appeared suitable, a site had to be found which was isolated and yet accessible for ant collecting on a regular basis. An area in Central Victoria some distance from a main highway where both Jumper ants and Bull ants abounded was selected. From November 1990, monthly trips were made to this site except between June and September when the ants were dormant.

Various methods of collecting ants were investigated and we found the safest and most reliable procedure was to use small portable vacuum cleaners and then freeze the collected ants with dry ice. The ants were stored frozen at low temperature until dissection.

Ant collecting was dirty, hot, thirsty and, at times, potentially dangerous work because some stinging was unavoidable. Sunstroke and close shaves with snakes were other hazards.

An essential part of the project was the establishment of In House References (IHR's). These preparations were used as internal reagents at CSL and were intended for skin testing patients participating in the clinical trial. They were also used as laboratory reagents for coating the solid phase in the important radioallergosorbent assays (RAST) and also enzyme immunoassay (EIA).

In January 1991, two IHR standards were dispensed and freeze-dried. One was Jumper ant venom sac extract and the other Bull ant venom sac extract. Over 800 ampoules of each of these containing 110 microgram of venom sac extract with 4 mg of mannitol as an excipient were prepared. These quantities followed exactly the formulation of the Bayer Albay bee venom for immunotherapy.

In February 1991 the project seemed to be going satisfactorily. Both IHR's had been dispensed, freeze-dried and proven to be sterile. On 25 February 1991, a meeting was held with CSL's Quality Assurance and Regulatory Affairs chiefs. This meeting was held to establish the parameters for stability studies and determine what tests had to be done by Quality Assurance and when.

In retrospect there were clear signs of problems ahead. The primary problem was not having established 'time-zero' parameters for the IHRs. However, it was felt if these were put in place as soon as possible then, when the batches for human immunotherapy were prepared, the established testing protocols could immediately be activated. The ampoules for immunotherapy were to contain 550µg of venom sac extract. For registration two sequential identical batches had to be prepared.

The second problem, which eventually caused the project to peter out by 1994, was under-funding. CSL was not inclined to meet the extensive cost of the preparation of the pre-registration documents nor were appropriately qualified CSL staff made available for overseeing the stages that would have to be reported in full in the registration documents.

The White-tailed and Other House and Garden Spiders

I was first alerted to the possibility that Australian spiders might cause significant skin damage in May 1978, when Mrs Joan Vivian nearly lost her hand. Joan, who lived near Swan Hill in Victoria, developed a gangrenous hand after a presumed bite by a garden spider the day before. After reviewing Joan's case in 1997, her surgeon, Alan McLeod, and I still believe that in all probability it was due to a spider bite. The details were published in *Australian Animal Toxins* in 1983 with several other similar reports. They represent the earliest accounts of the disease, necrotising arachnidism, possibly occurring in Australia.

In spring 1978, Megan and I, aided by Joan's family, surveyed the spider population around her home. Wolf spiders predominated but early anecdotal and experimental evidence implicating them was not validated by further experience. The identity of the spider that bit Joan is still unknown but I hold hopes of its eventual discovery.

Over the years three very common spiders have been accepted as causing skin damage. They are the White-tailed spider, the Black Window spider and spiders of the genus *Steatoda*. None of these have been proved to cause massive injuries. The bites usually go unnoticed when people are gardening or roll on the spider when asleep in bed and the patient may not seek medical advice until a day or so later.

Most of the known or suspected bites reported to CSL were minor and healed spontaneously. Many lesions probably had nothing to do with spiders; indeed, a few were later proved to be self-inflicted. Although necrotising arachnidism looked to be a significant and widespread Australian problem it was the occasional near-fatal case that brought home the urgent need for research. For example, in 1988 I felt helpless at the bedside of a critically ill man with a gangrenous arm, probably due to a spider bite. I nearly fainted as I tried to give helpful advice to the assembled staff while at very close quarters to the ghastly injury. The man survived and a sample of his serum taken some years later offered faint evidence that the Black Window spider might have been involved.

However, no solution is on hand as yet to the most difficult research I have experienced. In ten years there have been no breakthroughs, only negative findings. Indeed, we could fill several editions of the apocryphal *Journal of Failed Experiments!*

The aim of the project was to provide an antivenom to arrest significant necrotising arachnidism. Firstly, we had to determine the species and sex of the spider responsible for the bites causing the most concern. We could then proceed in an orderly fashion to obtain sufficient venom for research and antivenom production, and investigate methods of early diagnosis to aid prompt antivenom therapy.

One possibility was to develop venom detection kits suitable for diagnosing bites due to three or four common household and garden spiders. Rather than adopt this approach, however, we decided to screen the sera of several hundred patients for the presence of antibodies to the common spider venoms. A number of these sera had been taken from seriously ill patients, sometimes weeks after the onset of their illness and it was felt that a demonstration of a rising antibody level to a particular spider venom might be a rewarding approach.

We had successfully detected antibodies of various types against snake venoms and snake venom components in the sera of herpetologists and snake handlers months, even years, after they had been bitten by snakes.

Although the amount of spider venom introduced by the spider would be miniscule compared with the average snake venom, it was felt the sensitivity of enzyme immunoassays might prove sufficient for our purposes. Therefore, using the methodology that had successfully detected antibodies, we explored the serum samples of hundreds of patients, many of whom gave convincing stories of having suffered a spider bite which produced significant local reaction.

Initially we thought everything was working satisfactorily. High titres against White-tailed spider venom gland extract were found amongst a number of the samples, whereas normal sera gave uniformly low results. Unfortunately, when the number of controlled sera were increased it was proved that some non-specific binding was going on between components of White-tailed spider venom glands and normal human IgG. Despite exhaustive technical manoeuvres and consultation with experts, the problem has not to date been overcome though we determined that the chiton present in the venom gland extracts is responsible for the binding of IgG. The degree of binding varies from one serum to another.

Apart from these apparently unique technical difficulties, it is not known whether these common spiders introduce enough venom to provide either a humoral or a cellular response (or neither). Another factor might be that there is an allergic basis for the reactions because it is feasible that a person could suffer a succession of bites by the common White-tailed spider over a period of time.

To date, basic research on these common venoms has produced only bland results. In the meantime, as research continues, the existence of necrotising arachnidism in Australia is now fully accepted. We shall now leave this still-to-be-solved puzzle and return to CSL where some stage furniture is about to be moved around.

Departure of Dr Neville McCarthy

In December 1989 the Minister for Health (Mr Peter Staples) overruled the CSL's Commissioners' recommendation to re-appoint Neville McCarthy. He was to continue as acting Managing Director until 30 June 1990 when Dr Brian McNamee would succeed him. In the meantime Brian McNamee would join CSL as soon as possible to 'learn the ropes'.

Looking back on my relationship with Neville it's hard to see how things could have been different. He wanted to be the wise and acknowledged leader of a radically commercialised CSL. I wanted to contribute to scientific knowledge and aid my medical colleagues 'out there'. I found him scientifically and medically shallow. He no doubt found me outspoken and enthusiastic about my work to the point of arrogance. I considered he put profit ahead of humanitarian issues and on this point I had to chuckle when he and Alf Brogan called the history of CSL *Committed to Saving Lives*.

Neville McCarthy's farewell dinners

Two big farewell dinners were held for Neville McCarthy. On Tuesday, 19 June, a large function sponsored by CSL was held at the Windsor Hotel.

On Thursday, 21 June, an interesting dinner was held at the Cafe Tasma. It's interesting because Neville, with an eye to posterity, instructed that the speeches be recorded. The tapes and transcripts of this dinner were a great source of comfort to the writer who had not been invited.

The dinner at the Cafe Tasma was free to the eighty or so senior staff with the proviso that they donate '$20 towards a farewell gift to Dr McCarthy'. By the time David Doherty, who was CSL's Materials Manager, got to his feet to give an appreciation of McCarthy he was facing a sea of flushed and happy faces.

To his audience's surprise David gave McCarthy a highly entertaining roasting, intermittently sticking a knife in and twisting it. David's preamble set the pace:

> The events of 1974, some of you may remember, there was a Cyclone in Darwin that wiped out the town, the worst floods in history in Brisbane wiped out that town, Vince Gair was appointed as the Ambassador to Ireland and stuffed that up, Gerald Ford was appointed President of the United States, the worst President of the United States ever, Think Big won the Melbourne Cup, nobody backed it, Richmond won the VFL Premiership and Dr McCarthy started at CSL.

Of Neville's 1979 birthday party David said:

> I went to your fiftieth birthday celebration. You may remember that ... I made some most unsatisfactory comments about your rare visits to Australia and things like that, and as a consequence of that I said I was moving closer to getting stuck into your travel claims from an audit point of view. I got moved out of Audit in two weeks.

And later on:

> We once went out to lunch on a Christmas Eve, and you got hold of Joe O'Brien who was our boss at that time, I was with a very small crew of Mick O'Brien and those sort of people, so we hadn't drunk very much, and you got hold of Joe O'Brien and said 'This is absolutely disgraceful, these people are two hours late back from lunch, and I want two hours of their salary docked' which was duly done. I want to tell you Dr McCarthy, we were five hours late back from lunch.

David then related how McCarthy had asked him to draft a policy on publication of papers by CSL.

So, I thought here's an opportunity to convince the M.D. how good I am, and I assembled a list of people who had been producing papers, and I went to talk to the bloke at the top of the list, Struan Sutherland, and he had just published a paper which was unknown to me, which you didn't bother to tell me, was about Antivenoms that had an advertisement from CSL in the middle of it which John Blizzard put in and I walked into the office and I said to Struan, 'I want to talk to you about published papers, and the things we do,' and he said, 'Get out, McCarthy sent you, that's the finish of you,' threw me out of the damn office. Anyway, I did submit an Audit report which Struan helped me to put together. (I liked that story but the reader should bear in mind that David was about twice my size!)

You are keen on making quotes and telling us what famous people said what, I've got two which I think are relevant. One by you, which I think will go down in CSL folklore, and that is 'Betty, get me some more overseas travel forms,' and the other one is by Nancy Doherty, who is my wife, who said, 'When's that bastard McCarthy going to give you some more money.'

I think we can genuinely say that your coming to CSL changed the Organisation. That's unfortunate, we were going pretty good before you got there.

After warning McCarthy that if he spoke for longer than seven minutes all persons would get up and walk out David then reverted to normality and said, 'Enough rubbish, we will miss Neville McCarthy, and I am sure the Organisation is a better place for you having been here.'

After making the presentation to McCarthy he sat down to long and tremendous applause.

To McCarthy's credit he appeared to take this roasting on the chin. His speech was marginally longer than David's and he explained that he had no regrets about coming to CSL and turning down overseas job opportunities. He seemed to have a fixation on me as he mentioned my name four times, whereas only four other current CSL employees were named (once) in the whole address. Early on he stated, 'You mentioned my friend Struan, the most curious and ironic thing about Struan was that he was most reluctant to publish when I came to CSL, I encouraged him and almost forced him to publish, and to popularise the work he was doing, and I was certainly repaid in full subsequently.' (This was news to me as I was in full publishing stride before I ever heard of Neville.)

Neville then explained how he had been able to exercise some choice in the gifts of appreciation that he and his wife were receiving.

We were fortunate on Tuesday evening in being presented with a very fine commemorative present to receive a Salad Bowl and a pair of Salad Servers and, lo and behold tonight, we have another Salad Bowl and a pair of Salad Servers, not by accident, let me assure you, not by accident. This particular set, with its Water Jug and its Gum Leaf design, is intended particularly to fit the 9ft Messmate timber dining room table at Flowerdale, whereas the plated silver wire one of the other night will sit on the Georgian type dining room table at Kew.

This statement qualified Neville as a script writer for the television programme 'Keeping Up Appearances'!

The last word of the night went to David Doherty who said, 'Drink as much as you can, you are not paying for it.' They did.

Within a week or so copies of the complete tape and transcript were circulating at CSL for the benefit of the other one thousand employees who had not been privileged to contribute to the salad holding and dispensing equipment presented to Neville McCarthy.

From Sunday, 1 July 1990, a new McCarthy-free era opened at CSL. Most of my opponents had wandered off into the sunset with more than adequate pensions. Things had to get better, and they did, for a while. Then the tide turned against me and I had no option but to leave CSL, but not without a struggle!

Dr Brian McNamee

The appointment of Brian McNamee as Managing Director, with a goal to oversee privatisation, was a wise choice by the Federal Government. Twenty years my junior and almost boyish with enthusiasm, he was clearly a business man, having steered steadily away from the practice of medicine since graduating in 1979. He looked fit and healthy like his tennis-playing brother. Almost the entire staff assembled in the canteen to hear him say that he was a straightforward local boy and that from now on everything would be open. I was loudly applauded when moved to observe that some of us had waited a long time to hear such a statement. Although CSL was clearly going to move towards privatisation, Brian radiated a 'nice guy' image and most of us left the meeting reassured and optimistic.

At the time Brian McNamee arrived, some money had reached me via a Commonwealth Grant to activate an Ant Venom Project (Page 313). A freeze on appointment of scientific staff was in place but this Government money allowed the hire of scientific staff on an hourly basis. It was an expensive way to conduct research, but there was no alternative.

In the first few months Brian McNamee was encouraging and helpful. Once a month he would come to my attic for coffee and we happily discussed a wide range of subjects. One word from him released grant money for essential equipment and, after a seven-year drought, a laboratory dedicated to venom research was being equipped. There was even a possibility that my former research team could be reassembled.

It was wonderful to have staff speaking freely at senior management meetings. After years of silence I was fairly bubbling over with excitement.

One thing that worried me was Brian McNamee's first impression that antivenoms were essentially non-profitable. I thought this negative approach was false and wanted an external party to analyse the cost and other aspects of antivenom production in depth. Accordingly I sought, and got, an early audience with the new Director.

Whilst not knocking venom research, Brian said that in fact antivenom production was an expensive millstone around CSL's neck and the Government should be more supportive. He said that one of the great costs was the provision in each state of a twenty-four-hour service to distribute CSL products. I gave him a little burst on how antivenom usage could be rationalised on a state-by-state basis, and each state should determine exactly how much antivenom should be held, where it should be held and the amount used in previous years carefully analysed. I told him Marjory and I were currently concluding a survey of antivenom usage in Australia which could be very helpful in this matter.

Furthermore, I pointed out that it was not necessary for CSL to provide a twenty-four-hour service in each state, as the policy adopted with the storage of overseas antivenoms would be quite satisfactory. In the case of overseas antivenoms, which are held to supply private collectors, zoos, etc., a large hospital in each capital city has been designated the holder and the appropriate antivenoms are held in its pharmacy.

On the subject of the Farm at Woodend, I expressed the view that it was in CSL's interest to explore new improved ways of developing antivenoms and cited Funnel-web antivenom which is produced in rabbits locally at CSL and was selling for a handsome $600 an ampoule. Furthermore CSL should develop more in vitro testing systems to ensure optimal harvests from the snake antivenom horses and perhaps reintroduce a pharmacology research laboratory.

I also broached the possibility of making globally designed antivenoms, giving as an example the possibility of a latrodectus antivenom for world-wide use.

Too late I realised that the opponents to antivenom research had had greater access to McNamee. I pressed on, however, innocently believing that in time a compromise would be reached.

Dr Ian Gust

At a meeting of senior staff in August 1990, I pushed for greater interaction between CSL researchers and relevant outside experts. 'For example,' I piped up, 'the CSL virology workers should be regularly picking the brains of people like Ian Gust.' 'Good point,' said Brian McNamee. In all probability this had nothing to do with the appointment of Ian Gust as Research and Development Director to CSL from 12 November 1990. Gust was effectively to close down venom research at CSL and nearly finish me off at the same time.

I thought I knew Ian Gust well at the time of his appointment. He was five years younger than me and in our casual meetings over the years we had got along well. We had a few things in common: twice we had both hit the headlines over troubles with superiors at our respective establishments. In the early seventies Ian had had a great run-in with Dr Noel Bennett, the then Medical Superintendent of Fairfield Infectious Diseases Hospital in Melbourne. Noel was enforcing his authority over Ian regarding absences from meetings without permission and not wearing a tie in the laboratory, etc. In this first blue I gave public support to Ian Gust's cause because he seemed to be the underdog.

Gust and I also shared the same mentor: Dr Jock Frew (later Sir John Frew). In the early eighties Jock put both Ian and me on the vaccine sub-committee of the NHMRC. After some of these meetings, Ian and I would compare notes on our respective battlegrounds and proffer each other advice. He was becoming increasingly involved with WHO and kindly sent me documents of interest such as reviews of deaths in the Pacific region due to marine toxins.

Several months before his appointment to CSL, Gust was in the headlines again. This time it was about a dispute with the Administrator of Fairfield Hospital over the control of donated funds and other budget matters. His departure from Fairfield coincided with the peak of the unsuccessful campaign to save that wonderful hospital from closure.

When Ian joined CSL in November 1990 I welcomed him as a friend and colleague, as well as a fellow veteran of unsympathetic and sometimes obstructive beaurocracy. I soon discovered our agendas were miles apart!

Ian's job was to re-organise Research and Development and to pinpoint projects which would lead to profitable products. He had control over a $16 million budget, a staff of eighty or so and almost unlimited business-class travel opportunities. He was also in a supreme position to help, hinder or ignore antivenom and venom research.

After frequently seeking my advice on people and their backgrounds Ian and I had our first in-depth chat about antivenom and venom research. This started off promisingly enough with his desire 'to tap my vast experience and make up for lost time, etc., etc.' However, within five minutes I realised the conversation was one sided and I was having no chance to refute some of his broad generalisations which were frankly 'up the creek'. With horror, I recalled Charles Guthrie coming out of his office the day before and giving me a very smug smile. The day before that, I had been transferred from Guthrie's control back to Research and Development. When Gust left, I was more perturbed by his attitude, than by the fact that my 'vast experience' had been largely ignored.

Next day, Brian McNamee came in for our scheduled monthly coffee and wide roaming discussions. My chain of command now was Ian Gust and then McNamee. Early on he said, 'How are you getting on with Gustie?' When I replied that frankly I was quite worried about the relationship his face fell. At that moment the phone rang with the Chairman of CSL on the line. I politely waited outside my office until the conversation had ended, at which stage Brian said an urgent matter had come up and left abruptly. Brian McNamee never visited my office again.

From 1991 onwards, I thought the best way to survive was to demonstrate my usefulness by my output of research and publications. The problem was that I was allowed no CSL scientific staff and had to rely on the enthusiastic but largely inexperienced young science graduates who were hired on an hourly basis from an agency. I lacked a right-hand person who could supervise their work, knew the ropes of CSL and could link up with other departments. On the positive side, there were tangible benefits for some of these young scientists. For example, Sony Varma and Mary Papadopolous, who were both excellent workers, went on to bigger and better things.

There were no slackers in our little group and for the next two years a prodigious amount of work was done. Apart from the Ant venom project (page 313), which was our bread and butter, we established a wide range of enzyme-linked immunoassays to examine the sera of envenomed patients. Of particular interest were sera from possible victims of Whitetailed spider bites. Exciting as it was, this project proved extremely difficult (see page 317). Very successful work was done in developing tests to reduce the use of experimental animals in antivenom manufacture and testing. The examination of the sera of snake handlers who had suffered a number of snake bites produced some astounding findings. For example, one man developed antibodies to all the main species of snakes in the world. Although weak, these antibodies may in time help scientists working towards a global antivenom or venom detection kit.

During this time, Robert Premier was the only CSL scientist who interacted with and advised me on a regular basis. Apart from some good friends amongst the 'workers' in quality assurance and Chris Kapouleas, who performed some analytical chromatography on our behalf, we were somewhat isolated from the other CSL facilities.

Staff quickly learnt not to debate an issue with Ian Gust in public. Dr George Harris, whom he had inherited as Assistant Research and

Development Director, stood up to Gust, who decided George had to go. His departure left only two other medically qualified people in Research and Development, one of whom was Dr John McEwen who was skilled in the important field of medical registration. He shared many of my views and he left shortly before I did.

Reviewing the written exchanges between Ian Gust and myself, and occasionally Brian McNamee, in 1992 and 1993 I am struck by the hopelessness of my situation. McNamee failed to intervene, despite a number of requests to do so as my 'next level Manager'. No-one came out of this dispute particularly well. Gust and my past experiences ensured everything would be well documented. On one memorable day we exchanged six memos!

It might seem unbelievable, but not once in the two years after I moved into my new office and laboratory did either of my two seniors visit us to see what went on, offer advice or encourage staff, etc. My budget was about $0.5 million per annum over this period and we were considered a fairly dynamic little group which welcomed visitors.

What I thought was to be a golden finish to my time at CSL was turning sour. Even the most minor of requests tended to be blocked by Gust.

For example, I requested that Ian Gust allow my two hired scientists to be paid 'holiday pay' for Christmas Day 1991 because they had worked so industriously on ant venom collection and dissection. He refused this request so I gave them both a personal cheque.

Performance Management System Agreements

The core of the problems between myself and Gust, and ultimately CSL, related to the Performance Management System Agreements. These were generally known as Work Plans and were introduced into CSL in June 1991. The draft plan I submitted was almost unrecognisable when returned by Ian Gust for my signature later in June 1991. Some added goals were unachievable without access to at least one experienced CSL scientist and it made no mention of my routine consultant duties of advising doctors and answering extensive mail, etc. Some minor modifications were made but when I received the final work plan for my signature, there was no way I was going to sign the document. It was put in a drawer in my desk awaiting further negotiations. In December 1991 Gust realised that mine was the only work plan that had not been signed and sealed as scheduled. He was furious, as until then he had assumed I had succumbed.

On 16 January 1992 I wrote as follows to Ian Gust: 'I really feel that a third party should be called in to arbitrate on the matter of my hesitancy to sign the Performance Management System Agreement. I have been advised most strongly against signing an agreement which does not include what is seen as my day to day routine duties.' He replied the same day '... my view is that this is primarily a matter between you and me.

Should our opinions differ substantially—which I think is unlikely—the final arbitrator would have to be the MD.' The matter remained unresolved until readdressed with vigour by Gust in May 1992.

Since I was back running a laboratory again it seemed sensible to catch up with overseas trends. CSL had not sponsored any of my overseas trips for some twenty years. Any overseas experience I had I funded myself or with direct assistance from WHO. Ian Gust, who had been airborne for much of the previous year, gave only limited financial help to this proposal.

On 21 January 1992 I asked Gust to consider funding visits to a number of overseas centres in June. An early indication of support was requested and a rough draft of some of the centres to be visited was enclosed.

He replied requesting more details and suggested the ten centres I wished to visit could be completed in fourteen days! In March he informed me that CSL would fund one economy round the world airfare with expenses for two working weeks to visit colleagues and laboratories in June. It cost me $7000 to visit the centres not covered by CSL and on some of the more ghastly economy flights I thought dark thoughts about Gust and his frequent business-class trips.

May was not a pleasant month. My mother died, a fact of which Gust was aware, but he proffered no condolences. I had a bleed from my guts which was rather startling but found watching my own colonoscopy on a screen quite fascinating. Fortunately no cause was found. Then the travel agency that my wife and I were using, went into receivership, which added to the fun. In retrospect I realise that my handwriting was already compromised at this time by my neurological disease because my wife was kindly filling in all my arrival and departure forms.

Ian Gust used great pressure to get me to sign a modified but still unpalatable work plan before commencing this overseas trip. After a series of meetings with him I won a few minor concessions and signed a work plan which I was far from happy about on 4 June. However, it did acknowledge my current research and consultant duties. In other words it covered 80 per cent of what I was 'normally doing'. As I signed it, I wondered if the plane might hit a mountain and it would all be irrelevant.

The plane did not hit a mountain, but when I arrived back from overseas on 22 July and saw a memo from Gust dated 16 July I almost wished it had! It started with the statement: 'As you know, Commonwealth contract funding for the Jumper ant venom project expired on 30 June 1992' (the first I knew about this) and it went on to inform me that if either of my two scientific staff resigned before 31 December they would not be replaced unless Commonwealth funds had been obtained. He concluded: 'I am optimistic that discussions with the Commonwealth re special interest activities will be fruitful and will keep you informed of progress' (they weren't and he didn't).

Unbeknownst to me the Public Service Union had come down like a ton of bricks on CSL for long-term use of staff from labour hire companies. The Union regarded this procedure as exploitative and not in 'best practice'.

The outlook now was rather grim. I was locked into a fairly impossible work plan and access to experienced CSL staff was minimal. I could not hire staff whose specialised skills I might need for only a matter of weeks. On the other hand, 'LabStaff' the company which provided both Mary Papadopolous and Sony Varma were reasonable and agreed to reduce their commission as time went by so Mary and Sony's hourly rate could be slightly increased. One way or another, I was able to employ Mary and Sony right up until I left CSL in June 1994. They knew that, if they left, our research was completely sunk, since I was forbidden to replace them. It is good to report that in 1998 they both have interesting jobs which make good use of their talents and pleasant personalities.

On 4 August 1992 a friend at CSL rang me up to ask for my comments on a fax that five months earlier had gone to the Department of Health, Housing and Community Services in Canberra about my future at CSL. Ignorant of the document, I asked for a copy. This fax certainly raised my eyebrows. It requested financial assistance from the Government for the transfer of the Venom Research Unit (my unit!) out of CSL. The fax was written by Graeme Kaufman, General Manager of Finance and Planning, to Mick Roche, a Deputy Secretary in the Department of Health, Housing and Community Services.

In reference to antivenom research Kaufman said, '... there is a question of the significant costs incurred by CSL in research on antivenoms (with no prospect of a commercially viable outcome) and the consultant and advisory role performed by Struan Sutherland. These costs should be recognised as a CSO, with funding provided by the Commonwealth. In line with our proposal regarding the National Blood Group Reference Laboratory, these functions should be relocated away from CSL.' (CSO is Community Service Obligation.)

In December 1992 I submitted a summary of the Ant venom allergen project as requested by Mr Ken Sanderson of the Projects Section of the Federal Health Department (DHH and CS). In this report I stressed that it was essential that an appropriately qualified individual should be made available to bring all the specialised documentations together.

On 21 December 1992 I was awarded the AMA Diners Club Public Health Award ($10 000). I consequently received a memo from 'good guy' Gust congratulating me on the 'Pacific Health Award' and 'hopefully, the publicity surrounding your award and the recent *MJA* articles will assist our discussions with Government on continued funding'. (Later, despite repeated requests, no evidence of such discussions was forthcoming.)

On 24 December 1992 Graham Mitchell was appointed as Gust's

assistant. Graham had been Director of the Royal Melbourne Zoo and before that he had been at the Walter and Eliza Hall Institute. An affable, very talkative but honest fellow, his appointment almost seemed to make Ian Gust redundant. During the next eighteen months I only had one hour of talking hard science with Graham. Although he was unfamiliar with my field I could not fault his intelligent suggestions. It is a pleasure to record that on the few occasions I needed decisions from him he was scrupulously fair and helpful. I felt that he acted as a buffer between the excessively autocratic Gust and the nervously vulnerable researchers. Sadly for some of the staff, Graham left CSL several years later.

For most of 1993 and 1994 I spent every Sunday morning writing a weekly article for the *Australian Doctor*. Most of these were on dangerous Australian animals, which ranged from magpies to the hopefully non-extinct Tasmanian tiger. I really enjoyed trying to write in a 'popular' style. These articles, named 'Creature Feature' proved an excellent way to provide up-to-date information to doctors and paramedicals. Plans are well advanced for them to be published in 1999. The science editor, Guy Nolch, has joined me as joint author.

The first six months of 1993 passed peacefully enough with neither Gust nor I feverishly seeking each other's company. However, things soon hotted up! The seeds for our major clash were sown innocently enough in February, when I sent a memo to CSL's wealthy Pharmaceutical Division requesting help with expenses so I could speak at the Townsville hospital on the way back from a Medical Conference at Mt Isa in August. My fare, etc., to Mt Isa was to be paid by the organisers (which included the Royal Flying Doctor Service). Dr Stephen Silk replied he would be pleased to assist and was given details of the likely costs, so he could incorporate them in next year's budget. Having been able to fund the Townsville diversion I then accepted the invitation to attend the Mt Isa meeting on 16 July 1993. However, visiting Mt Isa was not that simple.

On Friday, 18 June, Stan McLiesh, the powerful General Manager of the Pharmaceutical Division, said he would like to come over to have a chat. This was quite an event, since Stan was perhaps the second most important person at CSL after McNamee. When Stan left I was in a state of elation. The gist of his conversation was that he believed that CSL had not made good use of me and that he wanted me more involved in all aspects of antivenoms. He would even sponsor more travel if that was what I wanted. 'It's time to make amends,' he said.

On 21 June I sent Stan a memo entitled 'The production and testing of antivenoms—some good news'. In this document I detailed the advantages of the new assays we had developed which used no animals, gave fast results and were far cheaper and more accurate than the current animal assays. This work was an excellent example of the advantages to CSL of an active venom research unit.

Fired up by Stan's apparent green light, on the morning of 1 July I gave Kevin Healy, Head of Quality Assurance, a draft of the methods we had developed for improved antivenom testing. I can't say Kevin reacted to the documents with ecstasy. I have subsequently felt that he was responsible in part for the slowness of CSL to implement a reduction in animal usage in regard to antivenom production and testing.

That afternoon I received a bombshell in the form of a memo from Stan McLiesh through Ian Gust. There was a dramatic change in tone. I was admonished for not following protocol and told that in future all communications should be forwarded through Ian Gust. My travel arrangements took on a different slant. I was to become an almost full-time one-man travelling circus visiting small and large states. Stan indicated that one of his marketing people would draw up a draft programme for this multitude of seminars and forward it to me through Ian Gust. A copy of the memo was sent to Brian McNamee.

And it was time for a new draft Work Plan. After some initial skirmishes Gust and I got down to business. However, the Plan he had prepared had no reference to the activities which occupied at least 70 per cent of my time! On 29 June he demanded details of these activities so he could complete the assessment of my performance the next day.

I provided these details plus copies of my timesheets which had basically remained unchanged for two years. These included an estimation that 50 per cent of my time was charged to corporate services not R and D. How I spent 80 per cent of my time was outlined in a typewritten diary which was never challenged by Ian Gust. This fairly impressive document detailed hundreds of after-hours calls concerning many other CSL products as well as antivenoms.

On 1 July 1993 Gust came into my office for the first time since I had moved into it two years earlier. He handed me an envelope marked 'Confidential', gave me a chilling look and left without comment. I toyed with opening it, but decided it might be a strategic mistake so I returned it with a memo suggesting that I had the impression it would only inflame an already unpleasant situation. I pointed out that, in regard to counselling, there were certain niceties to be observed. One was to allow me to select a third person to be present. I forwarded him an article on counselling which had appeared in the POA Journal *The Professional*, December 1991. I sent a memo to Brian McNamee requesting that a third party who was suitably qualified and acceptable to both parties be asked to arbitrate. I also suggested that this needed to be done promptly to avoid an unpleasant situation which could only worsen—possibly to CSL's detriment.

The next afternoon I had another compulsory and unproductive meeting with Gust about the Work Plan.

On 5 July I received a memo entitled '1993–94 Travel' from Ian Gust indicating that he would not authorise any travel in the current financial year until my Work Plan was agreed to and overall travel plans considered.

The Mt Isa meeting was looming up on 16 July and the organisers now requested an after dinner speech as well. This was a bit of a joke since stress was already putting me off my food.

By now I was running out of time so I wrote to McNamee protesting about the threat to my speaking commitments. I enclosed a recreation leave form for McNamee to sign to enable me to accept an invitation to talk on envenomation at the North-West Medical Conference, Mt Isa, and to be an after dinner speaker at the conference dinner; to address a meeting of physicians at the Townsville General Hospital on 19 July; and to visit James Cook University on 20 and 21 July regarding supplies of Box jellyfish venom for antivenom production.

McNamee washed his hands of the whole affair: 'I am returning your application to take recreation leave. The matter should be resolved with your manager.' To my mind this was very poor management. If you have two dogs who persistently fight, you don't deliberately put them in the same backyard. I felt that under no circumstances should I sit down with Ian Gust again without an appropriate third party present. Having had a number of sessions with him, which had been increasingly unpleasant, I wanted to avoid the pointlessness and stress of repeat performances.

On 6 July I went to see my old mate Dr Peter Sutherland because by this time I was finding it increasingly difficult to write. The first word or two in a sentence would be easy but then my writing would become smaller and smaller until it almost stopped due to a painless cramp. Although impressed by a demonstration, Peter allayed my fears that it was early Parkinson's Disease and laughingly suggested I get a bigger pen. Peter and I thought it was probably stress induced but because of the strong family history of Parkinson's Disease he referred me to a neurologist.

On 7 July I received a memo from Brian McNamee acknowledging that CSL's Performance Management System allowed for 'the next level manager to become involved in the objective setting process'. He said it was necessary for people to solve their differences through discussion and/or negotiation and stated 'this should be the case particularly when the process involves senior management at the highest level of the company'. I promptly replied enclosing a page of CSL's 1993 Guidelines to Work Planning and Performance Review. This gave details of the role of the next level manager and stated, 'Another role would be that of arbitrator where agreement cannot be reached between the employee and the immediate manager.' My memo included the question 'Since you are my next level manager, surely you have a legal obligation to arbitrate as I have repeatedly requested?'

On the afternoon of Friday, 9 July, I spent over two hours alone with Ian Gust, who in my opinion tried every trick in the book. I nearly succumbed, but was saved by his reaction to a comment I made. As mentioned earlier Sir John Frew had been a mentor to both of us. In exasperation I said, 'I wonder what Jock Frew would think of you for saying something like that.' He looked at me and snapped, 'Jock Frew is dead and so is Bill Lane and so you can forget them.' The comparison of these two with my adversary at the time was striking.

My attitude had suddenly changed. I told myself not to give in—survive the meeting and have a good think. For the last fifteen minutes I was reasonably relaxed. At the end of the meeting, during which no agreement had been reached, we parted amiably enough. I half agreed to meet with him again on Monday morning. I had no intention doing this but I didn't have a clue what to do!

When I returned to my office Gust's secretary had again delivered a confidential letter for my attention. I thought 'what the heck' and opened it. In it, Gust described my overall job performance as unsatisfactory and, as a direct result, he believed the question of my continuing employment by the company had to be considered. He also enclosed a spectacular work plan for the coming year, should I still be at CSL. After sitting down and studying this work plan I said to Marjory, 'I don't know of any three people in the world working together who could possibly fulfill this work plan.'

That night I made the decision to somehow go public. The person who could fix the situation was McNamee, but it appeared he would only do this if his own interests were threatened. His apparent administrative failings had to be spotlighted but I had no desire to be sacked for doing so. The problem was an internal matter entirely of CSL's making, and media involvement could be damaging to all parties.

In the end I adopted a novel approach and circulated an open letter to Brian McNamee around CSL on Monday, 12 July. As far as I knew there was no staff rule which prohibited the circulation of such letters! However, it was dangerous to write and distribute such a letter for, if one fact was wrong, it could be disastrous to the writer.

The letter which is reproduced below is far from polished but represents a weekend's work. Reading it again some years later I would agree it's a mite too strong in places. Desperation was driving me and it was no time to mince words. There also seemed little opportunity to seek wise and moderating counsel.

On Monday, 12 July, at 8.55 am I gave the letter to Brian McNamee's secretary and placed several copies in the library and the canteen. Within a minute CSL's photocopiers were helping to extend distribution ... I held my breath.

AN OPEN LETTER

TO DR BRIAN McNAMEE, MANAGING DIRECTOR, CSL LTD

FROM DR STRUAN SUTHERLAND, MEDICAL CONSULTANT,

REGARDING MATTERS RELATING TO HIS CONDITIONS OF EMPLOYMENT

Dr B. McNamee
Managing Director
CSL Ltd
45 Poplar Road
Parkville 3052

Dr S. K. Sutherland
6 Wallace Grove
Brighton 3186
11th July 1993.

Dear Dr McNamee,

Thank you for your memo of 8th July, 1993. I wish to bring the following matters to your attention:

1. **My workplan for the current year.**
On Friday 9th July I spent a ghastly two and a quarter hours being brow-beaten by Dr Ian Gust. No agreement was reached on the workplan. Since my repeated and rightful requests to have an arbitrator present at such meetings have been repeatedly refused, I think I am within my rights for the time being to avoid further such experiences.

Dr Gust reiterated a number of times and with considerable relish that he is not prepared to approve my recreation leave to attend the conferences this week in Queensland until agreement has been reached on this workplan. I will return to this point below.

2. **My performance review to July 1993.**
After leaving the meeting with Dr Gust I found his review of my performance in my In-tray. Having read this document most closely it is my intention to challenge its soundness and validity. I vehemently believe it presents a distorted view of my statements, grossly trivialises achievements and

amplifies faults or failures.

It is well documented that I have protested about the imposition of workplans which do not take into account the consultative activities I perform at CSL and after-hours. Requests for advice from doctors, pharmacists and Poison Centres have steadily increased over recent years and occupy at least 50 per cent of my normal working hours. Records of calls are meticulously kept. Over 2000 doctors have sought advice over the last 12 months.

In June 1992 on the eve of my departure overseas, and at a time of considerable stress for other reasons, I was coerced to sign the workplan.

Prior to signing it I had done two things: I had sought advice on the legality and flexibility of the agreement and, most importantly, I had succeeded in changing the stated "Primary Purpose of the Position". It was changed from "Research into development of new anti-toxins and anti-venoms of therapeutic value" to "Research into development of new anti-toxins and anti-venoms of therapeutic value and to undertake consultant duties for and on behalf of CSL". Without the establishment of this broader definition I would not have signed this agreement.

I should point out that I have not seen the June 1992 Annual Review of my performance by Dr Gust as required by CSL's Performance Management System. The requirements are that I should see the report and have an opportunity to comment upon it before signing it. Why was I not shown it and why is a copy of it not held by CSL's Human Resources Department as is also required?

It is also relevant to observe that over a two year period neither Dr Gust nor yourself (my "next level" manager) have visited my department to see what goes on, offer advice, encourage staff etc.

3. Refusal of granting recreation leave to attend a conference this week at which I am an invited speaker as well as the after-dinner speaker.

This Mt Isa conference will be opened by the Queensland Premier, Mr Wayne Goss, and is quite important especially to the Royal Flying Doctor Service.

Dr Gust is adamant that leave will not be granted unless the workplan is agreed to. I have appealed to you to intervene in this and other matters but you washed your hands of the matter and returned the leave application to me.

From an institutional point of view this is worse than the pressure applied prior to my overseas trip last year. It appears this time that you, as my "next level" manager are condoning this pressure which almost amounts to blackmail. (Your memorandum of 8th July, 1993.)

4. If recreation leave is not granted should I attend the conference in Mt Isa?

Morally and ethically I feel I must. Apart from the expectations of the organisers, the air fares they have paid are not refundable. If I don't go, the very least I would have to do would be to reimburse the organisers from my own pocket.

5. If this matter gets further out of hand the consequences may be most unsavoury.

It is well known that both Dr Gust and I have had, at a number of times in our careers, significant and persisting problems with our superiors. Putting two such people together may or may not work. In this case it has not and indeed, as I have repeatedly pointed out to you, the situation is worsening.

> I don't believe anyone at CSL should be as badly treated as I am. Such a poor example of "people management" could hardly inspire the confidence of potential shareholders of CSL.
>
> My preparations for the conference and for the Townsville meeting which immediately follows have been severely disrupted by the events of the last week or so. I request I be left in peace in the short time left to complete my preparations and also finalise staff activities in my absence.
>
> Finally I will stake my reputation on the quality and the usefulness to CSL of the research my group has done in the past year. This research, particularly that under Project 274 (Antivenom development) will significantly impact on all aspects of antivenom production. I get no support from Dr Gust for this work - just the opposite.
>
> I will welcome an inquiry by an appropriate party into the matters raised.
>
> Yours Sincerely,
>
> Struan K. Sutherland
> MD, DSc, FRACP, FRCPA.

By midday I was reading a memo from Ian Gust in which he said he would appreciate a copy of the letter delivered to Dr McNamee that morning. (He must have been the only person at CSL without one!) His memo continued, 'Given the progress I thought we had made on Friday and the difficulty of concluding the outstanding issues before you leave, I am prepared to approve your recreation leave, on the clear understanding that we need to agree to a work plan by the end July.'

Gust then sent out a memo to all members of the R and D staff stating that he understood that an open letter from Dr Sutherland to Dr McNamee was being circulated which addressed a number of issues. He stated that, while some differences remained, he was hopeful that these could be resolved by the end of the month and that in the meantime he had approved my application for leave.

Brian McNamee also responded in a conciliatory fashion but still maintained that it was not appropriate for him to get involved. He requested that any further correspondence with him on these matters should be on a confidential basis!

With the scores somewhat evened I farewelled my loyal, but traumatised, staff and set off to show CSL's flag in Mt Isa and then Townsville.

When I returned to CSL on Friday, 23 July, the battle resumed with Gust and I exchanging memos. I got in first stating the extraordinary draft work plan he had sent me was unacceptable for a variety of reasons. It set an amazing variety of Herculean tasks, the appropriateness of some of which I found hard to understand. There was no provision for my current National Interest projects which offered economic and other advantages to CSL. I pointed out that I could not fulfill this work plan; indeed it was a potent recipe for an accelerated burn-out. (I had secretly filled in a questionnaire on 'Burn-out' published by a reputable newspaper. To my alarm, my score suggested I was in the post-combustion range.)

I told Gust that I was adamant that the Ant venom project must be continued by an appropriately qualified CSL scientist. He responded with a comment that he was pleased I had reviewed the draft work plan and said clearly that there were several issues which must be resolved before agreement was reached. (I would have thought every issue.) He set some alternative dates for meeting the following week and said he still regarded it as unnecessary and inappropriate to have a third person present.

On 22 July Brent Jones, CSL's Director of Corporate Human Resources, was officially made aware by the public sector union that the matter had gone on long enough and a mediator should intercede immediately. John House, the AMA's Federal Industrial Officer, who had proved a staunch ally over recent months, further increased the pressure on CSL, via Brent Jones.

By a stroke of luck CSL's preferred policy was to hire necessary equipment. I stealthily filled out applications for both a photocopier and a fax machine. To the surprise of both Marjory and myself we shortly had installed not only a large and efficient photocopier but also the latest multi-skilled fax machine. These two beauties, which we tried to keep out of sight, ran hot for weeks. If Gust or McNamee had ever bothered to visit the office they would have seen the Sutherland/Davey Communication Centre in full action.

The Beginnings of the Australian Venom Research Unit (AVRU)

At 10.00 am on 30 July 1993 I was feeling quite jaunty. The third year medical students at Melbourne University had clapped with enthusiasm at the end of my lecture. They had laughed at the right spots, been delightfully attentive and asked some good questions. I no longer attempted to write on a blackboard. Two days before I had seen Bob the neurologist who assured me that I was free from Parkinson's Disease. He seemed frankly amused by my micro-writing, thought it was stress related and also advised me to get a larger pen. My wife subsequently presented me with a variety of monstrous pens, to no avail.

On that crisp sunny morning as the students dispersed like noisy parakeets I paused and looked around. Since it was too nice a day to go back to CSL immediately, I overcame my natural shyness of meeting strangers and decided to call on Professor James Angus. Earlier that year he had been appointed to the Chair of Pharmacology and I had just performed as one of his Department's external lecturers.

Accordingly, on this memorable morning I took the lift up to the fifth floor of the building which housed the lecture theatre. I told his secretary who I was and she rather offhandedly gestured to his office and said, 'He's in there.'

First impressions are seldom wrong. Although obviously frantically busy, Jim stopped in his tracks and made me warmly welcome. There was no doubting his sincerity when he thanked me for my contributions to the Department which he said 'fired up the students'. This was a good start and what was to be a mere courtesy call turned into something more significant. Almost as a joke I asked Jim if in time he might be able to spare a metre or two of bench space for me. Five minutes later we were inspecting an empty laboratory which in a few months would become AVRU's first home.

Jim Angus is a rare human being. He has enormous energy and enthusiasm for science but is also splendid at handling people. His interaction with students is a joy to observe and his Department was deemed the most caring Department in the Medical Faculty for 1996. Not all his staff may agree with this assessment but then everyone can't be the Professor. Speaking of Professors, Jim thought a personal chair for me was a distinct possibility. Although one was in the air at another University, working with Jim was clearly a more attractive proposition. After four years of sharing the highs and lows of AVRU's development my admiration for Jim Angus remains unabated.

On 1 September, a rather relaxed Brent Jones cruised into my office to sound me out on voluntary redundancy. Sensing that I was in no mood to volunteer for anything, Brent explained that all negotiations would be between him and me in confidence. Furthermore, he had been empowered to be 'quite flexible'.

Looking positively languid, Brent suggested I take advantage of the offer before the scheduled privatisation of CSL the following June. I said I would think about it and agreed to make a list of conditions for consideration regarding what I always officially referred to as my 'involuntary redundancy'. It was agreed there and then that Ian Gust would keep his beak out of this matter. In the months of bargaining that followed Brent Jones behaved with immaculate professionalism.

A week later, I informed Brent that in principle I accepted the inevitability of my redundancy. Part of the deal was an offer of voluntary redundancy to Marjory. She was happy about this since it fitted in with her long-term plan that we both leave together. Then began a basically successful struggle to ensure maximum help from CSL in the establishment of AVRU.

I submitted to Brent Jones a list of matters requiring resolution prior to 30 June 1994. This included re-deployment of staff, post-privatisation consultant work, possible initial support from CSL or Government and most particularly CSL's attitude regarding current equipment.

Since my salary had remained static for a number of years and my pension was to be set at a percentage of my final salary, the following week I requested that my salary be appropriately updated to take into account inflation, etc. Furthermore, no bonuses had been received and discreet inquiries disclosed I was the only eligible person at CSL who had missed out! My request was later knocked back as being inconsistent with the fundamental principles of CSL's Performance Management System. (CSL could follow its own rules when it chose to!) However, I didn't pursue the matter further as some generosity was being shown towards the funding of the unit and I didn't want to sacrifice that.

It appeared essential for AVRU to tap into some of the Federal funding which traditionally came to CSL for venom and antivenom research and I needed to know the exact state of play between CSL and the Government so I could start negotiations for a small slice of the cake. On 11 October, I homed in on Graeme Kaufman seeking details of any response to his fax of 28 February 1992 to Mick Roche, Deputy Secretary, Department of Health, Housing and Community Services (see page 328).

I explained to Kaufman that if funds were available from the Government, I needed to know the amount and what conditions would apply. On the other hand, if they weren't I would like CSL to consider another approach. This fishing expedition produced no reply from Kaufman.

On 13 October I sent Graeme Kaufman another memo challenging his statements on the non-profitability of antivenoms. I outlined the economically and scientifically more attractive assays we had developed, which also reduced the use of experimental animals. It was important that all these facts be fully and openly documented.

Melbourne Cup Day 1993 found me in the Coroner's Court, Condobolin, central NSW. I had been subpoenaed to give an opinion on an unusual case of snake bite in which there was a possibility the patient's sudden demise was due to an adverse reaction to a CSL antivenom. The importance of research done at CSL was particularly relevant to this case and the reputation of its antivenoms emerged untarnished. It also allowed the Coroner to clear the fine medical and paramedical staff of the local hospital of any hint of incompetence. (Details of this case were published in *Australian Doctor*, 25 March 1994.)

Next day, when I got back to Melbourne a trifle fatigued, I faced two letters from Brent Jones which set me back. Apart from dismissing most of my requests, they also informed me that all current research projects were to terminate on 31 December 1993 with my three contract staff leaving on that date. The Ant venom project was specifically targeted for termination but I was also informed that 'the study on sera from spider victims can be completed by March 1994' (with no scientific staff!). Most telling of all was the statement: 'The development of in vitro assays for antivenom development is a long-term project with a low priority at this point of time.' This was my clearest indication that the reduction of usage of experimental animals was in fact of low priority by CSL, running against a world-wide trend to reduce the usage of laboratory animals.

Under specific questioning, Brent Jones admitted that Ian Gust was dabbling in the matter. I turned the heat up in all directions and either wrote or spoke to many influential outsiders.

By 17 November I was able to tell Brent that I was to accept an appointment as an Associate Professor at the Department of Pharmacology at the University of Melbourne. Further, the Head of the Department, Professor James Angus, strongly supported the establishment of an Australian Venom Research Unit in his Department. I informed him that the appointment was basically self-funding and vigorous attempts would be made to attract grants and donations. I formally requested an indication of CSL's attitude to a possible transfer of my laboratory equipment (which had been paid for by direct Government grants for venom research). Furthermore, I asked for some corporate support to help in the establishment of the Unit.

Meantime in Parliament the Australian Democrats were attacking the Federal Government over its proposal to privatise CSL. On 23 November their spokesman on science, Senator John Coulter, criticised financial, scientific and humanitarian aspects of the sale. He later asked the Minister some specific questions about venom research.

By 25 November we were beginning to get somewhere. Jim Angus and I had agreed on the aims of the Unit and how it was to function. CSL decided to provide corporate funding of $20 000 for two to three years and lend to the University any equipment which was not required in other areas of CSL. It was agreed that my contract staff would be released one at a time so that I could keep two till 31 March and one until 30 June 1994. Notifying Jim Angus of this next day I observed 'with further external pressure, CSL may well be more generous ...'

By 1 December it was clear the Ant venom project was in real danger because of lack of resources (page 317). I approached the Australasian Society of Clinical Immunology and Allergy to stress the importance of this project to their members and bring pressure on the Federal Government and/or CSL. CSL ignored the approach and three years later the Society attempted to pick the project up again.

Later, in a desperate attempt to save it, I wrote to the three hundred patients waiting for ant venom therapy and also to their attending doctors. This resulted in an avalanche of appeals to Dr Carmen Lawrence, the then Minister for Health. No favourable response was forthcoming.

On 19 November 1993 I had written to Mick Roche seeking the Department's views on how venom and antivenom research might be continued.

By 2 December he still had not replied but nor had I been blasted by CSL for contacting him in the first place. I therefore sent off a 'hurry up' fax which got a prompt response from an assistant who said he would get him to reply as soon as possible. (He did on 16 May 1994 suggesting I apply to the National Health and Medical Research Council (NHMRC) for funding!) However, the assistant's faxsheet had the names and phone numbers of the Department's Big Wigs and I spotted the number of Dr Tony Adams, the Chief Medical Advisor to the Minister. Right, I thought, I will go to the top and speak to a senior medical colleague. I rang the number and left a message.

On 13 December I had a long conversation with Tony Adams. He astounded me by saying he had no idea venom research was closing down at CSL. He said a lot of kind things about my work and the importance of it in the Australian scene. I promised to send him some background information on current projects and we agreed to keep in touch. In high hopes I dispatched documents to him which supported the Ant venom work in particular. If he did put in a word for me, it was unfortunately quite ineffective.

I had a further conversation with Dr Adams on 10 January 1994 after which I sent him additional material including a note on antivenom costing. These two conversations must have had a profound effect on Dr Adams because, when appearing before the Senate's Estimates Committee on 22 June 1995, in response to a question from Senator Herron, Dr Adams said about me, 'He was talking to me nearly every day about the possibility of getting support ...' (This appears to be yet another case of me kidding myself that I had found an influential ally.)

In January 1994 I was firing memos around CSL about matters which I felt might cause future problems. These included concerns about the performance of the Venom Detection Kits, pricing of antivenom and CSL's failure to reduce animal usage. My research notes give the impression of business as usual. A surprising number of new venoms, including platypus venom, were obtained. On 13 January, I reported the successful capture of March flies and some other insects needed to investigate an allergic death.

We processed hundreds of patients' sera and stored them at -75°C. These samples and their case histories represent a gold mine for the venom or antivenom researcher. The venom samples processed at this time have also proved their worth. For example, in November 1996 we could compare the venom glands from Red-back spiders collected in Japan with stable samples of our local Red-back venom prepared in 1994.

There are lighter moments: on 17 January 1994 my diary records that I was stung by a bee at 1.05 pm when walking around the perimeter of CSL with Alan Coulter. The bee flew up my trouser leg and I had to wait a long ten minutes before I could get back to remove my pants, the offender and its sting. The latter two were then photographed.

On 18 January there was great excitement when we thought we had linked the venom of the Black Window spider with a case of massive and near fatal skin damage, using the Western blot technique which appeared to show the patient's serum contained antibodies to the venom of this particular spider. (See page 317)

By 1 February after hours of negotiation, Brent Jones and I had made real progress. To my surprise, CSL offered a once only financial grant to the Unit of $150 000. Always an optimist, I pointed out that the University would take 12.5 per cent of this for overheads. To my further surprise, CSL agreed to upgrade the grant to $168 750.

Agreement then had to be reached on the timing of the transfer to the University, what equipment could be 'loaned' and who was going to pay for the transfer. As the reader might expect I considered all the equipment vital and fought off any would-be contenders.

Graham Mitchell, Gust's assistant (see page 328), was most helpful in all these matters. In the end the grant from CSL was reduced by $10 000 to pay for some of the items we wished to take. This meant that we could move everything in the laboratory, from filing cabinets to deep freezers, over to the University.

On 15 March I signed an agreement with CSL to leave on 30 June 1994. CSL was to pay the University $80 000 within seven days and the remainder on 30 June.

On 22 March I sent a fax about the Ant venom project to Dr Tony Adams, Chief Medical Advisor to the Minister. I outlined the critical situation that this project was in and how I believed prompt action might bring it to a successful conclusion. Because of pressure from some of the 350 patients awaiting therapy I requested that, if no further help with this Ant venom project would be forthcoming, could I have a ministerial statement to this effect. Mr Mick Roche replied to this and my earlier request on 13 May. (The letter was dated 13 May and was faxed to me on 16 May.) He apologised for the delay and said that the Department, like myself, was waiting to see the outcome of the negotiations with the University of Melbourne. This was news to me. He also thought now that everything was 'sorted out' I would be in a position to continue my antivenom work.

On 5 April 1994 Kim Beazley, Minister for Finance, announced the public float of CSL would take place early in May (see page 346). I thought that, before this event and my move to the University, I should formally present Brian McNamee with a list of my concerns regarding venom/antivenom research. In my innocence (or stupidity) I believed that AVRU would be the obvious recipient of any relevant contract work that CSL deemed necessary, since CSL had decided to close down its own venom/antivenom research unit.

The letter to McNamee is reproduced in full as it highlights difficulties which might result from the close-down of venom research. (It became known as 'McNamee's serve'.) I wanted it to be constructive but at the same time care had to be taken that nothing was brushed under the carpet.

Research
and Development

9 May 1994

Dr Brian McNamee
Managing Director and Chief Executive Officer
Corporate Management
CSL Ltd

Dear Brian

RE: CESSATION OF VENOM AND ANTIVENOM RESEARCH AT CSL

I would like to be certain you are fully aware of some of the implications of the cessation of venom and antivenom research at CSL.

* Humanitarian issues and the reduction of animal experimentation aside, it would seem inexplicable that the work on in vitro assays under development specially for the monitoring and development of antivenoms have not been seized upon because of their economic advantages over the cumbersome animal assays. To cheaply monitor antivenom production and accurately pinpoint optimal harvest times is a great advantage. I attach documents forwarded to Mr Graeme Kaufman on 18 January 1994. More information is available if you require it.

* In regard to the ant venom project, we are ready to process and dispense material for the secondary stability studies and also the clinical trial. There are now over 350 patients on the ant venom allergy registry. These citizens who are of all ages have suffered one or more severe general reactions after ant stings. Some have been fortunate to survive the anaphylactic reaction and it is inevitable that deaths will occur.

* The cessation of work to produce a source of Box Jellyfish venom to replace the now exhausted supplies from the late Dr Barnes in Cairns may result in a shortage of Box Jellyfish antivenom within the foreseeable future. This work which was commenced after an approach by the Production Department was to get into full swing during the recently concluded Box Jellyfish season. A request made on 9 September 1993 for confirmation of support for this work (enclosed) went unanswered. After expressing my concerns about this matter in November last, I was informed "the sourcing of Box Jellyfish venom can be managed between the pharmaceutical and the logistics division". I am now told that current stocks are now near depleted with no guaranteed or acceptable source in sight.

Phone: +61 3 389 1911   CSL Limited A.C.N. 051 588 348   Fax: +61 3 381 1247
45 Poplar Road Parkville Victoria 3052 Australia

* CSL has inadequate stocks of beaked sea snake venom (*Enhydrina schistosa*) which is essential for the manufacture of sea snake antivenom. No reliable source is currently in sight or known.

* The investigation of sera from real and suspected cases of necrotising arachnidism has become a major project because of the frequency and widespread nature of the problem. I have little doubt that in time a rapid method of diagnosis and appropriate antivenom or antivenoms will be required in all parts of Australia. We have just shut down this project, leaving over 250 patients and their doctors in limbo.

There are two other points which are relevant:

* For years, CSL has been a source of attractive, up to date and informative literature on the use of its antivenoms etc. Last week, for the first time, I had no CSL literature to hand out to 180 medical students prior to a lecture. The failure to make use of such a good opportunity to promote CSL Ltd amongst CSL's future customers is a matter which I think should be addressed at a high level.

* There is a very negative attitude towards antivenoms which has infiltrated down to middle management who for personal and/or career motives tend to avoid the most basic collaboration, even though it would be clearly of benefit to CSL. The laboratory workers are keen to help and collaborate, but this is actively discouraged. Without doubt, the morale of all involved in antivenoms is extremely low. I often hear them say that they feel their work is not appreciated.

I would like CSL Ltd to adopt a positive, forward looking attitude and offer some suggestions for your consideration.

1. Although antivenoms are currently only marginally profitable, CSL is stuck with them for the moment and it would seem sensible to adopt a more positive approach rather than keeping them at arm's length. We should turn the situation to CSL's advantage and attempt to make them more profitable, expand their market and raise the morale of those associated with them. Would you, therefore, consider planning a meeting at CSL of all staff involved in all aspects of antivenom and antitoxins? Such a meeting has never taken place before, to my knowledge, and I am certain, apart from promoting cohesion, a number of sound, practical suggestions would be forthcoming. I would welcome the opportunity to discuss this proposal with you further.

2. If I obtain some external funding for the continuation of the ant venom project, I would like to be assured that I may investigate the possibility of contractual arrangements with CSL.

3. If asked, I am willing to help CSL obtain some of the rarer venoms. I am most interested in the problem of box jellyfish venom and, indeed, spent all of last Saturday at CSL, trying to make some progress in this matter. Again, some encouragement is required from on top, so that with co-operation solutions to shortages can be promptly found.

4. The establishment of the Australian Venom Research Unit at the University of Melbourne may allow the continuation of some of this work. It is good that CSL has found itself able to assist in the establishment of the Unit and I hope some of the collaborative work currently being done with some CSL departments will be encouraged to be continued.

Would you please consider encouraging some frank and open discussion on these issues with offices of appropriate level before I leave CSL, in order that matters of agreed mutual and on-going interest may be selected and prioritised.

Yours sincerely,

STRUAN K SUTHERLAND
MD DSc FRACP FRCPA
Medical Consultant

After I had left CSL Stan McLiesh replied on behalf of McNamee. His letter dated 13 July 1994 was superficially conciliatory. However, he quoted Kevin Healy, the Head of Quality Assurance, who was negative about the assays I was establishing to reduce experimental animal usage. Most of the issues I had raised were ignored and little progress has been made on them since that date.

On 9 May at the 8.30 am Research Meeting Ian Gust announced that CSL had 'arranged my appointment at the University and had been more than generous' in the establishment of AVRU. I was not impressed!

I was rather amused at this time to see an empty room near our laboratory slowly filling up with useful store items and bits and pieces 'donated' to the Unit by various Departments. These presents consisted of old but still functional equipment which we regularly used. It was thought prudent to screen it from the sight of those who passed by. For the sake of their careers, friends who contributed to this collection will not be named.

To give the Unit extra impetus, I contacted key people across Australia who would have an interest in the Unit. These ranged from herpetologists to intensive care specialists. In January 1994 thirty-two letters were sent. They contained background information about the Unit and requested the correspondents to sign an enclosed invitation to formalise their link with the Unit. All but one responded immediately and warmly to the invitation, the exception being Dr Julian White of Adelaide (see page 362). Later a number of international authorities also agreed to become collaborators.

Feverish activities went on in CSL's venom research laboratory right to the end. Ever optimistic, we ploughed on with stability studies on the ant venom standards and special plates which tested antivenoms without the use of experimental animals. In retrospect, we were wasting our time. For the couple of months before packing up many venom standards were prepared and dispensed. Advantage also had to be taken of some of CSL's testing facilities. Hundreds of vials were filled with a treasure trove of venoms ranging from those of rare snakes to extracts from poison glands of male platypuses.

The Red-back spider venom standard prepared at this time was later important both for antivenom testing and research. The haemolymph (serum) of Red-back spiders also proved useful for investigating the appearance of Red-back spiders in Japan in October 1996. Although by various wiles I had managed to keep Mary and Sony engaged full-time until the close-down, it was a desperate race to process all the venom and serum samples in time. For instance, I spent two weekends in May processing Box jellyfish tentacles literally from dawn to dusk. This was intended to supply a reliable source of reserve toxin for CSL's antivenom production unit as well as for venom research. The work has also benefited the Unit's current research into dangerous jellyfish.

I was allowed to transfer to the University five large freezers which, to my mind, held national treasures. They housed rare venoms and forensic samples awaiting the development of tests to detect specific venoms. I must admit that in the past the smaller freezers often contained a crayfish or two and garden produce (like stewed plums) I had brought in for the staff!

The Sale of CSL Ltd

The highly successful public float of CSL took place in May 1994, a month before my scheduled departure. I believe that in time the privatisation of this asset will be seen as a classic example of the downside of micro-economic reform.

Many people who were opposed to the sale of what had been a national institution recalled with irony the assurance that the then Prime Minister R. J. Hawke had given when visiting CSL ten years earlier. Hawke had departed from his prepared speech to say the following: 'It is appropriate this day that I give you one guarantee; and I give you, the employees of CSL a guarantee; and I give the people of Australia a guarantee—it is very appropriate that it be given today—CSL will not be sold off to private enterprise.'

In 1992, as the momentum towards privatisation increased, I had written a summary of my concerns from the point of view of a researcher. This was published as an editorial in the *Medical Journal of Australia* on 12 December 1992. It was entitled 'Privatisation of CSL—wither venom research?' The accidental misspelling of whither went largely unnoticed. In retrospect it was prophetic, most of the projects I thought to be at risk have withered! Others have not even had a chance to sprout.

As a prelude to the sale of CSL the splendid new blood fractionation complex at Broadmeadows, which cost $209 million, was to be opened by the Prime Minister, Mr Keating. However, the ceremony was actually performed fairly briskly by the then Minister for Health, Senator Graham Richardson, in February 1994. A little later Senator Richardson told a meeting of medical scientists that he was 'passionate about medical research'. This reassured them greatly. The only problem was he resigned a week later!

In June 1995 a discussion paper entitled 'The Privatisation of CSL' was published by the Australian Institute in Canberra. It concluded that from a taxpayer's viewpoint, the privatisation of CSL was a disastrous transaction. It predicted that six years after the sale, and each year thereafter, taxpayers will be $45 million worse off.

The sale of CSL was also the subject of a performance audit which was published as Audit Report Number 14 1995/96 by the Auditor-General. The report found the gross return on the sale of CSL was $299 million, total sales costs were $9.2 million (3 per cent of total proceeds).

However Brian McNamee and his senior executives certainly reaped great financial rewards in the years after privatisation. According to *The Age*, 28 June 1997, McNamee held 600 000 options which showed at that stage a paper profit of more than $3.3 million. The same paper reported on 19 January 1998 that CSL was out of Tiger snake antivenom. This first such shortage in sixty or more years could hardly have come at a worse time—mid summer.

Leaving CSL

Graham Mitchell asked me what type of farewell I wanted. I said 'low-keyed to nil' but that I'd like to give a final seminar. This was set for Tuesday, 28 June, and gave me the opportunity to review my work at CSL and acknowledge all my co-workers. It was also a good chance to present the aims and aspirations of AVRU. To my surprise and chagrin two weeks before the seminar I was informed that it was to be purely internal in nature and no outsiders were to attend. Two days before the seminar Graham rang me to say that Management had agreed to let my wife and Jim Angus come.

The theatre was packed and my talk had been well rehearsed. If the audience expected something controversial they were disappointed. I gave a deadly serious chronological account of activities, some of which had started before a few of the listeners had been born! One friend later kindly described the seminar as being delivered with 'quiet dignity'.

Forty minutes of lively and good-humoured questioning followed. Gust, who had remained silent throughout the whole session, sent me a memo for the record saying how much he enjoyed the seminar.

On 30 June Gust and I performed in public at Marjory Davey's farewell in the CSL library. Marjory, who had been out of action since breaking her ankle in early April, was in good form. I was embarrassed when she announced she thought I was a fine boss who had been very badly treated by certain people. To the wonderment of the gathering, Gust and I chatted amiably over coffee for some time. The surprise mirrored that scene when years earlier I had socialised with Norman Coles during a break in the disciplinary hearing. On this occasion, though, I was closely studying Gust for what I hoped was to be the last time!

The success of the move to the University in July was due to the hard work of Mary, Sony, and the Laboratory Manager, Mr Ian Macfarlane.

Anyone who has shifted a laboratory will appreciate the complexities. For example, the precious contents of five deep freezers had to be removed and packed in dry-ice for a couple of days until the relocated deep freezers were proved fully functional. On top of this there were twenty-eight years of records which only I was familiar with. Somehow we managed to pack all the equipment, filing cabinets, etc., into an area less than half of that we had occupied at CSL.

My last act was to send notes of thanks to seventy-two CSL people who had helped the venom research group. By 13 July the move to the University was complete and I braced myself to leave. I recalled a comment made to me by Leo Davis almost thirty years earlier. Leo was a tall gentle virologist who had led the team which produced the Salk poliomyelitis vaccine in 1956. In a private moment he admitted to loving CSL and assured me it would grow on me. It did, and I loved many of its people, quirky buildings and surrounds.

There was no note or farewell handshake from Brian McNamee. No gift from the Board, just a cheery wave from the guard as he swung the gate open. Perhaps I have wasted years, I mused. No promotion since 1974, excluded from lush corporate packages and now squeezed out by the new boys McNamee and Gust. I consoled myself with the thought that, although there are things in my personal life I would do differently, I was proud of what I and my colleagues had done at CSL.

CSL had been good to me. Now, aged fifty-eight, far older than your average associate professor, I had a new and challenging job. I also had independence. 'Don't stuff this opportunity up Stru,' were my thoughts. 'It's your last chance!' As we shall see, things worked out quite nicely.

Australian Venom Research Unit, July 1994 to December 1998

11

Optimism: Disposition to hope for the best or to look on the bright side of things under all circumstances

(*Shorter Oxford English Dictionary*)

Prologue: The Aims of AVRU	349
Raising Funds	350
Relations with the Federal Health Authorities	352
The Government of Victoria Saves the Day	357
Relations between CSL and the Unit, July 1994 to June 1998	358
New Blood, Fresh Hopes and a Happy Ending	364
Growing Down	367

Prologue: The Aims of AVRU

Advisory and Public Health
- *To provide the best available advice to medical practitioners and Poison Information Centres on the management of bites and stings by venomous Australian terrestrial and marine creatures.*
- *To increase public awareness of the dangers of venomous creatures, and the first aid measures which may have to be used.*
- *To collect data from Australia and the Indo-Pacific region on diseases caused by envenomations and to be a Resource Centre for such information.*
- *To work closely with the World Health Organisation in matters of antivenom standardisation as well as patient care.*
- *To contribute to the training of medical and paramedical staff.*

Research
- *To undertake basic research into venoms of importance in Australia and the region.*

- *To establish and maintain a National Reference Collection of venoms and venom components.*
- *To collaborate with scientists working in these areas.*
- *To develop and improve antivenoms in association with a production company or companies.*
- *To seek venoms with novel activities which may have potential as either new therapeutic agents or investigational tools.*

Raising Funds

The hardest part of getting to the university each day was remembering to turn right when I reached Flemington Road. Old habits die hard and many times, deep in thought, I turned left and found myself steering towards CSL! As the days went past I became happier and happier at the university. It was inspiring to observe how the staff combined what was at times a very intense teaching load with research activities. I had always enjoyed interaction with students and it was great having closer and ongoing contact, especially with the honours students. A splendid way to start the day's work was to ride up in a lift packed with students. The things one hears!

People could not have been more helpful: indeed I felt pampered by the Laboratory Manager, Ian Macfarlane, and his assistant Jenny Steen. Furthermore, Jim Angus's plans for a new home for the Pharmacology Department were progressing well and included fine facilities for AVRU. So, all in all, this old buffer took to university life like a duck to water.

There were some downsides as well. I could not afford to offer either Sony or Mary full-time work. Also the lines of communication I had established both inside and outside CSL were now disrupted. As I was no longer in contact with the CSL branches that were in most states the guaranteed arrival of serum samples and precious venoms proved nearly impossible. Although Peter Hobbs of CSL faithfully provided requesting doctors with my telephone number, all too often others did not.

I found it harder to gather details of important cases, including deaths. Although I was still a consultant to CSL, I could not make as much impact as I had when I had the full authority of CSL behind me. Since CSL had represented an arm of the Federal Government my requests in those days had added clout. The other side of the coin was that it was my job, as a public servant, to diligently handle inquiries concerning my field. These came, as they still do, from a myriad of sources, ranging from the general public to the Minister himself. The resulting decrease in surveillance was later addressed in a 1998 submission on the effects of the privatisation of CSL to a House of Representatives Standing Committee. The problem was compounded by increased privacy provisions which suit lazy bureaucrats to the ground!

If AVRU was to succeed it had to prove its usefulness as an advisory service and attract money for research. The CSL money had to last as long as possible. I put myself on half pay, which sounds like a great sacrifice but in fact, when combined with my pension, approached my former (rather unsatisfactory) CSL salary. I became a compulsive targeter of likely fund sources or persons of influence. Some secretarial assistance was available but, without a full-time secretary, the maintenance of the twenty-four-hour advisory service was more onerous. As my handwriting deteriorated the writing of faxes and posting of information to doctors became at times a nightmare. A dramatic improvement occurred in May 1995 when Ms Joanne Cook became Departmental Secretary. Like Marjory Davey in tolerance, skill and efficiency the proportion of Joanne's time allotted to the Unit was its best investment yet. When Joanne was promoted in 1997 her shoes were ably filled by Vanessa Tresidder.

The single biggest hurdle facing AVRU was to convince 'people in power' of the importance of venom and antivenom research.

With the tacit approval of the Federal Department, CSL had closed down its venom and antivenom research unit and, after the 1994 float of CSL had raised the expected $300 million, neither the Government nor CSL relished accusations of neglecting 'National Interest' problems, for example venom research. Furthermore, McNamee and Gust had not only played down the importance of venom and antivenom research but also, with good reason, were not kindly disposed towards the writer. Neither of them was likely to offer private or public support for this type of research and I feared an occasional aside from either of these two powerful figures could neutralise months of attempts to obtain Government, corporate, or private support. International connections such as WHO might be thwarted or delayed by a deprecating underhand murmur. On the other hand, I was hardly running the university branch of the Ian Gust/Brian McNamee fan club!

Despite the efforts described below, both CSL and the Federal Government showed little if any interest in the future of the Unit. I got two things wrong from the start. First, I thought it would be possible to bring pressure on the Federal Government so that both it and CSL would support the Unit. For over two years I coerced many organisations and individuals to write or talk to Dr Carmen Lawrence. My files bulge with copies of letters sent by practically all Medical Societies, Lifesaving Associations, RSL, CWA, Girl Guides and Scouts, Farmers Federations, etc, etc. A battery of public servants must have been engaged churning out responses. Some of the replies made me splutter with anger and I could only regain my equilibrium by communicating the true facts as I saw them to the dolt involved.

The second big mistake I made was in my methods of fundraising. I sent personalised letters containing a package of information to the chairpeople of major Australian companies. After finding out the correct name of the current chairperson, fifty drug companies were also approached.

Some early successes, for example the Commonwealth Bank gave us $10 000, falsely suggested that this was the best method. Replies often included nice comments about the Unit but later, when half promises were followed up, it was usually found that the company believed the Federal Government should be the first source of funding, not their shareholders. Nevertheless, some companies such as Kodak have been regular donors. An anonymous donor has given the Unit $15 000 to $25 000 a year, regularly as clockwork. Eventually we discovered that the most successful way to raise funding from corporations was to target a small number and lobby the person or persons controlling the purse-strings.

The long-term survival of the Unit could be ensured with ongoing support of both Federal and State Governments as well as CSL. As we shall see, persistence has yet to be fully rewarded

Relations with the Federal Health Authorities

In April 1994 I wrote the first of many letters to the then Minister for Human Services and Health, Dr Carmen Lawrence. As with the other letters I was sending to people like the Prime Minister and assorted politicians of all parties, I enclosed a three-page summary of the aims and aspirations of the Unit.

The letter to Carmen provoked a flurry of replies by various officials to long dormant letters. She eventually replied on 16 June via her parliamentary secretary, Dr Andrew Theophanous. Andrew, like all the officials, expressed the view that the Unit should apply for a grant through the National Health and Medical Research Council (NHMRC). Finally after exchanges with various members of her staff, a reply from Dr Lawrence herself was received on 2 August.

In her letter Carmen summarised the various knockbacks already transmitted to me by her officials but to my surprise told me that CSL had received $2 million to subsidise antivenom production in the current year and would receive the same amount in 1994–95.

By this time the full force of the impact of the closure of projects had hit me. On one hand, I would have to explain to doctors and their patients why the relevant research had stopped or why postmortem samples could not be analysed, etc. On the other, I was aware of some unhelpful misinformation being shuffled around by non-medically qualified people in the Health Department. My same-day reply to the Minister is reproduced in full below. It captures the flavour of the time and succinctly summarises my jaundiced view.

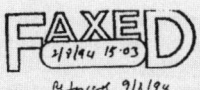  

THE AUSTRALIAN VENOM RESEARCH UNIT
Department of Pharmacology
Tel: +61 3 344 7753 Fax: +61 3 348 2048

THE UNIVERSITY
OF MELBOURNE

Professor J.A. Angus
HEAD OF DEPARTMENT

FOR THE URGENT ATTENTION OF DR LAWRENCE, PLEASE

Dr R.J. Summers
DEPUTY HEAD

2 August, 1994

Honourable Dr Carmen Lawrence, MP
Minister for Human Services and Health
Parliament House
CANBERRA   ACT   2600

(06) 273 4146

Dear Dr Lawrence

Re: The Australian Venom Research Unit

I was very pleased to receive a direct communication from you today in the form of your letter of 26 July. A copy is attached.

Since I am married to an Anglican priest, I should know better than to argue with a Minister, however some of the statements made in your letter cannot go unchallenged.

It may be safely assumed that departmental officers prepared the letter for your signature and therefore my criticisms are not directed at you personally.

I will respond to your letter paragraph at a time.

Para 1. The response from your department over the last 10 months or so has been far from impressive. The documentation makes sad reading.

Para 2. It is not appropriate for the NHMRC to provide routine day-to-day funding for the Unit, especially, if some of the functions previously carried out at the Commonwealth Serum Laboratories (now CSL) are to continue. One might ask if the National Blood Reference Laboratory which was moved from CSL at Parkville to Sydney last year relies on NHMRC grants. Like the Venom Research Unit that laboratory has some public health functions as well as background research activity.

DEPARTMENT OF PHARMACOLOGY, THE UNIVERSITY OF MELBOURNE, PARKVILLE, VICTORIA 3052 AUSTRALIA
TELEPHONE: +61 3 344 5674 FACSIMILE +61 3 347 1452

Para 3. The cessation of work on the ant venom allergy project is tragic. Of concern is that despite my regular summaries your letter refers to this project as *jack jumper ant bite anti venom*. The project has <u>nothing</u> to do with anti venoms; and ants sting, not bite, anyway. The aim of the project is to prepare a purified ant venom for the investigation and immunotherapy of highly allergic patients. This misconception surely casts doubts your advisers understanding of the problem.

Para 4. Your letter assures me if positive outcomes are generated by the unit then CSL will be able to undertake product development. Considering the current finances of the Unit and CSL's recent record in product development, any positive outcomes appear quite unlikely.

Para 5. The subject of the cost of antivenom production at CSL is extremely important. It is my opinion that the cosy agreements on costing between your Department and CSL, which have been made on a regular basis, have resulted in the failure to improve the efficiency of manufacturing, marketing and also the reduction of experiment animal usage.

To put an extra 2 million dollars in last year and further 2 million dollars in now in an attempt to control costs is very difficult to justify. For a number of years there has been a desperate need for a close examination of all aspects of antivenom production at CSL. You state by maintaining the viability of CSL antivenom you would "expect prices to hold steady". With my intimate knowledge of CSL I would suggest this expectation be modified.

Finally, it is somewhat disturbing that some of the submissions that I and others have forwarded to your Department have either not been carefully considered or have drawn routine and repetitive bureaucratic rebuffs. The need for Venom Research will not go away nor will stop gap measures solve the problems of rising antivenom costs.

I would welcome your comments as a matter of urgency and would be keen to provide additional information should it help your deliberations.

Yours sincerely

STRUAN K SUTHERLAND
MD DSc FRACP FRCPA
Foundation Director and Associate Professor

It was the first I had heard of CSL receiving $4 million to subsidise antivenom production. It came smack in the middle of CSL's privatisation and was certainly a closely-guarded secret. Naturally I was fascinated to observe how this $4 million would be spent. Since CSL would also gain revenue of at least $1.5 million per annum from antivenom sales I thought some money should flow to antivenom research.

On 2 August 1994 (the same day I received a reply from Dr Carmen Lawrence) *The Bulletin* ran an article by Greg Roberts on the privatisation of CSL. It reported very favourably on the Unit and its aspirations.

At the end of this article Brian McNamee was reported as saying in reference to my research, 'he believed the Government should conduct an independent review to consider if it should renew funding for the research'. The article concluded as follows:

> A spokeswoman for Federal Health Minister Carmen Lawrence, said medical research priorities are 'appropriately determined' by the NHMRC—a view which sits oddly with her recent 'request' to the NHMRC to commit $5 million to Breast Cancer research this year. Sutherland replies that the earliest he could obtain an NHMRC grant is 1996, and that in any event, NHMRC grants will not be available for much of the Unit's work, such as the services which provided advice on venom to medical officers.

The greatly loved Dame Phyllis Frost was amongst many notables who found Dr Lawrence an unsatisfactory correspondent. Dame Phyllis, who believed she had had the misfortune to be bitten by a White-tailed spider with devastating effect, took up the cause with a vengence. In her first letter to Dr Lawrence dated 17 August 1994 she wrote, 'I am not a woman who protests unduly when circumstances are not to her liking.' Several letters later, Dame Phyllis wrote to me on 21 November 1994 and let off steam about Carmen as follows, 'Of all the dilatory politicians, I think it would be hard to find someone to beat her because she does not get around to the task of acknowledging letters received. (The Lord protect me from too close an association with very many politicians ...).' The RSL's Bruce Ruxton, who had also suffered from a spider bite, didn't mince words either, but like Dame Phyllis made no apparent impact.

Over the next twelve months or so, further communications with Dr Lawrence went unrewarded. A request for an audience received a delaying reply addressed to 'Ms Sutherland'!

On 1 November 1994 Carmen Lawrence sent a cautionary and negative reply to my fax of 2 August. She noted various points such as my views on the increasing price of antivenoms. Interestingly a month later in NSW two suspected cases of Tiger snake bite arrived at Cooma Hospital within thirty minutes of each other. Cost cutting may well have contributed to the hospital holding only one instead of the recommended four ampoules of Tiger snake antivenom. One of the patients, Dr Bill Edmonds, died.

This tragedy attracted a great deal of publicity. I was greatly moved by an account by Dr Kate Blackmore, the victim's partner, of the circumstances which led to his death. Because of my concern about certain aspects, I appeared as a voluntary expert witness for Dr Blackmore at the Coronial Hearing in Sydney in May 1996. The Coroner and I were not impressed with each other and the four barristers representing the interested parties gave me a hard time. Kate was the only one in the court to know that my illness was by now affecting my speech. Care had to be taken not to slur my words and avoid ones that I stumbled over. (I could hardly ask the court to 'listen more clearly!')

Fortunately, I was on the way back to Melbourne when the Coroner, Mr West, gave, in my opinion, his watery findings. Had I been present, I would have protested that he had ignored many improvements in the management of snake bite victims introduced by me and others over the previous thirty years. I would even have willingly received a charge of contempt of court.

Dr Mike Ragg, who at the time worked part-time in emergency medicine, wrote an excellent article on the Edmonds' case for *The Australian* magazine of 20 October 1996. It set the record straight and also produced some astounding information, presumably from CSL.

> As part of the arrangement to allow privatisation, the Commonwealth Government and CSL signed a memorandum of understanding that CSL would continue to supply existing antivenoms, with the government ensuring CSL's profitability in that market. Initially, the Commonwealth tried to give the states $2 million a year to buy antivenoms, but the deal fell through due to mutual distrust. So now the Commonwealth gives CSL $2 million a year to underpin the antivenom market. The deal works, but it relies on a number of factors for its continuance. The Commonwealth has to offer CSL enough for it to consider it financially worthwhile. If CSL is ever sued over its antivenoms, that price will jump dramatically, and it is plausible that we could see the situation with antivenoms, as with vaccines in the US, where nobody wants to take the risk of making them.

Personally, I think it is most unlikely CSL could be sued if a reaction occurred to its antivenoms. The risks are well documented and are minimalised if proper precautions are taken (see page 235). On the other hand, if these precautions are ignored then the person administrating the antivenom may, under certain circumstances, be the target of legal action.

By this time I had forged links with a number of MPs in the conservative opposition. Senator Robert Hill, then Shadow Minister for Education, Science and Technology, had issued a splendid media release in support of the Unit in December 1994. Most encouraging letters had been received from Mr Tim Fischer and Senator John Herron. At the same time the Democrats spokesman on Science, Senator John Coulter, continued to use the cessation of venom research as a prime example in his campaign against the proposed privatisation of other government instrumentalities.

This show of support by the Federal Opposition prompted me to target their Shadow Minister for Health, Dr Michael Wooldridge. When I hot-footed over to see this 'proper doctor' in Box Hill he proved an interested and sympathetic listener who reminded me of a slightly obese but contented fox. Unfortunately on the way back I realised he had made no firm commitment to support AVRU.

In December 1995 the Federal President of the AMA, Dr David Weedon, called for Federal funding of the Unit. His press release coincided with the publication of an article in the *Medical Journal of Australia* by Dr Ria Leonard and myself disclosing an increase in deaths due to snake bite. (Ria had become my part-time assistant the previous October.) During the ensuing publicity Ria and I published a long letter in *The Age* to which Dr Lawrence declined to respond. I wrote to Dr Wooldridge requesting he make opposition policy known regarding ongoing support for the Unit.

In March 1996 when Michael Wooldridge became Minister for Health and Family Services my letter of congratulations was off like a flash. I requested access to an appropriate departmental officer for advice and guidance in the preparation of a submission to the Government. I responded promptly to a reply which came from a First Assistant Secretary six weeks later but had no response to this letter.

To date it appears the Federal health authorities do not give a hoot about improving the lot of envenomed patients in Australia. Nonetheless pressure will be maintained on Canberra and every opportunity for constructive publicity seized. After all, there is a fair population of snakes, spiders and dangerous ants in and around Canberra just waiting to highlight the importance of venom research!

The Government of Victoria Saves the Day

But for support from the Victorian State Government, the Unit would have ceased to exist within eighteen months. In retrospect, my approach to the State Government on several fronts had been well planned. In April 1994 I sent details of the Unit to my local MP, Alan Stockdale, who just happened to be the Victorian Treasurer. I also sought a meeting with him which took place at 7.15 am in his electoral office round the corner from my home. The Treasurer arrived at 7.14 am, unlocked the office, ushered me in, put the lights on and, after the meeting, reversed the process and drove off to his much bigger office in the city. Mr Stockdale gave me a good hearing and I admired his magnificent eyebrows. He said he would write to Marie Tehan, the Victorian Minister of Health.

(I enjoyed this interview better than the one I had with Dr David Kemp, my Federal MP, a week earlier. Kemp had opened up by saying what a great guy Ian Gust was and that he was all in favour of the sale of CSL! To get into his office I had to pass a giant portrait of the then opposition leader Dr John Hewson. I was not a fan of Dr Hewson and felt I was sinking a bit low!)

From the outset Mrs Tehan was sympathetic and, by August, discussions relating to the operations and future directions of the Unit were under way. I have no doubt that Sir Gustav Nossal's influence was crucial at this stage.There was also a long standing mutual respect between myself and a number of Health Department medical officers. After a lot of hard work, supervised by Dr Chris Brook, a funding and service agreement was signed between the Department and the University. Mrs Tehan allocated $100 000 towards the operations of the Unit.

In January 1995, Mrs Tehan wrote to Dr Carmen Lawrence requesting a matched Commonwealth contribution to the Unit. She also indicated that in April 1995 she would put forward proposals to the AHMAC (Australian Health Ministers Annual Conference) for ongoing joint Commonwealth/State funding.

The arrival of the Victorian grant triggered the search for some part-time staff and the printing of the Unit's fundraising brochure. Best of all, it made the Unit feel more secure.

In a powerful news release of 12 December 1995 Mrs Tehan came out to bat on the Unit's behalf against Dr Lawrence, and announced a further $100 000 to allow the Unit to survive during 1996–97.This second grant enabled important research to start on a regular basis again and for me to look for a successor. (A further similar grant was received in 1997.)

Relations between CSL and the Unit, July 1994 to June 1998

Stan McLiesh, General Manager of the Pharmaceutical Division of CSL, appeared supportive of the Unit from the start. He was a fast thinking, resilient man whom I had seen survive countless re-organisational changes at CSL. In fact over thirty years the title of his job had changed at least twenty times! Now he controlled CSL's largest and wealthiest division. Antivenoms for human use were under his control, but venom detection kits and veterinary antivenoms were not.

A six-month consultancy agreement was signed between Stan McLiesh and Jim Angus in June 1994. Under the terms of this agreement, the Unit (i.e. me) undertook various duties. These included the twenty-four-hour seven days a week Advisory Service to doctors and pharmacists, etc., throughout Australia. Stan provided a mobile phone and a pager which I carried at all times. Peter Hobbs of CSL kept things working smoothly: forwarding letters requiring responses and antivenom usage reports to the Unit on a regular basis. My practice of recording details of all inquiries continued and over the next two years almost one-fifth of all the calls received were referred directly or indirectly by CSL. Today stronger links between the Poison Information Centres in all states and the Unit result in CSL being increasingly by-passed.

Before leaving CSL I had been careful to prepare the ground for AVRU to undertake some specific product-related activities efficiently and economically. These included some projects in which I had considerable investment and which I was confident would succeed. The contract work would allow at least part-time employment of my staff and maximum use of equipment at AVRU.

The most important of these activities was the expansion of a new assay which could be used for testing antivenoms without the use of animals. Over the years I had felt an increasing compassion towards the fate of laboratory animals. I fully agreed with those who held they should be used only when there was absolutely no alternative. Fortunately, our developmental work went smoothly and by July 1993 we had developed an enzyme immunoassay which could quickly and cheaply analyse the likely effectiveness of antivenom. It worked against important individual components as well as the whole venom.

Such an assay would be a godsend to all aspects of antivenom production, ranging from horse immunisation, purification processes, standardisation and stability. Hundreds of samples could be examined in a morning in place of the cumbersome guinea-pig test which had severe limitations and took days to perform. Moreover, the information obtained from the guinea-pig model, which had been used at CSL for sixty years, was limited. Guinea-pigs received various venom and antivenom mixtures and the strengths of the antivenoms were determined by the number of animals that survived. This crude assay gave little indication as to the guinea-pig's reactions regarding clotting disturbances, kidney function, muscle damage, etc.

The collaboration of CSL's Quality Assurance Department in this work was vital as it was important that all antivenom samples being tested in animals should also be tested in our assay. In this fashion, we hoped to show the relevance of the assay and in time have permission from the Regulatory Authorities to phase out the use of animals. To this end, I documented all progress and meticulously kept CSL management informed of developments.

We had ploughed ahead with the work before leaving CSL and fortunately the Production Department had flooded us with spare duplicates of the samples going for animal testing. We also had a large number of antivenom samples for long-term storage and retesting. Coating the plates with all the different venom components had been quite a complicated procedure. We therefore did a large batch which was freeze-dried and the plates individually sealed. Stability studies done at six-monthly intervals were satisfactory and we were confident of the practicality and validity of the assay.

By using these plates we demonstrated that we could assay an antivenom in less time than it took to prepare and dose the guinea-pigs. Also we didn't have to wait seven days for the results like we did with guinea-pigs!

By March 1994 CSL's biostatistican, Bill Finger, confirmed good correlation between our assays and the guinea-pig assays.

In May 1994 in my letter to Brian McNamee (page 343) I stressed the commercial implications of this assay, as well as its humanitarian aspects. With the encouragement of Stan McLiesh I met with Darryl Mills, his Technical Manager, to consider possible contracted R and D work. On 31 May I submitted to Darryl a quote for continuation of these antivenom assays and development of others, including one for Red-back spider antivenom. Believing the issue of animal usage was vital for CSL I broached it with Sir Gustav Nossal during our meeting on 16 June. He was highly supportive and, after this meeting, I rang Brian McNamee to seek his permission to send Sir Gustav a copy of the as yet unacknowledged letter I had sent him on 9 May. Brian sounded a bit off balance by this request and said he was not too keen on the idea but then acquiesced.

On 13 July Sir Gustav wrote to me saying among other things, 'I have been in touch with the powers that be at CSL, and they, in turn, passed the buck to our regulators who continue to place their reliance on the more old-fashioned, rather than the newer, techniques for quality assurance and the like. Nevertheless, I believe I have made a tiny bit of impact, and you are certainly not without friends in that quarter.'

As a result of intense correspondence Stan McLiesh arranged for me to meet Kevin Healy of the Quality Assurance Department and his offsiders at CSL on 12 August. After two and a half hours of mind-numbing obstructionism I withdrew. Time and time again the possibility of contract work was ruled out because they were either starting the research I was proposing or were considering it. Healy was clearly opposed to any concrete co-operation and relied on a mixture of petty objections and evasion. Lack of indepth knowledge of venoms and antivenoms helped too. Recalling this meeting some years later I realise that if there had been a large club on the table I would have gladly used it. As it was, I let off steam the next morning (Saturday) by faxing my side of the debate to Stan.

Stan's reply was conciliatory and so I hired staff to open up the laboratory to perform stability studies, etc., on the plates to keep the project on schedule. This turned out to be quite an expensive exercise but it seemed worthwhile at the time. By CSL's Annual General Meeting in November 1994 I was quite browned off by them. No contract work had been forthcoming and both venom and antivenom work was at a standstill. I therefore appealed to a higher authority. I wrote to Mr Colin Harper, Chairman of CSL, as follows:

THE AUSTRALIAN VENOM RESEARCH UNIT
Foundation Director: Associate Professor Struan K Sutherland
Department of Pharmacology
Tel: +61 3 344 7753
Fax: +61 3 348 2048

THE UNIVERSITY
OF MELBOURNE

Confidential   16 November 1994

Professor J A Angus
HEAD OF DEPARTMENT

Dr R J Summers
DEPUTY HEAD

Mr Colin Harper
Chairman
CSL Ltd

Dear Mr Harper,

   Re. CSL and Venom Research

   At the shareholders' meeting yesterday I was tempted to rise in response to your answer regarding the above matter. I do not dispute the facts of your prepared statement but consider it gave an incomplete view of the situation to the shareholders.

   As Chairman I would like you to be acquainted with the following additional information.

   1. The agreement with CSL regarding initial funding and loan of equipment was confidential. The breaking of this confidentiality in May of this year by CSL (via the Minister) had an adverse effect on fundraising for the Australian Venom Research Unit.

   2. For over 12 months I have been seeking evidence that CSL actively sought funding from the Commonwealth Government for the continuation of venom research. All I have found is a 1992 fax. from Mr Graeme Kaufman to Mr MJ Roche, a deputy secretary in the Department of Health, Housing and Community Services which, amongst other matters, advocates my relocation but makes no request for funding. Without specific requests it is understandable that the Commonwealth did not proffer funding. I would be very grateful if you could help in the provision of any documentation on this matter as it may help current negotiations with the Federal Government.

   3. As the shareholder at the meeting stated, the PR value of venom and antivenom research done at CSL has been considerable. This is despite an apparent policy by management to play down this fact. Later, some shareholders expressed their surprise that the new CSL promotional video did not even mention antivenoms.

   4. Antivenoms are not a commercial liability and indeed are moderately profitable. Their profitability could be increased if appropriate measures were adopted.

DEPARTMENT OF PHARMACOLOGY, THE UNIVERSITY OF MELBOURNE, PARKVILLE VICTORIA 3052 AUSTRALIA
TELEPHONE +61 3 344 5674 FACSIMILE +61 3 347 1452

> Since it is particularly relevant to the above, I enclose a copy of my letter to Dr Brian McNamee dated 9 May 1994. There seems nothing ethically wrong in drawing your attention to the contents of this letter which raises a number of important issues. Although airing these matters would have aroused interest and perhaps even sympathy at the shareholders' meeting, little resolution of the issues was likely to occur.
>
> Since some of the problems outlined in my letter to Dr McNamee have since become even more urgent, prompt and decisive action is needed. Perhaps a short meeting to discuss the matter would be useful.
>
> I look forward to your reply in due course.
>
> Yours sincerely,
>
> *S.*
>
> STRUAN K SUTHERLAND
> MD DSc FRACP FRCPA

I received a somewhat pompous and unhelpful reply from Colin. Two days after receiving his letter I was asked to come out to CSL to discuss 'various matters'. At a meeting on 13 December 1994 at CSL Stan McLeish advised me that since I would not 'last forever' they intended to appoint a second consultant to handle antivenom inquiries, etc., and payments to the Unit would be appropriately reduced. My heart sank even further when Stan McLiesh named Dr White as the likely appointee. The appointment of Julian White as a CSL consultant set the stage for occasionally conflicting advice to emanate from CSL. Moreover, Stan knew that in all states the most experienced intensive care physicians had signed up as collaborators with the Unit. Appropriate cases were referred to these experts who stood in for me from time to time. When the Unit was being founded, Julian was the only one out of eighty people who declined a formal invitation to become a collaborator.

I disagreed with him about his attitude towards premedication before the administration of antivenoms. For years CSL had recommended a small dose of subcutaneous adrenaline before the administration of most antivenoms (see page 239 and Appendix).

Over the last few years Julian White had been campaigning against the use of adrenaline premedication on the grounds that the patient's blood pressure might rise and, since snake bite patients often have clotting defects, they would be more likely to have a brain haemorrhage. In a very thoughtful article (*Medical Journal of Australia*, 3 January 1994) Dr James Tibballs reviewed the evidence for and against adrenaline premedication. He pointed out that subcutaneous adrenaline did not elevate the patient's blood pressure and there was no evidence of increased brain haemorrhage when the recommended premedication was used. It is important to point out that intravenous injections of concentrated adrenaline may be quite dangerous to snake bite victims. The intravenous route was never recommended for premedication.

By June 1995 I was pretty browned-off with CSL. There was no sign of contract R and D work, yet a number of CSL scientists were still asking for various favours such as help with the new Venom Detection Kit, and freeze-drying Funnel-web venom for antivenom production, etc.

Jim Angus accompanied me to a meeting at CSL on 26 June. The requested agenda included possible contract R and D work, in particular, assays to reduce animal usage. Nothing major resulted from this meeting but I fancy it may have given Jim new insight into my problems at CSL. I have not been back to CSL since.

By November 1995 I was fed up with both CSL and Carmen Lawrence. I was finding lecturing very exhausting and writing very difficult. On the other hand, things were very sweet with the Victorian Health Department, the University and the world in general. With CSL's Annual General Meeting approaching, I decided to have one more go and wrote to the Chairman. The following are some extracts from my letter.

1. The disruption of the work my group was doing to reduce animal usage at CSL in both antivenom production and testing is of particular concern. Some shareholders are aware not only of the commercial implications of this work but also the ethical and humanitarian aspects. It is important to encourage progress in this area as it is Government policy to reduce animal usage, and there are financial benefits. Apart from some new antivenoms, one would like to see improvements being implemented in antivenom production and shelf-life.

I strongly believe in the potential and ongoing value of this work. As the months pass, the greater is the possibility that much of this work may be wasted due to the time-expiry of reagents and disruption of stability studies.

I need to know the exact source of the funding provided by CSL to establish the Unit and whether it was part of the $4 million provided by the Commonwealth Government. It would also be helpful to know what proportion of the $4 million has been spent on in-house venom/antivenom research by CSL Ltd. This information will facilitate current financial negotiations between the Unit and the two Governments.

This drew a brush-off from the company secretary which left me no better informed. I gave up on CSL for the time being. With luck, others will have more success.

New Blood, Fresh Hopes and a Happy Ending

The first State grant in 1995 signalled the end of my one-man band status and increased my job satisfaction no end. Dr Ria Leonard became a part-time associate for some months from October 1995. In December we published a joint paper on deaths from snake bite in *The Medical Journal of Australia* which highlighted the frequency of rapid death following Brown snake bites. The mechanism of some of these deaths remains uncertain and investigation of this snake's venom became one of the projects the Unit's fundraising brochure has featured since 1995. Others include the search for an antivenom to the toxin of the Blue-ringed octopus and exploring the venom of the White-tailed spider.

The brochure also includes a jellyfish, the Irukandji, which has to date been the main research focus of the Unit. The ultimate aim is to prepare an antivenom to this tiny but very dangerous tropical jellyfish. Stings by this creature may produce a unique syndrome, the main features of which are muscle cramps, sweating, anxiety and very high blood pressure. At least one hundred cases occur in Queensland waters each year and doctors have long wanted an antivenom for the seriously envenomed patient.

We were lucky in 1996 to obtain the skills of Carolyn Wiltshire who put no foot wrong as she got the Irukandji project up and running. In the summer of the same year, Jim Angus arranged a vacation scholarship for an honours student, Ayse Berke. Apart from laboratory work, Ayse greatly improved the Unit's filing system. After the vacation she did a couple of hours voluntary work a week for the rest of the year until starting her PhD studies. Ayse was vivacious and charming and loved her 'venomous critters'. Her death in a road accident in May 1998 was tragic. She was only twenty-four.

After the second State grant the University advertised widely for a Deputy Director of the Unit. Dr Ken Winkel accepted the appointment as my potential successor and commenced part-time in July 1996. Tall, thin and at times extremely intense, he was just completing his PhD over the road at the Walter and Eliza Hall Institute of Medical Research. I took to Ken instantly. I thought we were lucky to get him because, apart from his ideal background, he had plenty of 'get-up-and-go'. There was a whiff of an entrepreneur about him. Born in Brisbane in 1967, he had graduated in medicine collecting a B. Med. Sc. on the way. Although quite young he already had a long list of publications the first of which caught my eye. It dealt with the diet of a marsupial mole which I thought was cute and vaguely relevant to his proposed career path. His more recent papers were at the cutting edge of immunological research and were highly technical. He had spent the last three years exploring the activities of mouse dendritic cells and I confess my old brain cells found it a struggle to understand this complex field.

While doing his PhD studies Ken worked regularly in the Emergency Department of the Royal Melbourne Hospital. Jim Angus and I encouraged him to maintain his appointment as a Senior Medical Officer which has benefited all parties.

I wanted a second doctor to share my thoughts and experiences while I was indoctrinating poor Ken. Apart from needing someone to help Ken with the advisory service, I was uncertain how long I would be at AVRU. Therefore I did not want to put all my eggs in the one basket, namely Ken's head. A properly chosen third person could help with the process of formulating policy and treatment protocols, and also be curator of half of my idiosyncratic filing system. We advertised for a research officer, and again were lucky.

Dr Gabrielle Hawdon joined us on 9 September 1996. She was born the same year as my daughter Susie which suggests that 1965 saw the arrival of some particularly stubborn though charming females. Gabrielle, who is shorter than Ken, also has both a medical degree and a B. Med. Sc. To date her appointment has been part-time while she is continuing as a medical fellow in intensive care at St Francis Xavier Cabrini Private Hospital. Gabrielle is also our resident medical statistician having completed her Master of Public Health (Epidemiology and Biostatistics) in 1997. I can talk to Gabrielle openly about any subject and, though we are often moved to argue a point with vigour, I think we have a high regard for each other's opinions. She and Ken continue to find gaps in my knowledge of recent medical advances during the forty years since I was officially a student. There are many fields I have neglected. They are very tolerant of my occasional appalling ignorance.

We got a great fillip when the University appointed Dr James Tibballs (see Chapter 7) as an Honorary Senior Associate to the Unit in late 1996 just when Ken began to work full-time. Jim, who is the extremely able Deputy Director of Intensive Care at the Royal Children's Hospital, has put a great deal of effort into helping the Unit develop. My respect continues to grow for this delightful man with his cheering and infectious chuckle. Ken and Gabrielle hang on his every word which is nice to observe as Jim is good for at least ten more years of venom research.

Early in 1997 Professor David Warrell of Oxford University accepted our second honorary position. This was a feather in the Unit's cap, David being probably the foremost authority on snake bite in the world. His appointment enhanced the standing of the Unit internationally and will encourage more collaborative research.

Ken and Gabrielle were fast learners and quickly appreciated how complex some problems were. To help their understanding I delivered a series of mini-lectures to an audience of two on the main subjects. After each, a lengthy interrogation would begin! Between these sessions they would delve into the published and unpublished records in the Unit as well as scanning computer based data. Initially Ken took on snakes and snake bite whilst Gabrielle studied venomous arthropods. They then swapped subjects and later both moved into marine animals.

I really enjoyed these sessions. They were like mini scientific conferences but without all the hoo-ha. I felt a sense of progress, of building. Occasionally I would find myself murmuring words of encouragement similar to those used by my long dead mentors.

We pored over the antivenom usage reports which were regularly received from CSL. Gabrielle undertook the collation and analysis of these reports and presented her findings at an international meeting of toxinologists in London in April 1998.

After a few weeks they could both handle the daytime calls, provided I was at hand for backup, and since February 1997 have shared responsibility for the twenty-four-hour advisory service. We all carry pagers but nowadays they rarely need my opinion which means, when a particularly interesting case comes on my pager, I have to try not to interfere and restrain my curiosity until the next Monday morning meeting at the Unit.

Relinquishing the on-call duties was a great relief to me as sometimes I was becoming hard to understand, especially when tired or on a mobile phone. I have had glowing reports about my two young colleagues' conscientious handling both of calls for advice and questions at scientific meetings. They have also adopted my practice of obtaining follow-up details of important cases to check how their advice worked. The wealth of clinical records held by the Unit steadily increases and is watched over by Vanessa Tresidder, the Unit's admirable administrative assistant.

In February 1998, Dr Anna Young, a protein chemist, joined the unit. Anna who had just completed her post-doctoral studies in France took over the Irukandji project from Carolyn Wiltshire. In the same month the dynamic Steven Pincus started work with us as the first of two trainees in a novel scheme. The Australian College of Emergency Medicine for the first time approved an advanced training position in emergency medicine involving six months full-time laboratory research. (Bringing Steven in, at no cost to the Unit, was a spin-off of Ken's appointment at the Royal Melbourne Hospital.) Anna and Steven prove a good team as they work on aspects of necrotising arachnidism.

I can take little credit for the recent achievements of the Unit. The staff have worked, sometimes fanatically, to meet deadlines for a host of potential sources of funds. Every week seemed to bring knockbacks from one 'dead cert' and another paperwork challenge to put in the pipeline. Despite these setbacks there was a certain air of exhilaration about the Unit. We were getting more and more recognition. For example, Ken was invited to Japan for consultations following the discovery of Redback spiders in Osaka. This visit has led to several collaborative projects. Another good sign is the regular requests for consideration by overseas scientists. To date, we can only accommodate those who can cover their expenses, like Professor May Win who obtained a WHO fellowship specifically to visit AVRU in 1997.

By the beginning of 1998 Ken and Gabrielle had published a number of papers from the Unit and were becoming well known amongst other venom researchers and the medical profession as a whole. Their media statements and appearances were increasing as were requests to give lectures, sit on committees and write articles, etc. However, we only had enough funds to last to July and by May 1998 the Unit was facing a bleak future.

Despite the efforts of the past four years no funds had been forthcoming from the Commonwealth Government. We were ignored even though we could show that our advisory service was used by doctors in all states. No notice seemed to be taken of the logged inquiries from Commonwealth Departments like Defence, Tourism and Foreign Affairs. In February, the president of the AMA wrote to the Minister for Health and Family Services requesting recognition that the Unit conducted work of national importance and it should be funded accordingly. We had heard nothing further and were not optimistic in the short term.

Although Gabrielle and Ken had good relationships with CSL no research contracts had eventuated. Since they were now both CSL consultants I kept in the background believing in time research links would be forged. (If the Unit was still in existence!)

Our one hope was the Victorian Government but even that quarter had gone strangely quiet. Earlier the Unit had put a number of lengthy submissions to appropriate bodies putting the case for a special grant for equipment and staff. Competition for funds had never been fiercer and I had a sinking feeling that I might have misled my troops with false hopes.

Suddenly the sun came out and shone brightly. The said troops' paperwork and enthusiasm had won the day. On 29 May the Victorian Department of Human Services announced an equipment grant for the Unit of $178 000 and $100 000 a year for salaries for the next two years. The Unit was saved!

These funds, hopefully combined with other grants and donations, will give my young friends a chance to continue their research. They can buy the latest tools and devices and hopefully wear them, but not themselves, out. I wish them luck!

Growing Down

Well, there you have it. AVRU is in keen, capable and enthusiastic hands. Although its future is far from secure, it has the right staff, unwavering support from Jim Angus and, since October 1997, fine new headquarters. Ken and Gabrielle have become familiar with both my records and the way I think. They have challenged me to justify many of my views.

This has been an exhausting but exhilarating process which led to consensus on all manner of things from knotty research problems to fine tuning patient care.

I'll end this tale on 15 June 1998, as I leave for home after attending the weekly AVRU meeting. Present were Ken, and Gabrielle, Jim Tibballs had come over from the Children's Hospital. Steve Pincus, whose six-month stint with the Unit finishes next month, was also there, as was Anna Young. As a scientist, Anna was as usual, outnumbered by medical people. If I excluded Jim, I estimate the average age of those gathered would be about thirty-two years. Younger than my children!

In the midst of healthy debate on possible causes of a marine sting that had occurred recently, I was suddenly struck with how fortunate I was to have the chance to 'pass the baton' on to these talented people. I feel certain the advice they give doctors will be sound, practical and, best of all, up to date.

Jim Tibballs and I are well into the revision of *Australian Animal Toxins* and I look forward to working on it tomorrow at home. My study is bliss, away from the hurly-burly and complete with Pidey the cat and our log fire.

Bob the neurologist was right in that my bodily degeneration would be relentless. The bright side is that, to date, it's a wee bit slower than predicted.

Its nice how things work out. When I get home, although I walk like a marionette, a toddle along the beach is in order. Life is excellent!

Fine.

Appendix

First aid for snake bite in Australia

with notes on first aid for bites and stings by other animals including spiders

Snakes and Snake Bite	370
Prevention of snake bite	370
Special notes regarding the protection of children	370
Snake bite when far from civilisation	370
Some facts about snake bite	370
Rational first aid	371
First Aid Using the Pressure-immobilisation Procedure	371
Additional notes on first aid for snake bite	373
First Aid against Other Australian Creatures which	
May Cause Death	373
Other Land Creatures	373
Sydney Funnel-web spider	373
Red-back spider	373
Other types of spiders	373
Bees, wasps and ants	373
The Australian paralysis tick	374
Marine Creatures	374
Blue-ringed octopus and Conus shells	374
Box jellyfish or sea wasp	374
Other types of jellyfish	374
Stonefish and other stinging fish	374
Some Notes on Antivenoms for Doctors and Paramedical Staff	375
Administration of antivenoms	375
Quantity: how much snake antivenom should be held	
by hospitals?	376
Metropolitan and Regional Centres	376
Small Centres and One Doctor Hospitals, etc.	376
References	377
General References	378

Snakes and Snake Bite

Prevention of snake bite
Most cases of snake bite can be avoided by following these simple rules:
★   Leave snakes alone.
★   Wear sturdy shoes and adequate clothing in 'snake country'. Do not wear sandals or thongs.
★   Never put a hand in hollow logs or thick grass without prior inspection.
★   When stepping over logs, carefully check the ground on the other side.
★   Always use a torch around camps and farmhouses at night—most snakes are active on summer nights.
★   Keep barns and sheds free of mice and rats as they will attract snakes.
★   Keep grass well cut—particularly in playgrounds, etc.

Special notes regarding the protection of children
★   Never let children collect snakes.
★   If young children say they have had contact with a snake, please believe them. (Better to be safe than sorry.)

Snake bite when far from civilisation
Advice on this subject is often sought by leaders of bushwalking groups, scouts and individuals travelling alone in remote areas. Such groups should be advised on how to avoid snake bite before setting out (see above).
★   People travelling in isolated areas are far more likely to need medical aid following falls, heart attacks or other illnesses than for snake bite.
★   When possible a radio transmitter or mobile telephone should be part of the expedition's equipment.
★   If snake bite occurs, the current first aid measures offer a longer term advantage over the earlier methods.

Some facts about snake bite
★   Even though not all snakes are venomous it is safer, from a first aid point of view, to consider all snakes dangerous.
★   Sometimes only small amounts or no venom is injected, even though puncture marks are present.
★   At least 95 per cent of bites occur on the limbs. Perhaps 75 per cent involve the lower limb.
★   The venom is injected quite deeply. It was shown many years ago that little venom is removed by incision or excision.
★   Research has shown that very little venom reaches the blood stream if firm pressure is applied **over the bitten area** and the **limb is immobilised**.[1,2] (See References on page 377)

* Sea snakes: The pressure-immobilisation procedure is appropriate for sea snake bites.

Rational first aid
* Immediately apply a broad firm bandage around the limb and over the bitten area. It should be as tight as one would bind a sprained ankle. As much of the limb as possible should be bandaged. Bind from below upwards. Crêpe bandages are ideal but any flexible material can be used, e.g. clothing or old towels torn up into strips. **Panty hose are satisfactory.**
* Keep the limb and the victim as still as possible. Bind some type of splint to the limb, e.g. a piece of timber, spade or any rigid object.
* Lie the patient down to minimise movement.
* Bring transport to the victim whenever possible.
* Leave the bandages and splint on until medical care is reached.

Don't cut or excise the bitten area.
Don't apply an arterial tourniquet.
Don't wash the bitten area. The snake involved may be identified by the detection of venom on the skin.[3] If the snake can be safely killed bring it to hospital with the victim.

First Aid Using the Pressure-immobilisation Procedure

The following diagrams show the recommended first aid for snake bite to the lower limb.

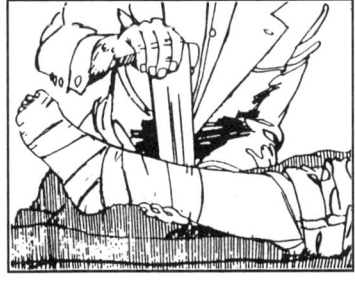

Apply a broad pressure bandage over the bite site as soon as possible (do not take off jeans as the movement will assist the venom to enter the blood stream. Keep the bitten leg still!)

The bandage should be as tight as you would apply to a sprained ankle.
Note: Bandage upwards from the lower part of the bitten limb. Even though a little venom will be squeezed upwards the bandage will be far more comfortable.

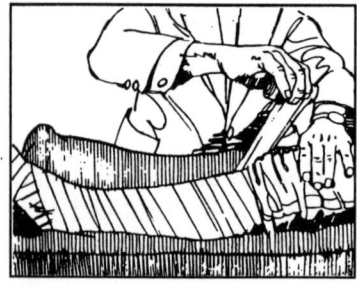
Extend the bandages as high as possible.

Apply a splint to the leg.

Research[4] stresses the importance of keeping the patient still. This includes the unenvenomed limbs! Bring transport to the patient.

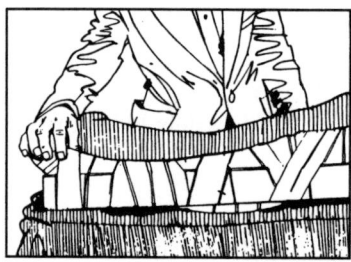
Bind it firmly to as much of the leg as possible

Bites on hand or forearm. 1. Bind to elbow with bandages. 2. Use splint to elbow. 3. Use sling.

If the bandages and splint have been applied correctly, they will be comfortable and may be left on for several hours.

They should not be taken off until the patient has reached medial care.

The doctor will decide when to remove the bandages.[5,6,7] If venom has been injected it may move into the blood stream very quickly after the bandages are removed. The doctor should leave them in position until he or she has assembled appropriate antivenom and drugs which may have to be used after the dressings and splint have been removed.[7,8]

Hospital Staff: Please note that first aid measures are usually removed soon after the patient is admitted.[8] Do not leave on longer than is necessary.

The first aid measures can always be quickly re-applied if deterioration occurs and left on until urgent antivenom therapy has been instituted.

Additional notes on first aid for snake bite
* Bites to the trunk: If possible apply firm pressure over the bitten area. Do not restrict chest movement.
* Bites to the head or neck: No first aid for bitten area.

First Aid against Other Australian Creatures which May Cause Death

The pressure-immobilisation procedure described on the previous pages is now recommended for use in the majority of other bites and stings with several exceptions discussed below. **Arterial tourniquets are no longer recommended for any type of bite or sting.**[5,9]

Other Land Creatures

Sydney Funnel-web spider
The pressure-immobilisation method should be applied as soon as possible and left in position until the patient is in hospital. Experimental evidence suggests that this venom may lose its activity if kept in the bitten limb.[9] Antivenom is available.[10]

Red-back spider
No first aid is recommended for these bites other than the local application of iced water. The venom works **very** slowly and if its movement is restricted, local pain may become very severe. Some 300 cases receive antivenom a year and no deaths have occurred since this treatment became available.[10]

Other types of spiders
A variety of common house and garden spiders frequently deliver bites, but usually very little occurs other than a little local swelling. Bites are best lightly washed with soap and water. Iced water may give relief from mild pain or itching. Medical advice should be sought if the bite site causes concern.[10] **Note:** If bitten by a spider always try to capture and preserve the culprit in methylated spirits for identification purposes, even if it has been squashed.

Bees, wasps and ants
These may cause an anaphylactic shock or death in allergic persons. In non-allergic persons iced water usually relieves the pain. In all cases bee stings should be scraped or pulled off as quickly as possible to prevent further injection of venom from the venom gland which remains

attached to the sting. In persons known to be allergic, the pressure-immobilisation procedure should be used and medical care sought immediately. Patients who have already suffered severe reactions should always have access to injectable adrenaline and know how to use it. Purified venoms are available for use in immunotherapy to desensitise people allergic to bees and wasps. Purified Jumper ant and Bull ant venom for this purpose is not yet commercially available.

The Australian paralysis tick
The tick should be carefully removed as soon as possible. Lever it out using a pair of curved scissors. If the patient is already ill the pressure-immobilisation procedure should be used, if possible, to inhibit the movement of any toxic saliva which has been expressed during the removal of the tick. Check carefully for other attached ticks! An antitoxin is available to treat cases of paralysis.[5]

Marine Creatures

Blue-ringed octopus and Conus shells
The pressure-immobilisation method is recommended. Prolonged artificial respiration may be required following a bite or sting.[11,12]

Box jellyfish or sea wasp
Pour domestic vinegar (never methylated spirits or alcohol) over the adhering tentacles to inactivate them as soon as possible.[13] Artificial respiration and cardiac massage may be required. An arterial tourniquet is not recommended but in severe cases the pressure-immobilisation procedure should be applied to the affected limb or limbs **after** the application of vinegar. An antivenom is available.[11,12]

Other types of jellyfish
The application of vinegar is also recommended for stings by the other dangerous box-shaped jellyfish such as the Morbakka and Irukandji. Current opinion is that all other jellyfish stings, including those due to Physalia, 'Blue bottle', or 'Portuguese man o' war' are best washed with sea water and then covered with iced water packs.[11,13,14]

Stonefish and other stinging fish
Do not attempt to restrict the movement of the injected toxin. Some stonefish stings respond to bathing in warm (not scalding) water. Pain relief from severe stonefish stings may require a regional nerve block. All stonefish stings require medical attention as do most deep stings caused by other fish. Often foreign material and bacteria are deposited quite deeply. A stonefish antivenom is available.[5,11,12]

Note: Even if the bitten or stung person is ill when first seen, the application of the pressure-immobilisation procedure in the cases recommended will prevent further movement of venom from the bitten or stung limb during transport to hospital.

Some Notes on Antivenoms for Doctors and Paramedical Staff

Detailed information on the use of antivenoms is packaged with the individual antivenoms. If you require additional advice, either contact the **Poison Information Centres** on **13 11 26 Australia-wide** which will put you in touch with a Consultant, or contact **CSL Ltd** on **(03) 9389 1911**. The Australian Venom Research Unit provides a twenty-four-hour Advisory Service **for doctors** and can be contacted via the Poison Information Service.

Australia is the only country in the world that has snake venom detection kits. A swab from the bite site, blood, or urine allows the doctor to select the type of snake antivenom which may have to be used. Less than one in ten cases of snake bite need antivenom because often the snake injects very little venom.

Antivenoms therefore should not be given unless there is evidence of significant poisoning. For example, in snake bite, signs of systemic poisoning such as nausea, vomiting, ptosis, etc., or positive laboratory findings such as a coagulation defect. Fang marks alone are not an indication of antivenom. Likewise, after a Red-back spider bite, if the only problem is moderate local pain, then antivenom is not indicated at that stage.

Administration of antivenoms (see refs 5, 6 ,7 ,8)
Australian antivenoms are established as amongst the safest in the world. Provided they are administered with appropriate premedication, there is no reason for them to be withheld, even if the patient has a past history of reaction to equine proteins. Such patients, e.g. snake handlers who have suffered reactions in the past, have had minimal or no problems with repeat antivenom therapy after appropriate premedication.

Most antivenoms are given by the intravenous route. Skin testing with antivenom for allergy to antivenom is unreliable and a waste of time. It may also delay urgent therapy.

The issue of premedication is controversial. Premedication with **subcutaneous** adrenaline is currently recommended prior to the intravenous administration of antivenoms. Adults should receive 0.25–0.3 mg of adrenaline via the subcutaneous route (0.005–0.01 mg/kg for a child). Adrenaline as a premedicant should never be administered intravenously, in order to avoid hypertension in the coagulopathic patient with the potential for bleeding. Similarly, it should not be administered intramuscularly, as this may also lead to hypertension, as well as to haematoma formation in the presence of coagulopathy.

An antihistamine may also be given parenterally, bearing in mind it may have sedatory effects and may also cause hypotension. If the patient has a known history of reacting to antivenoms, then corticosteroids should also be administered. Antivenoms which are to be given intravenously are preferably diluted 1 in 10.

Note: the antivenom requirements of patients will vary considerably. Some patients with minimal envenoming will require no antivenom, whereas others may require multiple doses of antivenom.

Both the incidence and severity of delayed serum sickness may be markedly reduced by the administration of prednisolone, 50 mg (adult dose), for five days after the administration of antivenoms.

Quantity: how much snake antivenom should be held by hospitals? The following is a suggested guide on quantity of antivenoms to be held.

Metropolitan and Regional Centres
* Adequate antivenom to cover two serious cases of envenomation by the major snakes found in that state, i.e. four ampoules of each monovalent antivenom.
* Except in southern Victoria and Tasmania, four ampoules of polyvalent antivenom should be held for the treatment of cases in which the snake has not been positively identified. In Victoria polyvalent antivenom is not required and a combination of Tiger snake antivenom and Brown snake antivenom can be employed when the identity of the snake has not been determined. In Tasmania, only Tiger snake antivenom is required.

Small Centres and One Doctor Hospitals, etc.
As a general rule, sufficient antivenom should be held to treat one serious case of snake bite. The decision whether two ampoules of polyvalent should be held or two ampoules of each of the appropriate monovalent antivenoms should be made by the local practitioner. If a small centre treats a significant number of snake bites per year, the quantity of antivenoms should be increased. Alternatively, if snake bite is rare in the area and a larger hospital can be reached within thirty minutes, then it is probably inappropriate to stock polyvalent antivenoms.

NB. The shelf life of antivenom is three years when stored protected from light at 2°C to 8°C. Antivenom should not be frozen

References

1. Sutherland SK, Coulter AR, Harris RD. The rationalisation of first aid measures for elapid snake bite. *Lancet* 1979;1:183–6.
2. Sutherland SK, Coulter AR, Harris RD, Lovering KE, Roberts ID. A study of the major Australian snake venoms in the monkey (*Macaca fascicularis*). 1. The movement of injected venom, methods which retard this movement, and the response to antivenoms. *Pathol* 1981;13:13–27.
3. Cox JC, Moisidis AV, Shepherd JM, Drane DP, Jones SL. A novel format for a rapid sandwich EIA and its application to the identification of snake venoms and enteric viral pathogens. *J Immunol Methods* 1992;146:293–4.
4. Howarth DM, Southee AE, Whyte IM. Lymphatic flow rates and first-aid in simulated peripheral snake or spider envenomation. *Med J Aust* 1994;161:695–700.
5. Sutherland SK. *Australian Animal Toxins*. Oxford University Press, Melbourne, 1983.
6. Sutherland SK, King K. Management of snake bite in Australia. *Royal Flying Doctor Service of Australia Monograph Series* No.1. 1991.
7. Hawdon GM, Winkel KD. Could this be Snakebite? *Aust. Fam. Physician* 1997;26:1386–91.
8. Sutherland SK, Leonard RL. Snake bite deaths in Australia 1992–1994 and a management update. *Med J Aust* 1995;163:616–18.
9. Sutherland SK, Duncan AW. New first-aid measures for envenomation: with special reference to bites by the Sydney Funnel-web spider (*Atrax robustus*). *Med J Aust* 1980;1:378–9.
10. Hawdon GM, Winkel KD. Spider Bite: A Rational Approach. *Aust. Fam. Physician* 1997; 26:1380–5.
11. Williamson JA, Fenner PJ, Burnett JW, Rifkin JF (eds). *Venomous and Poisonous Marine Animals: A Medical and Biological Handbook*. University of New South Wales Press, 1996.
12. Hawdon GM, Winkel KD. Venomous Marine Creatures. *Aust. Fam. Physician* 1997; 26:1369–74.
13. Hartwick R, Callanan V, Williamson J. Disarming the box-jellyfish. Nematocyst inhibition in *Chironex fleckeri*. *Med J Aust* 1980;1:15–20.
14. Exton DR, Fenner PJ, Williamson JA. Cold packs: Effective topical analgesia in the treatment of painful stings by Physalia and other Jellyfish. *Med J Aust* 1989;151:625–6.

General References

Cogger HG. *Reptiles and Amphibians of Australia*. Reed Books, Australia. 1996.
Covacevich J, Davie P, Pearn J (eds). *Toxic Plants and Animals: A Guide for Australia*. Brisbane: Queensland Museum. 1987.
Edmonds C. *Dangerous Marine Creatures*. Reed Books, Frenchs Forest, NSW. 1989.
Gow GF. *Graeme Gow's Complete Guide to Australian Snakes*. Cornstalk Publishing, Sydney, 1993.
Mascord R. *Australian Spiders in Colour*. AH and AW Reed, Sydney. 1983.
Mirtschin P, Davies R. *Snakes of Australia: Dangerous and Harmless*. Hill of Content, Melbourne. 1992.
Shine R. *Australian Snakes: a natural history*. Reed Books, Australia, 1993.
Sutherland SK. *Venomous Creatures of Australia*. Oxford University Press, Melbourne. 1994.
Underhill D. *Australia's Dangerous Creatures*. Reader's Digest. Sydney. 1987.
Weigel J. *Guide to the Snakes of South-East Australia*. Australian Reptile Park, Gosford. 1990.

Literature on the treatment of snake bite and other envenomations and literature on the use of antivenom and hospital care of the envenomed patient is available upon request to Australian Venom Research Unit, Department of Pharmacology, University of Melbourne, Parkville, Vic. 3052.

Donations for venom research are welcomed by the Unit and are tax deductible. A brochure on the Unit is available upon request.

'Stop it at the start, it's late for medicine to be prepared when disease has grown strong through long delays.'

Ovid (43BC–AD17)

Index

The use of italics for page numbers (e.g. '*opp 18*') indicates an illustration opposite page 18.

Abra, Lyn 204, 218
Ackland, Norman 241, 288, 312
Ada, Dr Gordon 187
Adams, Dr Tony 341, 342
Age, The 60, 205, 234, 284, 298, 347, 357
Alston, Richard 275, 277, 278, 279, 281, 283–5, 299, *opp 307*
Angus, Professor James 338, 340, 347, 350, 358, 363, 364, 365, 367
ant bites, first aid 373–4
ant venom project 303, 313–17, 322, 325, 327, 337, 340, 342, 343, 344, 354
antivenoms
  administration of 375–6
  adverse reaction research 160, 235–41
  notes for doctors on 344, 375–6
  shelf life of 376
  quality of 266, 296, 302, 330
Argus, The 48
Atkinson, Dr Ron 200, 215
Attwood, Professor Harold 274, 281
Australian Animal Toxins xi, 236, 246, 287, 298–9, 302, 317, 368
Australian paralysis tick, first aid 374
Australian, The 213, 356
Australian Venom Research Unit 161, 337–46, 349–68
  CSL grant to 161, 342, 361
  donations to 303, 351–2
  relations with CSL 358–63
  relations with Federal Government 352–7
  State Government funding 357–8, 364, 367

Australian Doctor 329, 339
Australian Family Medicine 254
Australian Family Physician 254, 279, 280

Bailey, Commander 150
Baldo, Dr Brian 315
Barbaro, Mirella *opp 242*
Barclay, Mrs 118
Batt, Neil 302
Bazeley, Dr P. (Val) 167, 274, 302
Beazley, Kim 161, 342
bee stings, first aid 373–4
bee venom project 268, 310–13
Bendigo, Victoria
  All Saints Cathedral 32–4
  Camp Hill State School 22–6, 32
  Girton CEGGS 33–4, 40
  hawkers 30
  High School 40–6
  hospital 37
  motor vehicles 27
  picnics near 26
  sketch map *23*
  swimming pools 28–9
  World War II and 31–2
Bennet, Gordon 24
Bennett, Dr Noel 324
Berke, Ayse xiii, *opp 307*, 364
Birner, Jan 172, 265
Birrell, Dr John 104
Blackhouse or Window spider *opp 147*, 342
Blackman, Dr Graeme 310
Blackmore, Dr Kate 355

379

Blackwood, Dame Margaret 73, 74
Blamey, Field Marshall Sir T. 72
Blamey, Georgia 72, 73, 74, 77
Blamey, Ted 73, 74
Blamey, Terry 73, 74
Blamey, Tom 72
Blank, Miss 43–4
Blaskett, Dr Alan 303
Blizzard, John 321
Blizzard, Les 101
Blossom 119–20, 121–2, 124–7
Blue–ringed octopus *opp 147*
 dissection of *opp 211*
 first aid 374
 venom research 160, 175–8, 364
Boardman, Mr 76
Bob (neurologist) xi, 337, 368
Bolton, Commander Chris ix
Bowles, Miss Gwen 44–5
Box jellyfish
 antivenom 343, 345, 346
 first aid 374
Broad, Allen 220, 227, *opp 242*, 243–4, 253, 267, 269–70, 288, 311–13,
Brogan, Alf 167, 262, 263, 271, 276, 279, 284, 285, 293, 298, 320
Brook, Dr Chris 358
Brown, Dr Kester 197, 245
Brown, Dr Ron 76, 77, 78, 80
Brown, Malcolm 271, 282
Bull Ant research 314
Bulletin, The 295–6
Burke, Don 310
Burnet, Sir Macfarlane 63, 165, 231, 251, 255
Burnside, Cameron 217
Burnside, Julie 217
Burnside, Wendy 217
Bush, Dr Peter 298

CSL Ltd, emergency phone number 375
Calvert, John 41
Cameos of Crime 127
Camm, Mark 213
Camp Hill State School, Bendigo 22, 32, *opp 82*
Campbell, C. P. O. Coxswain 125
Campbell, Dr Charles 237
Carpendale, Richard *opp 146*
Catt, Dr Kevin 223
Chandler, Howard 220, *opp 242*, 251, 298
Chatelier, Charles 210, 211
Cocking, Mr 38–9
Cocking, Mrs 39

Coles, Dr Norman 180, 184, 186, 187, 201, 208, 257, 259, 261, 267–74, 277, 280, 283, 287, 288, 289, 292, 293, 296, 297, 299, *opp 306*, 348 Collins, Tony 223
Committed to Saving Lives 166, 304, 320
Commonwealth Serum Laboratories (CSL)
 chronology of events at 160–2, 180–1
 emergencies, phone number 375
 establishment of 166
 Immunochemistry Dept. 259–60
 Immunology research 160, 182–3, 251, 252
 In House References 317
 Performance Management
  Systems agreements 326, 331, 334, 339
 Pharmacology Laboratory 251
 public float of 161, 346, 350
 Radioisotope Laboratory 251, 259
 relations with AVRU 358–63
 relations with Government 167
 Venom Research Unit 163, 328, 346
Conus shells, first aid 374
Cook, Joanne ix, 351
Cook, Verna ix
Corrigan, Fred 39, 60
Corrigan's Brochitis Mixture 39
Coulter, Alan ix, 200, 220, 222, 223, 224, 228, 233, 241, 242, *opp 242*, 251, 274, *opp 274*, 288, 341
Coulter, Senator John 340, 356
Covacevitch, Jeanette 243
Cowling, Mr Jnr 32
Cowling, Mr Snr 32
Cox, John 220, *opp 242*, 286–7
Coxwell, Henry 14, *opp 51*
Crocker, Mr 43
Crotty, Dr John 146
Crowley, Frank ix
Culley, James 180
Curnow, Mr 92

Daily Mirror, The 215
Dalgleish, Bob *opp 115*
Daniels, Professor Peter 207
Darwin 141–8
Davey, Marjory ix, *opp 178*, 263, 267, 268, 269, 270, 271, 278, 290, 299, 304, 315, 323, 332, 339, 348
Davey, Vivian 205–8, 211, 257, 258, 259, 266, 267, 268, 271, 273, 274, 279, 291, 293, 296, 299, 302, 303, *opp 306*
Davis, Leo 348
Dennett, Xenia 173
'Dick' 306–8

INDEX 381

Discovery and Healing in Peace and War 281
D'Lisle, Lord 129
Doherty, David 320, 322
Doherty, Nancy 321
Drane, Paul 247, 248
Drewe, Robert 266, 296
Duncan, Dr Alan 160, 180, 197, 199, 209,
  212, 213, 214, 231, *opp 243*, 278, 294,
Duxbury, Dr Alan 173
Dyke, Corporal 81, 82

Edmonds, Dr Bill 355
Embury, Pauline ix
Evans, Senator Gareth 282, 293, 297

Fahey, Kevin 220, 251
Family Guide to Dangerous Plants and Animals in Australia 246
Feery, Dr Brian 256, 261, 287
Finger, Bill 360
first aid, pressure–immobilisation treatment 160, 196, 226–35, 258, 261, *opp 274*, 281, 371–2, 373
First Aid for Snake Bite in Australia 233
Fischer, Tim 356
Fisher, Dr Malcolm 181, 196, 203, 205, 210, 212–17, 274, 295
Forbes, Dr Jim 167, 263, 274, 275–8, 280, 283, 289, 291, 292–3
Freeman, Dr Shirley 177
French, Dr Eric 251
Frew, Sir John 274, 281–2, 324, 331, 332
Frickie, Graeme 284, 285, 287, 291–3
Frost, Dame Phyllis 355
fruit picking 70–2
Funnel–web spider *opp 147*, *opp 242*
  antivenom 160, 180–1, 183–90, 192–218, 231, 244, 295, 297, 298
  first aid 373
Funnell, Fred 309

Galbally, P. 284, 285
Galvin, Bill 42
Galvin, Mrs 42
Garnet, J. Ros 245
Gilbo, Merle 172, 180, 184, 186, 187
Glaisher, Professor J. 2, 14, *opp 51*
Gledhill, Captain J. 145
Godden, Anne ix, 246
Godden, Susie ix
Goellner, Maria 308
Goodman, Dr 27
Goodman, Russell 27, 312
Gorton, John 129

Graydon, Dr John 169, 221, 222, 251
Greenfield, Nerissa ix
Greenwood, Senator 282
Guilfoyle, Senator 205
Gulasekharam, Dr 254, 256, 262
Gust, Dr Ian 161, 283, 324–37, 338, 340, 345, 348, 351
Guthrie, Charles 266, 303, 325, 357

HMAS *Cerberus* 134–41
HMAS *Melbourne opp 146*, 148–57
HMAS *Melville* 141–8
HMAS *Perth* 131
HMAS *Voyager* 114–34, *opp 146*, 148, 165, *opp 210*
Hachiya, Michihiko 133
Hammersley, Corporal 81, 82
Hancock, Arnold 166
Hancock, Justin 166
Hancock, Marjory 166
Hanger, Dr Brian 309
Hardy, Barbara 275, 276, 289
Harper, Colin 360–2
Harris, Dr George 303, 313, 325
Harris, Rodney 202, 203, 223, 230, 233, *opp 242*, 268, 274, *opp 274*
Hartman, Dr Len 256, 288, 290, 313
Harvey, Dick *opp 115*
Hasluck, Sir Paul 270
Hawdon, Dr Gabrielle ix, 162, *opp 179*, 365–8
Hawke, Bob 302, 346
Hawkins, Mr 48
Healy, Kevin 330, 345, 360
Hepsie, Aunt 16
Herald, The 48
Herron, Senator 341, 356
Hewson, Dr John 357
Hial, John 31, *opp 83*
Hill, Senator Robert 356
Hince, Kenneth 234
Hinton, Merv 169, 257, 267, 268, *opp 306*
Hird, Dr Zana 250
Hobbs, Peter 350, 358
Holiday Hazards 247
Horsburgh, Rhonda *opp 242*
House, John 337
Howden, Dr Merlin 200, 215, 216,
Hudson, Lt Mike 117, *opp 146*
Hudson, Professor Bryan 274, 278, 296
human gamma globulin 171, 238
Hunt, Celia (née Glaisher) 2, 14–15
Hunt, Dr Frederick E. 2, 14–15, 17
Hunt, Frederick K. (born 1814) 2, 14

382 A VENOMOUS LIFE

Hunt, Frederick K. (born 1868) 2, 12–14, opp 50
Hunt, Mary C. 14
Hunt, May Rose 15, 16, opp 50
Hunt, Ralph 255, 264
Hurley, Dr Tom 276, 287, 289
Hurrell, Dr 298
Huyer, H. D. 276, 283, 285, 287
hydroponics 309–10
Hydroponics for Everyone 309

immunoelectrophoresis 172, 173, opp 211
Ipsen, John 111, 192, 281
Irukandji (jellyfish) research 364
Isaacs, Deputy President (Arbitration) 264

Jacobs, Dr Maurie 37
jellyfish stings, first aid 374
Johnson, Charlie 92, 93, 94
Jones, Brent 337, 338, 339–40, 342
Joyce x
Jumper Ant research 314–17, 327, 354

Kapouleas, Chris 325
Kaufman, Graeme 328, 339, 343, 361
Keating, Paul 347
Keatinge, Captain 143, 144, 145
Kellaway, Dr Charles 180, 184, 221
Kelly, Peter 294
Kemp, Dr David 357
Kennedy family 24, 40
Kennedy, Trevor 24, opp 83
Kennewell, Susie ix
Kerr, Alan ix, 29, 36–7, 67, 81, 82, 83, 88, opp 115, opp 179
Kerr, Dr Keith 36, 81, 109, opp 114
Kerr, Gwen 36
Keynes, J. M. 52
King, Arthur 72
Knight, Al ix, 246

Lafferty, Professor Kevin 251
Lamb, Len 247
Lancet, The 196, 233
Lane, Dr Bill 170, 178, 182, 183, 186, 187, 252, 253, 255, opp 275, 332
Lane, Leslie 294
Lawrence, Dr Carmen 161, 340, 351, 352, 353–4, 355, 357, 358, 363
Leonard, Dr Ria 161, 357, 364
Leslie, Doug 233, 274, 280–1
Lewis, Sergeant 81, 82
Lichtenstein, Dr 311
Lockwood, Douglas 142

Lockwood, Surgeon Rear Admiral L. 110, 113
Looker, Keith 4
Looker, Margaret (née Read) 2, 7, 8
Looker, William H. 2, 8
Looker, William R. 2, 7, 8
Lovering, Erin 201, 240, opp 242, opp 274
Lucas, Dr Ron 290
Lynch, Mr 32

McCarthy, Dr Neville 161, 182, 198, 208, 211, 213, 215, 233, 252–93, 296, 297, 298, 299, 303, 304, 319–22
McEwen, Dr John 326
Macfarlane, Ian 348, 350
McGarvie, Lesley 37
McGarvie, Richard 37
MacGregor, Adrian 298
Macintosh, Andrew 241
McKay, Peter 208
MacKellar, Michael 264, 273, 294
McKinty, Judy 234
McLeod, Alan 317
McLiesh, Stan 329, 330, 345, 358, 360, 362
Macmillan, Reverend J. ix
McMullan, Mr 276
McNamee, Dr Brian 161, 319, 322–3, 325, 326, 331, 332–6, 342–5, 347, 348, 351, 355, 360, 362
McPhee, Ian 264
Mallard, Rodney F., 'Harry' 305, 308–9
Malloy, Margaret (Megan Sutherland) opp 178, 190, 191, 192, 209, 233, 247, 269, 304, 318
mammoth ball 136
Marks, Deborah 28
Marks, Harry 28
Maslen, Geoff 234
Matheson, Laurie 148
Medical Journal of Australia, The 164, 165–6, 206, 210, 211, 240, 241, 294, 347, 357, 362, 364
Menzies, Sir Robert 132, 167, 302
Metherell, Mark 284
Metz, Dr Geoffrey 282
Meyer, Dr John ix
Microbe Hunters 67–8
Miller, Anne opp 242
Miller, Lt Colonel 142
Mills, Darryl 360
Minifie, Mrs 63
Mitchell, Graham 328–9, 342, 347
Monetary Puzzle, The 51–2
Money, Lindsay 265

INDEX   383

Moore, Bruce 272, 275, 283, 284, 287
Myers, A. J. 291

Nakajima, Dr Hiroshi 265–6
necrotising arachnidism 318, 344, 366
Nelly, Roger 43
Neville, Dr 36
Newman, Joyce ix
Newman, Mr Justice 308
Nihat, Najiye ix
Nolch, Guy 329
Northrop, Mr Justice 285
Nossal, Sir Gustav 251, 358, 360

O'Brien, Joe 320
O'Brien, Mick 320
Oxer, Dr Don 169

Papadopoulos, Mary 316, 325, 328, 350
Parkinson's disease xi, 331, 337
Pascoe, Mr (Coroner) 246
Pearn, Professor John ix
Pearson, Joan 172
Perkins, Dr Frank 258, 259
Pettigrew, Tom 94
Phillips, Dr Bryce 262
Pickett, Ismay 191
Pincus, Steven opp 179, 366, 368
Pockley, Peter 82
Poison Information Centres 375
Polacsek, Emma 173
Potter, Andrew 297
Premier, Robert 220, opp 242, 325
pressure–immobilisation treatment 160, 226–35, opp 274, 281, 371–2, 373
Professional, The 330
Property and War 52
Puffer fish 177

rabbit shooting 26–7
radioimmunoassay 223, 224
Ragg, Dr Mike 356
Rank, Sir Benjamin 98, 99, 100
Read, Captain G. F. 7–8
Reale–Key, Angela opp 242, 288–9, 293
Red–back spider opp 147
    first aid 373
    research 160, 239–41, 346
Reid, Dr Alistair 237, 252
Richards, Jack 83, 84
Richards, Valme ('Venus') 83–6, 88, 89
Richardson, Graham 347
Riley, Bishop 34
Roberts, Bruce 223

Roberts, Greg 354
Roberts, Ian 202, 203, 233
Roberts, Peter 205
Robins, Doc 45
Robson, Dave 205
Robson, Ross 275–7, 279, 280, 282–93, 299
Roche, Mick 328, 339, 341, 342, 361
Rogers, Dr 76
Russell, Professor Findlay 226–7
Ruxton, Bruce 355

Sanderson, Commander 145
Sanderson, Ken 328
Sandner, Dr E. 27, 38
Schiff, Dr Peter 170, 183, 203, 207, 208, 211, 256, 257, 259, 260, 268, 271, 291, 293, 302, 303
sea snake antivenom 344
sea wasp stings, first aid 374
Sharland, Dr Jim 27, 38, 109, *opp 114*
Sharland, 'Poppet' 91
Sharland–Normark, Pamela ix
Shellam, Geoff 220
Shelmerdine, R. T. 257
Sheumack, David 217
Shoddy, Peter 29
Short, Jim (Senator) 35
Siegfried, Slimy 72
Silk, Dr Stephen 329
'60 Minutes' programme 296–7
small–scaled snake 160, 243–4
Smith, Belinda 224–5
Smith, Dr Stephen ix
Smith, Mr ('Drip') 43
snake bite
    First Aid for Snake Bite in Australia 233
    first aid kit 370–3
    pressure–immobilisation 371–3
    research 160, 221–35, 239, 242–7, 251, *opp 274*, 364
    venom detection kits 160, 241–2, 363
Solomon, Wendy 4, 134, *opp 147*, 166
Souter, Gavin 270
South African Medical Journal 219
Spalding, Sister May 103, 104
spider bites, first aid 373
spider research 161, 239–41, 317–19
    Funnel–web antivenom 183–90, 192–3, 244, 295, 373
    Red–back spider 160, 239–41, 346, 373
Stack, Dr Ella 142
Stannard, Bruce 296
Staples, Peter 319
Steen, Jenny 350

Stockdale, Alan 161, 357
Stone, Gerald 296
stonefish, first aid 374
Strellon, Jack 34, 43, 46, opp 83
Stretton, Dr P. 164
Striato Nigral Degeneration (SND) xi–xii, 356
Sturges, Christine 180, 195–6, 198
Styles, Miss 45
Sukarno, President 132
Sullivan, Peter 208, 211, 261, 266
Sun, The 208
Sutcliffe, P. O. 130, 131
Sutherland, Barbara R. 2, 3, 8, 9, 18, 82, 90
Sutherland, Dr Charles 2, 6, 8, 12, 55–68, opp 114, 170, 311
Sutherland, Dr Diana M. 2, 3, 12, 18, 46, 57, opp 83, 192
Sutherland, Dr Peter 73, 77, 97, 331
Sutherland, Dr Rod 54, 60
Sutherland, Dorothy E. (née Hunt) 2, 12, 16–19, opp 19, 281, 327
Sutherland, Hector Anaeus (Angus) 2, 6, 8, 9, 53–5, opp 114
Sutherland, Hector McK. 2, 5–7, 8, opp 18
Sutherland, Ian McK. 2, 8, 51–3, opp 114, 116
Sutherland, John ix, 4, 9, 134, 164, 166, opp 178, 191, 192, 241, opp 242, 246, 248, 304
Sutherland, Margaret F. 8, 18
Sutherland, Margaret H. ('Megan', née Malloy) opp 178, 190, 191, 192, 209, 233, 247, 269, 304, 318
Sutherland, Marjory (née Minifie) 57, 59, 62, opp 114
Sutherland, Minnie (née Looker) 2, 7–9, opp 18
Sutherland, Neil ('Abdul') 73
Sutherland, Rosemary B. ix, 2, 14, 19, opp 83, 91
Sutherland, Struan opp 82, opp 83, opp 115, opp 146, opp 147, opp 242
  AMA Diners Club Public Health Award 328
  antecedents 3–19
  appointed to CSL 160–78
  Associate Professor, Melbourne University 340
  birth 3
  choirboy 32–4
  departure from CSL 161, 341–2, 347–8
  education 22–6, 40–6, opp 82, opp 83
  family tree 2

  FRCPA 256–7
  honorary at RMH 168
  Immunology Research, Dept Head 160, 182–3, opp 243
  influence of Uncle Charles 66–8
  James Cook Medal 302
  medical course 70, 72–8, 80, 83, 86, 89–106
  National Service 81–3
  Naval service 108–66, opp 210
  part-time jobs during medical course 70–2, 77, 78–9, 91–101, 108–10
  pressure–immobilisation demonstration opp 274
  prison visiting 304–9
  suspensions from CSL 161, 259–65, 265–93
  Triennial 1935 BMA Prize 228, 257
  visit to Hiroshima 132–3
  Whitley Award 302
Sutherland, Susie ix, 4, 134, 164, opp 178, 191, 192, 193, 202, 247, 304, 309
Sutherland, Walter 2, 3, 8, 9–12, opp 18, opp 19, opp 83, 191, 192
Sutherland, Wendy (née Solomon) 4, 134, opp 147, 166
Sutherland, William (Willy) 5, 8

Take Care! Poisonous Australian Animals 246
Tanner, Charles 243
Tehan, Marie 161, 357, 358
Telford, Allan 186, 187
Theophanous, Dr Andrew 352
Thomas, Dr Laurel 206
Thompson, Barry V. 83, opp 115
Tibballs, Dr James xi, 160, 162, opp 179, 180, 200, 231, 234, opp 243, 362, 365, 368
Tiberius, Claudius 142
Tilley, Miss 24–5
Toadfish 177
'Tom' 305–6
Torda, Dr Tom 204, 210, 212
Trehy, Liam 217
Treloar, A. S. C. 150
Tresidder, Vanessa ix, 351, 366
Trethewie, Dr Everton 177, 194
Trikojus, Professor 63
Trinca, Dr John 170, 171, 172, 221, 236, 240, 245, 253, 256, opp 306
Trinder, Tommy 58
Turnbull, Jan opp 242
Tyler, Mr ('Cactus') 41
Tyson, Dr Chris 268, 313

INDEX   385

USS *Peary* 144

Vaccaro, Maree 246, 247
Varma, Sony 325, 328, 350
Venomous Australian Animals Dangerous to Man 245, 246
venomous wildlife in Australia 220, 244
  ants 313–17, 327
  bees 310–13
  Blue–ringed octopus *opp 147*, 160, 175–8, *opp 211*
  Funnel–web spider research *opp 147*, 160, 183–90, 192–218, 231, *opp 242*
  puffer fish 177
  small–scaled snake 160, 243–4
  snake venom research 160, 221–35
  spiders 161, 239–41, 317–19
Vivian, Joan 317–18

Warburton, Dr Frank 176
Warrell, Professor David 162, 365
wasp stings, first aid 373–4
Weedon, Dr David 357
Weiner, Dr John 315
Welch, John St V. 262, 271
Wells, Captain David 112–13, 120, 121, 123, 124, 128, 132, *opp 210*
Wells, David 271, 272, 273, 275, 279, 283, 287, 294

Wells, Miss 25–6, 32
Wendt, Jana 296
West, Mr (Coroner) 356
Wheatley, Gordon 181, 214
White, Dr Julian 346, 362
White, Lee ix
White–tailed spider *opp 147*
  research 161, 317–19, 325, 364
Whitlam, Gough 270
Wiener, Dr Saul 180, 184, 185, 187, 193, 204, 239
Williams, Robyn 297
Williamson, Professor John ix
Wiltshire, Carolyn 161, 364, 366
Win, Professor May 367
Winkel, Dr Ken 161, 162, *opp 179*, 364–8
Wombey, Dr John 243
Wood, J. H. 52
Wood, Sir Ian 57, 113, 165, 274, 281
Wooldridge, Dr Michael 161, 356, 357
Worrell, Eric 204, 205, *opp 243*
Worrell, Robyn 204, 208
Wright, Professor R. D. ('Pansy') 189–90, 207, 274, 278, 298

'Xuereb, Dr J.' (Joe the Welder) 256

Yallow, Dr Rosalind 223
Young, Dr Anna 162, *opp 179*, 366, 368